Lecture Notes in Civil Engineering

Volume 41

Lecture Notes in Civil Engineering (LNCE) publishes the latest developments in Civil Engineering - quickly, informally and in top quality. Though original research reported in proceedings and post-proceedings represents the core of LNCE, edited volumes of exceptionally high quality and interest may also be considered for publication. Volumes published in LNCE embrace all aspects and subfields of, as well as new challenges in, Civil Engineering. Topics in the series include:

- Construction and Structural Mechanics
- Building Materials
- Concrete, Steel and Timber Structures
- Geotechnical Engineering
- Earthquake Engineering
- Coastal Engineering
- Hydraulics, Hydrology and Water Resources Engineering
- Environmental Engineering and Sustainability
- Structural Health and Monitoring
- Surveying and Geographical Information Systems
- Heating, Ventilation and Air Conditioning (HVAC)
- Transportation and Traffic
- Risk Analysis
- Safety and Security

To submit a proposal or request further information, please contact the appropriate Springer Editor:

- Mr. Pierpaolo Riva at pierpaolo.riva@springer.com (Europe and Americas);
- Ms. Swati Meherishi at swati.meherishi@springer.com (India);
- Ms. Li Shen at li.shen@springer.com (China);
- Dr. Loyola D'Silva at loyola.dsilva@springer.com (S-E Asia and Australia/NZ).

Indexed by Scopus

More information about this series at http://www.springer.com/series/15087

Chien Ming Wang · Soon Heng Lim ·
Zhi Yung Tay
Editors

WCFS2019

Proceedings of the World Conference on Floating Solutions

Springer

Editors
Chien Ming Wang
School of Civil Engineering
The University of Queensland
Saint Lucia, QLD, Australia

Soon Heng Lim
Society of Floating Solutions
Singapore, Singapore

Zhi Yung Tay
Singapore Institute of Technology
Singapore, Singapore

ISSN 2366-2557 ISSN 2366-2565 (electronic)
Lecture Notes in Civil Engineering
ISBN 978-981-13-8742-5 ISBN 978-981-13-8743-2 (eBook)
https://doi.org/10.1007/978-981-13-8743-2

This Springer imprint is published by the registered company Springer Nature Singapore Pte Ltd.
The registered company address is: 152 Beach Road, #21-01/04 Gateway East, Singapore 189721, Singapore

Conference Organising Committee

Prof. Chien Ming Wang (Chairman)
Dr. Sabet Divsholi Bahador (Vice-Chairman)
Mr. Charles Lim (Secretary)
Mr. Anil Thapar (Treasurer)
Dr. Zhi Yung Tay (Technical Programme)
Captain James Fong (Committee Member)
Mr. Dave Kinrade (Committee Member)
Dr. Hock Seng Koh (Committee Member)
Dr. Paul Ong (Committee Member)
Mr. Ivan Stoytchev (Committee Member)
Mr. Soon Heng Lim (Advisor)

Organised by

Society of Floating Solutions Singapore

Supported by

Conference Programme for Day 1

Programme—Day 1 (Monday, 22 April 2019)	
0800hrs	**Registration @ Foyer of Engineering Auditorium, Faculty of Engineering, National University of Singapore**
0900hrs	Welcome address *Prof. Chien Ming Wang (Chairperson of WCFS2019 Organising Committee)*
Session 1 Living: Moderator is Dr. Paul Ong	
0910hrs	Design guidelines for upgrading living conditions in wetslums *Koen Olthuis (Waterstudio.NL, The Netherlands)*
0940hrs	Repurposing jack-ups, semi-submersibles and super-barges into offshore and nearshore settlements *Joseph Lim (National University of Singapore, Singapore)*
1010hrs	**Morning Tea/Coffee Break @ Foyer of Engineering Auditorium**
Session 2 Infrastructure I: Moderator is Mr. Soon Heng Lim	
1040hrs	Floating infrastructure: Large scale public spaces on water *Gerard Ronzatti and Petar Lovric (Seine Design, France)*
1110hrs	Floating shipyard design: Concept and application *Cliff Ohl (Royal HaskoningDHV, United Kingdom)*
1140hrs	Floating bridges and submerged tunnels in Norway—the history and future outlook *Torgeir Moan (Norwegian University of Science and Technology, Norway)*
1210hrs	**Networking Lunch @ Foyer of Engineering Auditorium**
Session 3 Infrastructure II: Moderator is Dr. Koh Hock Seng	
1330hrs	Dynamics of super-scale modularized floating airport *Daolin Xu (Hunan University, China)*
1400hrs	Design and potential applications of floating structures in Singapore *Kok Keng Ang (National University of Singapore, Singapore)*
1430hrs	Hydrodynamic responses and loads of a model floating hydrocarbon storage tank system for concept validation and numerical verification *Chi Zhang (National University of Singapore, Singapore)*
1500hrs	Wave induced motions of floating mega island *William Otto (MARIN, The Netherlands)*

(continued)

(continued)

Programme—Day 1 (Monday, 22 April 2019)	
1530hrs	**Afternoon Tea/Coffee Break @ Foyer of Engineering Auditorium**
Session 4 Food and Environment: Moderator is Mr. Anil Thapar	
1600hrs	Fish Farming in Floating Structures *Tor Ole Olsen (Dr.techn.OlavOlsen AS, Norway)*
1630hrs	Technology-driven sustainable aquaculture for eco-tourism *Hoon Kiang Tan (Nature Resources Aquaculture Pte Ltd., Singapore)*
1700hrs	Floating forest: A novel breakwater-windbreak structure *Chien Ming Wang (The University of Queensland, Australia)*
1715hrs	Seasteading = More Floating Singapores *Joe Quirk (The Seasteading Institute, USA)*
1730hrs	**Presentation of tokens of appreciation by President SFSS Soon Heng Lim** **Day 1 Programme Ends**

Conference Programme for Day 2

Programme—Day 2 (Tuesday, 23 April 2019)	
0830hrs	**Registration @ Foyer of Engineering Auditorium, Faculty of Engineering, National University of Singapore**
Session 5 Design of VLFS: Moderator is Mr. Ivan Stoytchev	
0900hrs	Classification principles for very large floating structures *Chua Wee Ng (DNV GL, Singapore)*
0930hrs	Mooring systems for very large floating structures *Aditya Sankalp (DEEPBLUE Pte Ltd., Singapore)*
1000hrs	**Morning Tea/Coffee Break @ Foyer of Engineering Auditorium**
Session 6 Concrete Technology: Moderator is Dr. Zhi Yung Tay	
1030hrs	Durability of floating concrete platforms *Sabet Divsholi Bahador (eCreators-global, Singapore)*
1100hrs	DemoSATH Project—Demonstrating the technical feasibility of a disruptive concrete floating offshore wind solution *David Carrascosa Francis (Saitec Offshore Technologies, S.L., Spain)*
1130hrs	Design and construction of concrete floating pier in Golden Harbor, Incheon *Kwanghoe Jung (Hyundai Engineering and Construction, South Korea)*
1200hrs	**Networking lunch @ Foyer of Engineering Auditorium**
Session 7 Sustainable City I: Moderator is Dr. Sabet Divsholi Bahador	
1330hrs	An integrated floating community based upon a hybrid water system: Toward a super-sustainable water city *Toshio Nakajima (Waterfront Real Estate Co. Ltd., Japan)*
1400hrs	Floating clean multi-energy systems towards driving blue economic growth *Narasimalu Srikanth (Nanyang Technological University, Singapore)*
1430hrs	Floating wind turbines in Goto Islands, Nagazaki, Japan *Tomoaki Utsunomiya (Kyushu University, Japan)*
1500hrs	The dawn of floating solar—technology, benefits, and challenges *Thomas Reindl (SERIS, Singapore)*
1530hrs	**Afternoon Tea/Coffee Break @ Foyer of Engineering Auditorium**

(continued)

(continued)

Programme—Day 2 (Tuesday, 23 April 2019)	
Session 8 Sustainable City II: Moderator is Captain James Fong	
1600hrs	Seascape the landscape of Singapore, Repurposing land in a land scarce nation *Soon Heng Lim (President of Society of Floating Solutions Singapore)*
1630hrs	Floating solutions: The new meaning of mobility *Milica Simovic (University of Nis, Serbia)*
1700hrs	Driving innovations in support of greenery, biodiversity and clean energy initiatives in Singapore *Sze Tiong Tan (Building &* Research *Institute, HDB, Singapore)*
1730hrs	Potential of floating urban development for coastal cities *Karina Czapiewska (Blue 21, The Netherlands)*
1745hrs	**Presentation of tokens of appreciation and closing remarks by President of SFSS Mr. Soon Heng Lim**
1800hrs	**End of Conference**

There are post-conference site visits to the world's largest floating PV testbed at Tengeh Reservoir, Singapore and the Deepwater Ocean Basin of TCOMS on 24 April 2019.

Foreword

Dear Sponsors, Supporters, Speakers and Delegates,

The Society is grateful that you have made this rare event possible. Floating structures is a disruptive technology to address issues of land scarcity, degradation of coastal ecosystems, climate change, rising sea level and ocean acidification. The world is under an insidious threat. Sir David Attenborough, the world-renowned naturalist speaking at Davos 2019 warned that we could 'wreck (the natural world) without even noticing it'.

The emission of carbon dioxide into the atmosphere is changing the global climate. The main sources of that emission are power generation plants, marine vessels and road vehicles. The greenhouse gas traps heat, causing polar ice caps to melt which in turn reduces ability to reflect solar heat. Permafrost thaws adding methane to the atmosphere. The positive feedback loop may spin out of control making the world too warm and stressing flora and fauna. Instead of discharging waste heat into the atmosphere, it goes directly into the oceans, which absorb heat more efficiently (water has higher thermal conductivity and specific heat.) Carbon dioxide from factories and power plants floating in water can be bio-sequestered with algae. Hydrogen produced by offshore power plants (either nuclear or fossil fired) should be used to power ships and road vehicles (or batteries when mass is not a problem).

The acidity level of seawater has increased by 25% since the industrial revolution. That has harmed planktons, which feeds the bottom rung of the world's food chain, which supports all life in the ocean and on land. Reducing CO_2 emission would reduce that risk. Top soil is lost at an alarming rate due to deforestation and over cultivation. It needs to rejuvenate by leaving it to fallow. Teaching people to live offshore will help heal over worked land. Salt-tolerant crops are being researched and could feed millions in a few decades.

As floating structures are mobile, it is possible now to relocate used office buildings, residences, hospitals and factories to less developed countries for another cycle of productive life instead of demolishing it. This means building materials would be used more sustainably and their source material be less rapidly depleted.

Living over water also means not having to build highways and railways with beneficial effects on the planet.

Your endeavour to make floating structures as alternatives to landed ones is a noble undertaking. You deserve the title Combatants against Climate Change. I applaud you for your endeavour. I hope many will follow your example.

I take this opportunity to thank Professor Wang Chien Ming and his team for organising this event. Many hours of work went in over the past 12 months.

Singapore

Soon Heng Lim, B.E., P.E., FIMarEST, FSSSS
President of Society of Floating Solutions
(Singapore)

Preface

Singapore, once a third world country under British rule, is today the world's most expensive city (sharing the top position with Hong Kong and Paris) according to a 2019 EIU survey. It has achieved striking success in its land reclamation projects to support its urban and economic growth. 25% of its original 578 sq. km footprint has been reclaimed from the sea. Despite this, in the same time period, its population grew by 380%, creating an urban environment that makes it the world's densest country.

In 2013, a White Paper was released by the government essentially preparing this island nation for further population growth that would see it grow from the current 5.7 million to 6.9 million by 2030. The Ministry of National Development projected that Singapore would need to find space amounting to 56 sq km. However, land reclamation has become more challenging year by year, as sand mining is now seen to be ecologically damaging and banned by countries that had traditionally supplied sand to Singapore. Fortunately, it has about 700 sq km of territorial waters well sheltered by a cluster of Indonesian islands to the south. A 65-million TEU capacity mega port is being constructed which will see all port activities on the waterfront of downtown Singapore being diverted to the west.

We believe this opens up huge opportunities for creation of sea space floating solutions for the future. A group of multi-disciplinary professionals with this vision has come together to form the Society of FLOATING SOLUTIONS Singapore (SFSS) in November 2017.

In January 2018, the Society invited Professor Jacopo Buongiorno of the Massachusetts Institute of Technology to speak at the National University of Singapore. He delivered a talk suggesting that to meet Singapore's energy resilience needs and in view of the limited land area, a floating nuclear plant could be considered. We also had other speakers sharing other floating solutions with us on 1 March 2018. Dr. Øyvind Hellan and Dr. Arnstein Watn of SINTEF Norway and Dr. Masaki Takeuchi of Shimizu Japan spoke at our conference which is titled 'The Applications and Technology of Floating Structures' at the Singapore Institute of Technology. In August 2018, the Society organised a 1-day seminar on floating farms for a sustainable future. The seminar speakers include Mr. Peter van Wingerden (CEO and founder of Belandon) who gave a talk on farming on water, Mr. Leow Ban Tat (Founder and Managing Director of AME2) who spoke on floating fish farms, Dr. Shinjo Sato (Nihon University) who spoke on floating composite plant factory, Mr. Morten Lund Hoffman (Multiconsult Asia Pte Ltd) who gave a talk on floating structures in Norway and Mr. Gary Wong (Keppel Offshore and Marine) who spoke on repurposing offshore technology to unlock the potentials of sea space.

Continuing on this journey, the Society organised a World Conference on Floating Solutions with a view of elevating international awareness of the benefits of floating structures. This volume of conference proceedings contains 23 invited papers presented at the conference. The papers cover a wide range of applications of floating structures that include floating wind turbines, floating bridges, floating houses, floating entertainment facilities, floating hydrocarbon storage facilities, floating piers, floating fish farms, floating forest, floating wetlands, floating solar farms, floating shipyards, floating airports and floating cities.

We hope that the floating solutions described in the Conference Proceedings will inspire urban planners, architects, naval architects, structural engineers, contractors and academics to create, innovate and build even more awesome and eco-friendly floating structures that nourish the human body, soul and spirit.

Saint Lucia, Australia Chien Ming Wang
Singapore Soon Heng Lim
Singapore Zhi Yung Tay

Acknowledgements

The Conference Organising Committee acknowledges with much gratitude the generous sponsorships received from the following organisations and companies:

Platinum Sponsors

GLE&C
GROWTH & LEADERSHIP

Keppel Offshore & Marine

Gold Sponsors

WOH HUP
BUILDING WITH INTEGRITY

Waterpolitan Initiative
WATERFRONT REAL ESTATES CO. LTD.

CONTINENTAL STEEL PTE LTD

KEMSA

Springer

The Organising Committee also takes this opportunity to thank Surbana Jurong, Torrisi & Procopio Architetti and Nature Resources Aquaculture for advertising in this publication.

Contents

About the Editors

Prof. Chien Ming Wang is the TMR Chair Professor in Structural Engineering at the School of Civil Engineering, The University of Queensland, Australia. He is a Chartered Structural Engineer, a Fellow of the Academy of Engineering Singapore, a Fellow of the Institution of Engineers Singapore and a Fellow of the Institution of Structural Engineers. His research interests are in the areas of structural stability, vibration, optimization and very large floating structures. He has published over 420 journal papers and 9 books in the aforementioned areas. He is the Editor-in-Chief of the *International Journal of Structural Stability and Dynamics* and an Editorial Board Member of several journals including *Engineering Structures*, *Ocean Systems Engineering*, *Structures* and *International Journal of Applied Mechanics*. His many awards include the Minister for National Development's R&D Awards 2017, IStructE Structural Award for Sustainability 2016, Monash Civil Engineering Alumnus of the Year 2015 Award, Keith Eaton Award 2014, Lewis Kent Award 2009, the IES Prestigious Engineering Achievement Award 2013 and the US$1 million Grand Prize in the Next Generation Port Challenge.

Soon Heng Lim has a long career in the maritime industry since graduating from the National University of Singapore in 1968. He worked many years in Keppel Offshore and Marine and was involved in ship construction, conversion and maintenance. He has international experience in the feasibility study, master planning, design and construction of shipyards. He holds a degree in Mechanical Engineering and is a Fellow of IMarEST. He is an active Advocate of floating solutions to land scarce Singapore.

Dr. Zhi Yung Tay is currently an Assistant Professor in the Singapore Institute of Technology (SIT). From 2014 to 2016, he was attached to the Institute of Energy System (IES), University of Edinburgh as a Postdoctoral Research Associate where he worked on the EPSRC EcoWatt2050 Project as the lead hydrodynamic modeller. Prior to joining IES, he held a position as a Senior Research Engineer in Keppel

Offshore and Marine Technology Centre and as a Research Fellow in the National University of Singapore (NUS). He obtained his Ph.D. from NUS and specialised in the hydroelastic response of very large floating structure. His research interests are in the areas of hydroelasticity, marine energy devices and very large floating structures.

Design Guidelines for Upgrading Living Conditions in Wetslums

Koen Olthuis, Pierre-Baptiste Tartas and Chris Zevenbergen

Abstract The megalopolis of Dhaka, Bangladesh faces, in a larger scale, common issues with the rest of Asian ones and more particularly South Asian ones: a massive population growth leading to urban sprawl on flood-prone areas—more precisely on areas located along the shores or on a water body—mainly in form of slums; named wetslum in this article. However, even if this condition is expected to increase in the next decades, these wetslums have poor living conditions and lack access to basic services. Thus, the purpose of this paper is to propose an approach and guidelines to favor small scale project able to improve wetslum dwellers' living conditions. To do so, the specific issues of these areas which are a shortage of available space, a lack of services due to investment risks and the insecurity of land tenure, are identified along with the position of the different stakeholders involved in these areas. Then, guidelines are set up to bypass these difficulties. Consequently, the findings are used to establish eight constraint requirements that have to be addressed by a wetslum upgrading project proposition: geographic location in wetslum, flood proof ability, flexibility, transportability, standardization, affordability, safety and legality. These guidelines will then be used in a near future to develop a small scale proposition named City App.

Keywords Wetslum · Dhaka · Services provision · Improving living conditions · Flooding

1 Introduction

This paper focuses on Dhaka, Bangladesh, a megalopolis, located in South Asia, that faces the second fastest population growth in the world [1], resulting in urban sprawl on flood-prone areas, mostly in the form of slums [2, 3]. As a result, this city

K. Olthuis (✉) · C. Zevenbergen
Flood Resilience Group, UNESCO-IHE, PO Box 3015, 2601 DA Delft, The Netherlands
e-mail: koen@waterstudio.nl

K. Olthuis · P.-B. Tartas
Waterstudio.NL, Generaal Berenschotlaan 211, 2283 JM Rijswijk, The Netherlands

C. M. Wang et al. (eds.), *WCFS2019*, Lecture Notes in Civil Engineering 41,
https://doi.org/10.1007/978-981-13-8743-2_1

is prone to important water related disasters such as floods, windstorms, heavy rain and waterlogging [4]. This situation is likely to getting worse due to climate change. Indeed, Dhaka is the third most threatened city in the world by rising sea level in term of population [5]. Thus, Dhaka will be used as a case study in this paper. Indeed, South Asia, and Asia as a whole face important urban sprawl on flood-prone areas [4]. Moreover, the continent hosts half of the world urban population with no less than 28% of it residing in slums [6]. At the same time, according to a report from the World Bank [7], 38% of the world population lives in highly flood prone zones with 24% of them living in densely populated coastal areas. As a result most of the population and cities settled in flood-prone areas are in Asia which explains the fact that almost all[1] of the water-related disasters observed in the world between 1980 and 2006 are in this region [4].

Yet, this situation is anticipated to continue because of population growth and rising sea level [6, 7]. Indeed, South Asia is expected to double its urban population around 2030. However, most of this population is expected to settle in slums, leading to a massive slum growth likely to take place in flood-prone areas [6]. Thereby, the choice of Dhaka is also motivated by the fact that more than one third of its population lives in slums mostly located in flood-prone areas [2]. Indeed, in the world, urban poor are the most threatened by water-related disasters. This is due to the fact that most of slums are located in the riskiest areas where flood risks, especially prolonged flood are pairing with health hazard, contamination vulnerability, work disruption as well as increasing costs for basic needs, during flood periods, leading to the deterioration of life condition [8–10].

In addition, those slums are confronted to low construction quality combined to the important density in slums as well as to a lack of services, such as waste collection [7, 11]. For instance, in Dhaka no waste collection is made in slums due to access difficulties [12]. Moreover, this situation is likely to increase due to the fact that slums located on or along water bodies as well as on flood-prone areas tend to expand onto the water due to a lack of available space and a lesser risk of eviction [13, 14]. In addition, according to Arachchilage and Jayaratne [15] the level of service provision decreased for the settlements built along or on a water body.

Thus, this specific type of slums or part of a slum, located along the shores or on a water body, where exist a considerable correlation between flooding and living conditions will be defined as wetslums herein. However, a study made in 2015 [13] highlighted the fact that the dynamic and diverse nature of slums owing to their localization and environment is not taken into account in physical upgrading projects. For this paper, upgrading—more precisely physical upgrading focus on addressing physical issues—is defined as the implementation of functions in a part of slum to improve the level of services. The ultimate goal of such upgrading process is to match the service level existing in the formal part of the city surrounding that particular slum, i.e. integrate slums in cities.

[1] Indeed, between 1980 and 2006, respectively 97, 90 and 95% of the world population affected by floods, windstorms, waves and surges where living in this area [4].

As a consequence of the observations made by Olthuis et al. [13], wetslums, where the risks are the highest for the population due to their localization and environment, are less likely to be upgraded with the implementation of services owing to the fact that the 'classic' approach used in slum physical upgrading is not set up to cope with water-related disasters [10, 16]. Furthermore, these slums are quickly evolving and expanding while facing increasing disasters in areas with a shortage in available space [2]. In addition, this situation leads wetslum dwellers to turn to what Porio [9] calls a "*water-based lifestyle (p.438)*", where they get use to regular rise of contaminated water and adjust their daily life in accordance with no conscience about safety risks. Consequently, solutions that overcome this situation are needed and are likely to be small scale intervention, instead of large scale upgrading programs.

Therefore, the aim of this paper is to answer the following question: What approach may be proposed to favor small scale project able to improve wetslum dwellers' living conditions? To this end the identification of the specific issues faced by wetslums dwellers as well as the stakeholder's position will be necessary, taking Dhaka, Bangladesh as context. The end goal is to set up guidelines for future small scale project proposition, named City App, able to provide either technical (electricity, sewers, sanitation, water provision…) or social services (schools, clinics, recreational and cultural facilities…) on available spaces and in a short response time in those wetslums in order to improve the long-term living conditions.

2 Identifying Problems Faced by Wetslums Dwellers

Before being able to establish a program of demand for service upgrading in wetslums, it is necessary to define the current context and the specific issues faced by the urban poor living in flood-prone areas. Indeed, as reported by Olthuis et al. [13], the environment and the localization have to be considered in order to provide adapted upgrading projects. In wetslums, issues are mainly connected to floods. As a result it is necessary to understand the reasons of their construction and continuous development as well as the causes of the lack of services and the absence of improvement.

2.1 A Settling in Risky Areas Due to Land Shortage

Owing to a lack of available space in fast growing cities, people are forced to settle in high risk areas, especially poor, migrant, refugees and people searching work. According to Hassan [17], the apparition of this situation in those cities follows a recurrent pattern. First, land prices and urban pressure increase drastically and agricultural lands are sold to the poorest and the middle class to support the demand. As a result of this, there is a drastic reduction of available lands and a continuous land price increase forcing the poorest to move to dangerous ecological areas such as flood plains. However, once there, slums dwellers face eviction risk because cities and States regard these areas as too hazardous or develop infrastructural solutions to provide additional spaces for formal urbanization.

In Dhaka, slum population doubled between 1996 and 2006. In addition, the number of slum communities increased about 70% where the total population grew by 5% [3]. Furthermore, due to the process described by Hassan [17], most of these slums are developed in hazardous zones such as: extremely flood prone areas, open drain areas, dumb site or along the railways. Moreover, these settlements are mainly composed by low quality buildings making them highly vulnerable and fragile [16]. Thus, flood prone areas formerly wetlands, low-lying farms or even water bodies are turned into informal build-area, i.e. wetslums, at an unprecedented rate: 270% between 1975 and 2005 [2]. This phenomenon was mainly taking place between 1975 and 2000 [14]. However, a study from the Centre of Urban Studies et al. [18] demonstrates that population in those hazardous areas will increase more than ever before in the next decades. As a result, in Dhaka, wetslums' population is growing as well as their density and similar situations are observed in fast growing megacities around the world [10–13, 19].

Therefore, most of the slums located in fast growing megacities are developed in flood prone areas and their constant growth lead to an increasing density coupled to a sprawl on water bodies (Fig. 1), caused by a shortage in available land [13]. Because of this, wetslums are facing numerous water-related disasters, such as typhoons, floods, storm surges, heavy monsoon rain and rising sea level, which are, on account of climate change, more and more frequent and stronger [9]. In addition, the development of wetslums is pairing with the destruction and the reduction of cities' natural protection against floods [3]. As a result the main disaster encountered by wetslums dwellers is floods and more precisely long lasting floods [2]. Indeed, prolonged floods generated several health hazards, going from drinkable water con-

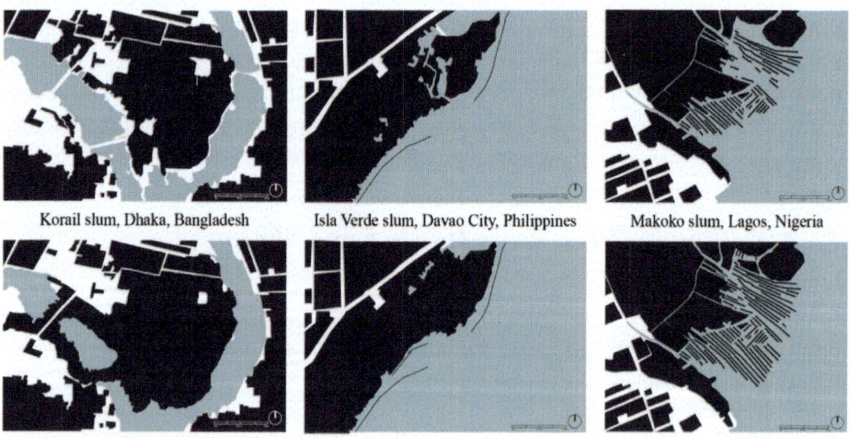

Fig. 1 Owing to a lack of space, slums located close to water bodies tend to expand on them. (Above: 2001; Below: 2014) [13]

tamination to mosquitos' infestation and this risk increase in accordance with flood duration which could last several months [8].

2.2 A Lack of Services that Worsen the Threats

Wetslums face an important scarcity of service provision that aggravate the already poor living conditions. For instance, wetslums' growing population increases waste productions which are accumulating onsite due to the absence of proper waste management and collection[2] [20]. Over and above, existing services usually do not integrate the environmental conditions and the local demand in their conception [21, 22]. Therefore, it could result in an aggravation of the risks during floods [23]. Indeed, sanitation and sewage can overflow when blocked by solid wastes; water taps may be contaminated and social services such as schools or clinics may not be accessible [24]. Moreover, these settlements are often excluded from public-sector resources which severely limit their access to basic services, such as drinking water, sanitation access, education and healthcare services [9, 25].

Furthermore, according to Adebayo and Iweka [26], slums upgrading projects generally do not integrate social services such as library, schools or community centers. However, such facilities have a strong impact in the long term, reducing illiteracy, unemployment and criminality as well as increasing human capital i.e. individual skills, knowledge and ability to work [27, 28]. Moreover, education is crucial to secure household's incomes and increase their safety because it provides better paid jobs and higher savings [25]. Thus, according to Braun and Aßheuer [2], schooling is as crucial as facilities preventing long lasting floods. Indeed, even a basic education is a key to develop financial capital which in result will improve access to other services and better building materials, allowing an improved ability to cope with floods.

Several reasons could explain the shortfall of services in wetslums. One of them is the cost. Indeed, developing specific structures in flood-prone areas is more expansive and requires an expertise in order to sustain floods and secure the investment [10]. In addition, most of the wetslums are seen as temporary by governments due to their localization which means that any investment could be wipe out at any time [13]. Furthermore, the constant growth in population and density lead to quick and constant evolution and expansion of the buildup area as well as a rapid under sizing of services provision [21].

Besides, another reason to this shortage is the fact that wetslums settlements are seen as 'illegal' and hazardous [22]. Because of this, urban poor basic rights to access services and to live safely are commonly denied while they contribute extensively

[2]As a matter of fact, waste collection issue is common to all slums and in a broader sense to all cities in the developing world [20]. For example in Dhaka, only half of the total wastes are collected [12].

to city's productivity and growth[3] [10]. On top of that, there is a perception that wetslums dwellers are incapable to pay for these services. Yet in most of the cities around the world, urban poor are paying more than the others [7]. Indeed, in the case of electricity provision in Dhaka, the prices are three time higher than the price for people with a legal access [23]. Moreover, before 2016, access to water was either controlled by *mastaan*,[4] through illegal connections, and sold at a high price or available for free from unreliable tap that were rarely functioning—sometimes just half an hour a day. Such situation is largely due to the absence of involvement from governments concerning services provision in wetslums [16]. Furthermore, in Bangladesh, low ranking officials collude with the *mastaans* [22]. Thanks to that support, these organizations will hampers any service provisions project in order to keep their business based on outrageous fees for services provision [29, 30]. Nevertheless, the implementation of a proper water supply system in Korail by Dhaka Water Supply and Sewerage Authority (DWASA) [30], proved that this issue can be overcome when, the government and the population support the project, and the *mastaans* are the first to benefit from it.

Thus, given the specific context of wetslums, the implementation of services cannot follow the pattern implemented in other slums upgrading projects [13]. At the same time planners have to find solutions to provide those services and connect wetslums to the rest of the city while proposing structures easily adaptable and able to answer the needs. According to Brillembourg and Klumpner [31], the best projects in these settlements are the one which are always in progress. However, such project should also be able to cope with evictions.

2.3 An Absence of Evolution Explained by the Existing Status Quo

Relatively speaking, the main risk face by wetslums dweller is not environmental hazards but eviction [7]. These evictions are officially motivated by the fact that it will reduce population's risk exposure. However, in several occasions, these evictions are made in order to turn the area in new districts or other urban projects [19]. In addition, according to Wendt [32], three other justifications are used by governments: clearing criminality hotspots, avoid health issues to spread to the rest of the city and improvement of city's attractiveness.

When there is no planned project for the evicted area, most of the wetslums dwellers re-build on the same localization, otherwise they move to another hazardous area [8]. For instance, in Dhaka, wetslums were first implemented on public owned

[3] Nevertheless, in Dhaka, slums dwellers could also face a barrier for job opportunity. Indeed, people living in slums are seen as unreliable because they are living in illegal settlements and to bypass this stigma they need an authorization from local-level leader to have a guarantor [22].

[4] According to Ahmed [16] mastaan are 'mafia' like organizations in Bangladesh that are working in slums built around political patronage and police support through corruption.

land, but after massive evictions in 2002, they are now settled on low-lying privately owned land destined for urban development [2]. Thus, according to the Centre of Urban Studies [18], about 7% of Dhaka's slums had faced eviction from their present location or were facing this threat. For Degert et al. [12], this threat is due to the refusal from Dhaka's government to recognize slums spreading added to a lack of legal representatives for slums dwellers. However, eviction is not the only approach that has been tried by Bangladeshi Government. Indeed,[5] resettlements, slums upgrading or even relocation approaches were experimented with various results [33].

Nonetheless, in poor countries such as Bangladesh, resettlements and relocations are actually unrealistic because of cost and shortage in available land [8]. Thus, in the case of Bangladeshi Government, the resulting approach is "*a* de facto *policy of either "doing nothing" or occasionally demolishing certain squatter settlements without any systematic plan (p.103)*" [25]. Moreover, wetslums dwellers are willing to move only if employment is provided; alternatively they will prefer the status quo even if loans and grant are granted. Additionally, wetslums dwellers not only prefers to stay where they are if recurrent flooding could be eliminated or reduced but also perceive no difference between flood-free areas and sectors facing annual flood [25]. Yet, evictions continue in Dhaka despite a High Court Division of the Supreme Court order, large scale protestations and the actions of the Coalition for Urban Poor (CUP) [8, 22].

Consequently, this insecurity of land tenure endure by wetslums dweller explain why stakeholders, including them, are reluctant to invest in these settlements [22], strengthening the status quo. Thus, no improvements are made to their houses to cope with floods. At the same time, investments from NGOs (Non-Governmental Organizations) are limited for population living in wetslums owing to eviction risk and urban land cost [16]. As a result, NGOs are hesitant to work in wetslums and rarely construct permanent facilities such as sanitation, schools and drainage [22]. Indeed, if an eviction occurs, it will result in financial loss, in regard of the invested capital as well as the investment in staff time and training,[6] and small NGOs do not have the capital to sustain it.

Therefore, rather than persisting in the status quo, there is much to gain by upgrading these wetslums [7]. Moreover, in Dhaka's case, upgrading slums instead of useless evictions appears to be more realistic [8]. Besides, such programs planned with the purpose to improve wetslums' life conditions and physical environment exist in the

---

[5]Concerning wetslums upgrading in Dhaka, the political will seems to change with the different governments. Indeed, different improvement programs were initiated by one government and the support of donor agencies at the beginning of the 2000s [8]. Then, eviction continued as reported by Ahmed [16] with the massive eviction that took place in Korail in April 2012. Finally, in 2016, Degert et al. [12] studied an upgrading program supported by another government at this same place.

[6]The second case study presented by Rashid [22] at the page 579, can be an explanation about this reluctance from NGOs. Indeed, this presentation of Agargaon's eviction showcases that Plan International lost all its investments: in running water and sanitation programs as well as healthcare and education ones, made in this 20 years old settlement.

city [23, 30]. Furthermore, Baker [7], report that "*studies show that slum dwellers gain more from slum upgrading than from relocation (p.77)*". On top of that, in a study taking place in Manila, Ballesteros [10], reports that there is an increasing involvement in favor of wetslums upgrading from local politician since population is willing to pay for the provided services. Ultimately, for Habib [34] and Degert et al. [12], the recognition by the government that slums settlements are part of the city where people live could allow dwellers helped by CBOs (Community Based Organization) and NGOs to manage themselves, initiating a quick improvement of the living conditions [30].

3 Establishing a Small Scale and Flexible Approach Towards Wetslum Upgrading

Now that wetslums' context and specificities are identified, the stakeholders' positions and demands toward wetslums upgrading is needed. Then, these information combine with the ones gathered above will be used to define a strategy and listing the constraints that a project located in a wetslum should have to address. Then, several guidelines will be set up for the future City App proposition.

3.1 Positions of the Stakeholders Involved in Wetslum Upgrading

In order to set up the proposed approach, it is necessary to identify the stakeholders involved in such programs as well as their position about it. Thus, several groups of stakeholders intervening in slum upgrading can be identified. In that sense, Alam et al. [29], pinpoint four groups of stakeholders:

- The key stakeholders who are leading programs implementation. This group includes the governments at the national or local level and the different governmental department and agencies [29, 34], in other words the political realm described by Boulding [35].
- The primary stakeholders who benefit from the programs. This group includes the wetslums dwellers, local committees and CBOs. CBOs and committees can be present at the settlement or at the city level such as Basti Basheer Odhikar Surakha Committee (Slum Dwellers Rights Protection Committee BOSC) in Dhaka, a slum dwellers committee organized at the city scale to discuss directly with local governments [36].
- The secondary stakeholders who are directly involved in programs implementation process. This group includes international and national NGOs, local agencies and

private development agency such as CARE [23] as well as international, national and local organizations.

- The tertiary stakeholders who support and advocate the work done by the others groups. This group includes international donor, funding agencies, international institutions and the private sector. Thus this group provides to key and secondary stakeholders funds, expertise, training and technical assistance for upgrading programs due to different success for such programs around the world [30, 37]. As a result, they may support the implementation of such projects but will not be directly involved in it.

Consequently, the proposal will be dedicated to upgrade the life condition of the primary stakeholders. Indeed, according to Rashid et al. [25], wetslum dwellers prefer to stay at their current location, mostly because of the existing social capital, which is quite strong in these settlements in addition to job proximity. Thus, they need access to basic services adapting to the demand and taking the population growth in account[7] [29]. Moreover, they need to be able to access functioning services during floods periods. However, even if these stakeholders are reliable, responsible and willing to pay to use services, they are nonetheless unlikely to participate in their acquisition [22]. According to Alam et al. [29], this is due to eviction risks and *mastaans* control over existing services which reduce even more their small resources. As a result, communities and CBOs are more likely to run and maintain a project than invest in it. Nevertheless, to do so, they will have to be involved in the process, through their consultation for instance [30, 34]. Hence, in order to maximize the impact of the proposal, these players should be involved in the proposal's realization and implementation process. Furthermore, there is neither accurate database nor real-time data yet necessary to know the type of service required and where it is needed [13]. Therefore, the involvement of the wetslums population, besides the benefit it brings, can provide these data.

Concerning the key stakeholders, a project proposition will have to adapt to their ambiguous vision toward wetslums upgrading. Indeed, as explain before, these settlements are seen as illegal and temporary [13]. Because of this and land ownership issues, bureaucratic regulations and approval systems are long and can be tedious as well as the negotiation to access the land needed to implement the projected services [38, 39]. Thus, the response time for new initiatives can be very long. Nevertheless, beside DWASA [30], different government bodies such as Dhaka City Corporation (DCC), Local Government Engineering Department (LGED) are involved in upgrading programs, with the technical and financial support from tertiary stakeholders [23, 34]. Moreover, Habib [34] reports, page 262, that "*The different government authorities have recently prioritised the need for slum upgrading, due to rapid urbanisation*

[7] Indeed, page 26, Alam et al. [28] report that upgrading projects are focused on "improving the current situation" and do not take into account the population growth. Because of this the urban poor see these interventions by key and secondary stakeholders as "unfruitful".

and the deterioration of law and order". Yet, these government bodies have insufficient financial and physical capacity to address services issues. In addition, even if some regulations were changed in order to allow the implementation of services in slums, government and land owners still have the right to evict these settlements [29]. As a result, governments are reluctant to invest in wetslums due to their temporary nature and the costs involved. This lead to the implementation of low cost programs with high maintenance costs [37]. Moreover, these programs are frequently inefficient, do not answer the population needs and could increase the local issues because of the absence of consideration about environmental risks [23]. Thus strengthen these stakeholders' reluctance to invest in such programs. Furthermore, this perception about wetslums temporary nature is even stronger for wetslums located in high value area owned by the government such as Korail. Indeed, this stakeholder will have substantial gain by selling it to let place to waterfront projects instead of upgrading the existing settlement [19, 40]. Over and above, according to Binte Razzak et al. [41] almost 80% of Dhaka's slums are located on privately owned land which implies additional cost for the government. However, this can change when a tertiary stakeholder showcase the benefit of investing in these areas[8] as well as by making it a condition of its involvement [30]. Thus, to match these stakeholders' demands, the proposition of a project should be authorized by local authorities, flood proof, low cost in its implementation and maintenance as well as match with their perception that wetslums are temporary or prove that it is necessary to improve the overall territory.

Regarding to the secondary stakeholders, it seems that they are the most likely to implement such equipment. Indeed, even if their involvement stays limited in Dhaka due to eviction risks, NGOs play the main role in services provision to wetslums [29, 41]. Thus, the land prices as well as the environmental and eviction risks that limit investment from these stakeholders in wetslums should be address by the proposal. Consequently, the proposition will have to be affordable, flood proof and be able to be preserved in case of eviction. Furthermore, NGOs are involved in different areas such as water and sanitation, health, education, child protection, skill-based training, income generating activities, environment and governance [23, 29], functions that the targeted projects will have to provide. However, the services provided by these stakeholders are implemented little by little without proper assessment, staffing and coordination between them [37, 41, 42]. In addition, most of the time the provided services do not address the wetslums dwellers needs, are frequently selective and the coverage is incomplete, with overlapping services [34, 41]. For Habib [34] and Alam et al. [29], most of this situation is due to stakeholders from the tertiary group: the foreign donors that finance them. Indeed, because of their financing, they have a word to say about the design and guideline of the activities made by the NGO, and their vision is usually not corresponding to the real needs. In addition, the NGOs activities are based on the short term and they could withdraw from their involvement in a

---

[8]Indeed, in the program lead by the Asian Development Bank and conducted by DWASA [30], the provision of water supply to slums, prevent the reoccurrence of technical and financial issues, unavoidable otherwise.

project when required by their donors [29]. Nevertheless, according to Rachid [22], due to the fact that primary stakeholders are willing to pay for the provided services and able to maintain them, certain NGOs can gain some independence from their donors thanks to the collection of fees for the services they provide. In addition, the initial cost can be recovered in few years [22]. Thanks to this result, the initial amount needed for such initiatives is backed by numerous tertiary players. Therefore, similar initiatives could encourage secondary stakeholders to provide the needed services in the long term while solving the coverage and overlapping issues. Furthermore, in order to collect fees through all the year, the provided service will need to be accessible even during flood periods.

3.2 Determine a Design Strategy for Wetslum Upgrading Projects with the Gathered Information

Now that the problems faced by wetslums and the position of the different stakeholders are identified, a strategy can be developed. To this end, it is necessary to determine the main objective of such projects, in other words identify their beneficiaries, their action and their purpose.

Thus, the main function of such equipment is to improve the living conditions through the provision of technical and social services to primary stakeholders, more precisely wetslums dwellers. Indeed, as reported before, there is an important lack of all kind of services in wetslums. Consequently, a project located in a wetslum will have to achieve its intended goal while addressing the different issues specific to wetslums that prevent such interventions nowadays. The latter are closely linked to the localization of these settlements—i.e. close to water, in flood prone areas or even on water, making them vulnerable to increasingly high and frequent floods and important health risks exacerbate by this lack of basic services.

In addition to flood risk, these settlements are overcrowded, dynamic in nature and face population growth while enduring a shortage of available land, driving to high land prices. As a result, beside the important cost to access to a plot, the guidelines set by NGOs cannot necessarily be achieved due to this lack of space and the need to preserve the facility from floods. For instance, NGOs such as WaterAid implement low cost services in high places to protect them from flooding, however they just consider the previous floods level to select the localization [29]. In such situations, the provided services are not necessarily where they are needed as well as still threatened by higher flood levels.

Besides that, the lack of available space leads the wetslums to grow onto water bodies [13] leading to the "*water-based lifestyle*" describes by Porio, page 438 [9]. As a result, with the wetslums' extension the services are increasingly distant from where they are needed. Therefore, the parts of a wetslum located and expanding along the shores or on a water bodies, faced a more important lack of services than other parts of the same wetslum, that are still affected by floods but that are located further away from this water body [15].

Furthermore, wetslums are seen as illegal settlement and perceived as temporary by key stakeholders, leading to a permanent risk of eviction. Moreover, the few existing wetslums service upgrading projects have neither long term vision nor proper assessment and investment, leading to low quality interventions expensive to maintain, vulnerable to high floods and lost in the case of eviction. For all these reasons, investing in wetslums is seen as risky, thus dissuading most of the stakeholders involved in upgrading programs.

Therefore, the resulting design strategy has to provide all kind of services close to where they are needed—i.e. along the shores or on a water body, while being able to cope with water-related disasters and continue to operate when such event occur. At the same time, it has to occupy a minimal space and integrate the dynamic nature of wetslums in order to fulfill its goal with a large impact. For these reasons, such approach should favor replicable proposition. Moreover, it has to be capable to handle eviction risk and thus make investment in wetslums appealing and cost effective for all stakeholders involved. However, to do so, it has to integrate the perception of the key stakeholders toward this type of settlements. In other words, the proposition has to be perceived as temporary. As a result, a project proposition in a wetslum has to be reusable and should be apt to relocate.

3.3 Developing the Design Strategy by Elaborating Guidelines

Hence, the resulting design strategy induces different constraints requirements (CR) that a project proposal, to improve living conditions in wetslums, would have to integrate. More precisely eight of them (Fig. 2): be **located** close to where it is needed (CR1) to provide the desired coverage; **flood-proof** (CR2) to cope with the environmental specificities of wetslums; **flexible** (CR3) in its design to adapt to the dynamic nature of such settlements; **movable** (CR4) for this same reason as well as to cope with eviction; **standardized** (CR5) to be easily replicable; **affordable** (CR6) to favor investment in these urban areas; **safe** (CR7) to protect the investment and the local population; **legal** (CR8) to facilitate its installation in a short response time. Furthermore, to guide the conception of such potential projects, it is necessary to detail these constraints requirements. Indeed, thanks to this specific solutions can be pinpointed and then incorporated in a future design strategy.

First of all, it must be located close to the population who will benefit from its services (CR1). However, its installation has to target the available spaces in wetslum. In addition, no destruction and no modifications of the settlement's structure have to be done and the rapid densification and expansion toward water bodies has to be taken into account. Indeed, according to the analysis of aerial images made by

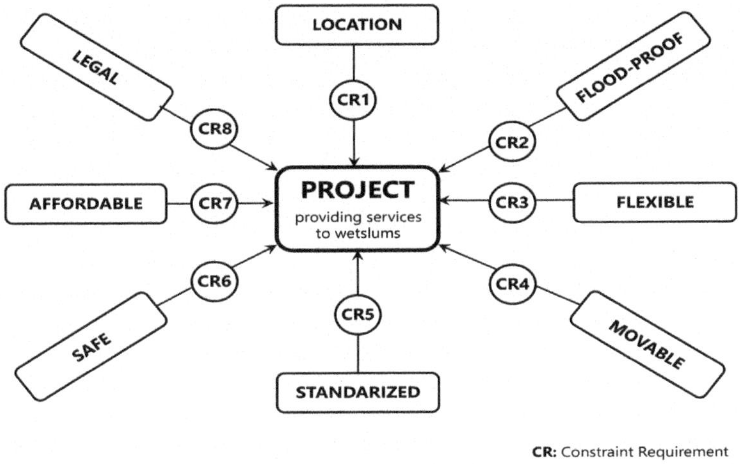

Fig. 2 Besides responding to the desired main function, a project proposition for wetslum has to address different constraint in order to be operational [by authors]

Koen et al. [13] for Korail slum in Dhaka, Makoko slum in Lagos and Isla Verde slum in Davos, water bodies are the only available spaces in wetslums. In addition, these spaces are the ones facing the most important lack of services due to their location [15]. Besides, as expressed by Olthuis and Keuning [43], the solution to space shortage is to build on water to provide more space without interfering with the existing functions.

Furthermore, such project has to be flood-proof (CR2) in order to allow it to sustain floods and different water levels, which are the environmental threat specific to wetslums, as well as to be installed on water bodies. To do so, stilts constructions or floating ones can be considered. However, the solution has to withstand all types of water-related disasters as well as the possibility of constant rising floods levels due to climate change. As a result, floating solution seems to be more fitted than construction on stilts. Indeed, the design proposals made by Altea [24] and Aman et al. [8] proposed floating design for wetslums, located on water bodies, owing to their better performance against these disasters.

In addition, the resulting system has to be flexible and adaptable (CR3) to host all kinds and scales of social and technical services. Moreover, its structure has to be easily transformable and replaceable by another one. For this reason, the different services should be plugged into a unique element that could be installed, transformed and replicated in a short response time as much as being easy to realize. At the same time, this element should allow the implementation of a combination of several similar or different services. Similarly, due to the shortage of existing technical services, the system has to be autonomous utilizing onsite conditions to achieve this. It should also use local materials and knowledge for its realization, in order to involve local population. Thus, it should be a modular solution based on prefabricates components while involving wetslums dwellers and use local input. For these reasons, the combination of several elements or structures should be considered.

Then, it has to be movable (CR4) to address eviction risks as well as its relocation to a more suitable plot when wetslums are expanding. This confirms the selection of a floating solution. Indeed, construction on stilts cannot be easily moved. Moreover, it has to be able to relocate to another wetslum. Under those circumstances, the system or the element that host the service preferably has to be transportable whole and intact. In that case, its weight and size has to be restricted to favor its portability and the system or the element containing the service should allow its shipping to the desired localization from anywhere.

Hence, the proposal has to be standardized and easy to reproduce (CR5) while offering a sufficient volume to host different types of services. Additionally, the system should use preexisting and standardized elements to favor its portability and reduce its cost. More specifically, these elements have to use existing reusable components, compatible to the existing infrastructures and transportation systems without any modifications, while being easy to combine together. Because of this, the elements have to allow modifications and transformations without interfering with their initial structure to avoid their dismantlement for transport. Thus, the resulting system may be composed of several elements, but the one containing the function should meet all the constraints listed above. As a result, refurbished shipping container appears to be the best match for this requirement of standardization as well as the ones of flexibility (CR3) and transportability (CR4).

Regarding to the safety of this kind of equipment (CR6), it has to preserve the utilities from degradation and water-related disasters as well as allow the system to operate normally when a flood occur. Here again, the selection of a floating solution (CR2) is appealing. In addition, the proposition should provide a shelter and a safe area for its users at any time and particularly during floods periods. Furthermore, it has to be robust to sustain handling during transportation and installation as much as giving it a long lifespan. For this reason, the technical elements composing the different services have to use proven technologies to be resistant and able to work in all conditions.

Concerning the cost, the proposed project has to be affordable (CR7) to match with the demands of the different stakeholders as well as to make it plausible. Here, all the previous constraints serve this objective. Indeed, the flood-proof ability (CR2) preserve the system from water-related disasters and allow it to operate normally throughout the year (CR6), the flexibility (CR3) allow the installation of different function that could be combined and are movable (CR4), thus coping with eviction risks while using standardized (CR5), reusable and robust elements design to have a long lifespan (CR6). Moreover, the proposition should have to use material and components easily available, recycled and mass produced with minimal modifications or not at all. Furthermore, its maintenance has to be minimal and easy as well as its installation and its construction to reduce the costs. At the same time, it has to encourage the local population to participate in its construction and management while leaving them the possibility to own it. In fact, one possible prospect is to make such system available through short term renting to encourage wetslums entrepreneurs and CBOs that cannot buy it to operate it thanks to renewable short term contracts. These contracts have to be flexible allowing them to easily renew or terminate them. Moreover, to preserve their saving in case of eviction, the contracts are canceled—i.e. they do not have to pay the rent. Ultimately, the overall design of such project should be easy to reproduce or even allow the possibility of its mass-production, at least in theory, allowing cost reduction and higher adaptability for service provision in wetslums in the long term.

Finally, these designs will have to follow different rules and regulations (CR8) to be authorized by the key stakeholders. To this end, its installation in the targeted wetslum has to be accepted and validated by local authorities and must be in compliance with the local regulations and laws regarding what could be applied for it. The purpose is to allow a common validation for the one standard system in the entire city allowing its deployment in a short response time. Moreover, the system or the transportable elements must be certified according to the international transport regulations. This last point supports the selection of refurbished containers.

4 Conclusion

Dhaka, Bangladesh is, like a large part of the megalopolis in the developing world and more precisely like a large part of Asian megalopolis, facing a fast urban sprawl on flood-prone areas. Such urban sprawl leads to the development of a specific type of slums, where a considerable correlation between flooding and living condition exist: wetslums. The specificities of these settlements, defined as wetslums in this article, shown that their establishment in areas vulnerable to water related hazard is due to a shortage in available land that occur in these fast expanding megalopolis. Moreover, because of the important population growth coupled with eviction risks, these wetslums tend to expand on water bodies, leading the dwellers to turn to a "*water-based lifestyle (p.438)*" [9] where they faced increasing floods event and intensity and health hazard.

In addition, there is an important lack of technical and social services in these buildup areas owing to floods and eviction risks that explain, for the most part, why just few investments are made by the stakeholders involved in these settlements. Indeed, the costs for implementing a service able to deal with floods in these areas are higher than in other ones, more protected. As a result, such costs are considered too high for settlements viewed as temporary and illegal by governments. Besides, the insecurity of tenure caused by eviction risks means that any investment made in wetslums can be wipe out at any time when the land owner, that can be private or public decide to retrieve its land. Additionally, even when services are implemented in wetslums, they are frequently inefficient, do not address population's needs and do not take into account environmental hazards. As a result, even if the position of the different stakeholders is different toward wetslums upgrading, they share a lack of motivation to invest in these areas. However, projects designed to improve onsite living conditions are considered by different studies [7, 8] as the most efficient solution to achieve this goal and dwellers are willing to pay for the provided services while being quite effective in maintaining and securing it.

Therefore, the proposed approach for a service upgrading project is to improve the living conditions in these settlements through the provision of technical and/or social services to the primary stakeholders by encouraging investment in it. The inclusion of social services, rarely integrated in slum upgrading programs, is motivated by the fact that they have a strong impact on the community's future. However, to achieve these goals, the future design proposition should address the eight constraints requirement identified in this paper to convince future investors. These constraints requirements are the following ones: the localization of the services which have to occupy available space; its flood proof ability to deal with floods and water-related hazards; its flexibility to answer the different local needs; its movability to cope with eviction risks and settlement's expansion; and its standardization, affordability, safety and legality to attract investment. Furthermore, attracting investment can then facilitate the implementation and the development several similar equipment, in a short response time, to any wetslums. To go further, this program of demands could also be used to design replicable structure that can host all the desired services.

Thereby, the result of the guidelines proposed in this paper is a system providing all types of services in wetslum, located on the shores or on water bodies, where the service provisions are the most lacking, (CR1) and thus build on a floating support (CR2) that sustain water-related disasters and allow it to run throughout the year (CR6) while permitting it to move to a new plot if requested (CR4). For this reasons, it has to cope with the local floating or boat regulation (CR8). In addition, it is a modular and flexible system (CR3) based on prefabricate, standardized, transportable and reusable components (CR4, CR5 and CR6). Consequently, in order to comply with all the demands, such system is likely to be composed by two elements: one element that host the requested service and another that constitute the buoyant part.

As a result, an overall concept consisting on providing plug & play floating services without interfering with the existing urban fabric, authorizing fast deployment, upgrade or modification of the requested service in order to respond to the specific needs of a targeted wetslum, as long as needed, may be the end goal of these guidelines. Therefore, the future development of such concept, named City App, will be developed with the approach and the guidelines established in this article.

References

1. United Nations. (2015). *World Urbanization Prospects: The 2014 Revision*. New York, USA: United Nations, Department of Economic and Social Affairs, Population Division.
2. Braun, B., & Aßheuer, T. (2011). Floods in megacity environments: Vulnerability and coping strategies of slum dwellers in Dhaka/Bangladesh. *Natural Hazards, 58,* 771–787.
3. Dewan, A., & Yamaguchi, Y. (2009). Using remote sensing and GIS to detect and monitor land use and land cover change in Dhaka Metropolitan of Bangladesh during 1960–2005. *Environmental Monitoring and Assessment, 150,* 237–249.

4. Adikari, Y., & Yoshitani, J. (2009). *Global Trends in Water-Related Disasters: An insight for policymakers*. International Centre for Water Hazard and Risk Management (ICHARM), Paris: France.
5. Strauss, B. H., Kulp, S., & Levermann, A. (2015). *Mapping choices: Carbon, climate, and rising seas*. Princeton, USA: Our Global Legacy. Climate Central Research Report.
6. UN HABITAT. (2016). PSUP team Nairobi. *Slum Alamnac 2015–2016: Tracking improvement in the lives of slum dwellers.* UNON: Publishing Service Station, Nairobi, Kenya, 2016.
7. Baker, J. L. (2012). *Climate change, disaster risk, and the urban poor. Cities building resilience for a changing world.* Washington D.C., USA: Urban Development Series, The World Bank.
8. Aman, J., Rahman, N. N., & Zahir, S. (2016). 'Beltola Lilies'—A solution of housing for lower income people and introduction of a module for flood areas. *Creative Space, 3*(2), 119–131.
9. Porio, E. (2011). Vulnerability, adaptation, and resilience to floods and climate change-related risks among marginal, Riverine communities in metro Manila. *Asian Journal of Social Science, 39*(4), 425–445.
10. Ballesteros, M. M. (2010). *Linking poverty and the environment: Evidence from slums in Philippine cities.* Philippine Institute for Development Studies, Discussion Paper Series NO. 2010–33. Makati City, Philippines.
11. Baker, J. L., Gadgil, G. U. (2017). *East Asia and Pacific Cities: Expanding opportunities for the urban poor.* Washington D.C., USA: Urban Development Series, World Bank.
12. Degert, I., Parikh, P., & Kabir, R. (2016). Sustainability assessment of a slum upgrading intervention in Bangladesh. *Cities, 56,* 63–73.
13. Olthuis, K., Benni, J., Eichwede, K., & Zevenbergen, C. (2015). Slum upgrading: Assesing the importance of location and a plea for a spatial approach. *Habitat International, 50,* 270–288.
14. Adikari, Y., Osti, R., & Noro, T. (2010). Flood-related disaster vulnerability: An impending crisis of megacities in Asia. *Journal of Flood Risk Management, 3*(3), 185–191.
15. Arachchilage, K., & Jayaratne, K. (2013). *Reducing the Vulnerability of the Urban Poor to Climatic Change: Experiences from Colombo and Dhaka.* The Faculty of Graduate and Postdoctoral Studies (Architecture), University of British Colombia, Vancouver, Canada.
16. Ahmed, I. (2014). Factors in building resilience in urban slums of Dhaka, Bangladesh. *Procedia Economics and Finance, 18,* 745–753.
17. Hassan, A. (1998). The informal city. *UNCHS (Habitat) Regional Symposium on Urban Poverty in Asia.* Fukuoka, Japan, October 27–29, 1998.
18. Centre for Urban Studies (CUS), National Institute of Population Research and Training (NIPORT) and MEASURE Evaluation. (2006). *Slums of urban Bangladesh: Mapping and census, 2005.* Dhaka, Bangladesh and Chapel Hill, USA.
19. Adelekam, I. O. (2010). Vulnerability of poor urban coastal communities to climate change in Lagos Nigeria. *Environment and Urbanization, 22*(2), 433–450.
20. Hasan, S., & Mulamoottil, G. (1994). Environmental problems of Dhaka city: A study of mismanagement. *Cities, 11*(3), 195–200.
21. McFarlane, C. (2008). Sanitation in Mumbai's informal settlement: state, 'slum' and infrastructure. *Environment and Planning A, 40*(1), 88–107.
22. Rashid, S. F. (2009). Strategies to reduce exclusion among populations. *Journal of Health, Population and Nutrition, 27*(4), 574–586.
23. Chowdhury, F. J., & Amin, A. N. (2006). Environmental assessment in slum improvement programs: Some evidence from a study on infrastructure projects in two Dhaka slums. *Environmental Impact Assessment Review, 26*(6), 530–552.
24. Altea, G. J. (2013). *Urban acupuncture & dwelling formability, regeneration of coastal informal settlement in Manilla.* Thesis under the supervision of Miller, D.; Prakash, V. and Hou, J., Washington University, St. Louis, USA.
25. Rashid, H., Hunt, L. M., & Haider, W. (2007). Urban flood problems in Dhaka, Bangladesh: Slum resident's choices for relocation to flood-free areas. *Environmental Management, 40,* 95–104.
26. Adebayo, A. K., & Iweka, A. C. (2014). Sustainable infrastructure upgrade in slum settlements of Lagos, Nigeria: The role of the architect. *Environmental Research, Engineering and Management, 2*(68), 53–60.

27. Rupa, C. K. (2015). *Urban informality and the need for spatial.* Housing and Urbanism Term-2, Housing and the Informal City, Architectural Association School of Architecture, London, UK.
28. Khan, A. I., & Seeley, J. (2005). *Making a Living: The livelihoods of the rural poor in Bengladesh*. Dhaka, Bangladesh: University Press Limited.
29. Alam, S. S., Alam, A. J., & Rahman, S. (2015). *Urban climate resilience, water and sanitation. Improving multi-stakeholders collaboration in Dhaka, Bangladesh,* Asian Cities Climate Resilience Working Paper Series 25:2015, IIED, London, UK.
30. Sharma, M., & Alipalo, M. (2017). *The Dhaka water service turnaround*. Manila, Philippines: Asian Development Bank.
31. Brillembourg, A., & Klumpner, H. (2013). Smart option for the informal city. *The Social Design Public Action Reader; SLUM Lab Magazine*, vol. 8, pp. 101–105.
32. Wendt, S. (1997). *Slum and squatter settlements in Dhaka: A study of consolidation processes in Dhaka's low income settlement*. Thesis under the supervision of Martinussen J. and Simonsen K., Roskilde University, Roskilde, Denmark.
33. Mohit, M. A. (2012). Bastee settlements of Dhaka City, Bangladesh: A review of policy approaches and challenges ahead. *Procedia—Social and Behavioral Sciences, 36,* 611–622.
34. Habib, E. (2009). The role of government and NGOs in slum development: the case of Dhaka City. *Development in Practice, 19*(2), 259–265.
35. Boulding, K. E. (1970). *A primer on social dynamic*. New-York, USA: The Free Press.
36. Banks, N. (2008). A tale of two wards: Political participation and the urban poor in Dhaka City. *Environment and Urbanization, 20*(2), 361–376.
37. Raju, A. A., & Rehan, T. (2018). On site upgrading of slum dwellers: Problems and prospects (A case on Railway Slum of Khulna City). *ResearchGate*. November 2016. Available at https://www.researchgate.net/publication/310480370. Accessed April 17, 2018
38. Rahman, M. M. (2002). Problems of the NGOs in housing the urban poor in Bangladesh. *Habitat International, 26,* 433–451.
39. Siddique, Q. I., Alam, A., Rahman, A., $ Jahan, H. (2000). *Case study on a slum improvement project in Dhaka Metropolitan City*. Local Government Engineering Department. Ministry of Local Government, Rural Development and Co-operatives, Dhaka, Bangladesh.
40. Alam, S. S., & Matsuyuki, M. (2017). Applicability of land sharing scheme to Korail Slum, Dhaka, Bangladesh. *Urban and Regional Planning Review, 4,* 151–167.
41. Binte Razzak, N. R., Chowdhury, S., Ohi, S. (2014). Assessment of essential public services in slums of. *Current Advance in Civil Engineering (CACE), 2,* 126–132.
42. Baker, J. L. (2008). *Urban Poverty: A Global View.* Urban Papers, World Bank, Washington D.C., USA.
43. Olthius, K., & Keuning, D. (2010). *Float!: Building on water to combat urban congestion and climate change*. Amsterdam, The Netherlands: Frame.

Repurposing Jack-ups, Semi-submersibles and Superbarges into Offshore and Nearshore Settlements

Joseph Lim

Abstract This research study is a collection of ideas proposing offshore and nearshore settlements responding to emergent need in the light of climate change. They are counterpoints to megacities with ecological footprints that are unsustainable. The continued depletion of natural resources leading to scarcity has resulted in displaced communities and prompted these research questions: *What if we floated on sea instead of consuming land inefficiently? And could we use wave energy instead of nuclear energy? Could we replenish food supply and regenerate marine eco-diversity? How would our lives be shaped by new offshore settlements? What would we use as structures for shelter, farming scaffold and recreation?* Floating cities emerged in the 1960s with Buckminster Fuller's Triton City and Kenzo Tange's Tokyo Bay Plan. Current manifestations include Vincent Callebaut's Lilypad, the Seasteading Institute and the mile long Freedom Ship housing 50,000 people. As an alternative to these examples, three types of vessels in the marine industry, namely, the jack-up platforms, the semi-submersibles and the superbarges are repurposed as small footprint habitable propositions to accommodate 20% of a projected global population of 8.1 billion people in 2050. Floating settlements are spatially conceived with food and energy estimates for housing, recreation, education at sea, post-disaster healthcare and resettlement for nearshore deployment.

Keywords Repurposed oil rigs · Floating settlements · Sustainable · Offshore · Nearshore

---

Research Team: Christopher Wijatno, Davis Wong, Chen Qisen, Roy Tay, Wang Yigeng, Sakinah Halim and Bek Tai Keng.

J. Lim (✉)
Department of Architecture, School of Design and Environment, National University of Singapore, 4 Architecture Drive, Singapore 117566, Singapore
e-mail: akilimem@nus.edu.sg

C. M. Wang et al. (eds.), *WCFS2019*, Lecture Notes in Civil Engineering 41,
https://doi.org/10.1007/978-981-13-8743-2_2

1 Introduction

The paper explores the idea of repurposing oil rigs and barges in the context of nearshore and offshore marine settlements. It suggests developmental options to cold stacking or decommissioning oil rigs with serviceable physical structures, or the immediate deployment of recently constructed oil rigs in a depressed market when operational costs will not yield returns.

The study identified emergent need in the context of climate change and rising global population, and proposed solutions to resilient forms of food production and supply, energy and waste water management by converting jack-ups, semi-submersibles and barges into specialised vessels. Specialised vessels were designed for accommodation, food production, waste management and medical care. These were then grouped into flotillas for post-disaster relief or 250,000 person marine settlement fleets. These autonomous settlement fleets which occupied 38 km^2 of sea space are sea-cities and an alternative to land-based mega cities accommodating 100,000 per square kilometer.

The study estimated that 1.6 billion people in the year 2050 can be accommodated in 6510 settlements of oil rig structures, spaced 240 km^2 apart over 54.25 million square meters of water. The space standards of each marine inhabitant—at 50 m^2 recreational area per person—exceeds 27 m^2 in London, is equivalent to Amsterdam with one settlement type reaching 115 m^2 per person equivalent to Vienna. Singapore by comparison, has 65 m^2 of recreational area per person.

2 Economic Cycles and Challenges to Mitigating Environmental Pollution from Disused Oil Wells and Rigs

This section introduces the economic backdrop to surplus oil rig production and the environmental concerns over decades of oil exploration, the accumulation of disused structures, the technical challenges to the decommissioning unused rigs and the uncertainty of environmental impact after decommissioning.

Oil rigs are decommissioned in sustained periods of falling oil price making drilling operations unviable. Time lags between supply and demand create large oil rig surpluses. For example in 2010–2011, worldwide orders for rigs tripled with one-year long waiting times for available oil rigs. However, oil rigs enter an oversupply market by the time their construction is completed over 3-year periods. Bloomberg Intelligence estimated $65 billion worth of offshore rigs still under construction in 2016 in a market with idling rigs. By 2016, oil prices declined sharply as China, Russia, India experienced a slowdown in economic growth, and North America extracted shale oil in response to the 2000–2008 spikes with the Middle East sustaining its supply volume. Rigs without contracts to drill are either "cold-stacked" (anchored without crew) to wait for a market recovery [5], warm-stacked with motors running or sold for demolition. Cold Stacking is estimated at $15,000 a day whilst warm stacking costs $40,000 a day. Because rigs are not designed for idling, operators prefer to bear the costs of warm stacking and not risk machines requiring total replacement over indeterminate periods of disuse.

With North Sea hydrocarbon reserves depleting, several hundreds of oil and gas rigs are about to expire their productive life cycles and are costing more to operate. The low oil price compounding the economic downturn results in one-third of oil fields operating at a loss [1].

Political uncertainty on the matter of Britain and the EU also has an impact on North Sea Operations, which in 2000 was one of the world's largest sources of oil, yielding six million barrels a day. At only 1.5 million barrels now, an estimated 600 production platforms in the North Sea are due to be decommissioned. The British sector alone comprises 470 platforms and as many other offshore installations, 10,000 km of pipelines and 5000 wells [2]. The British industry anticipates more than 200 decommissions until 2025.

The decommissioning of offshore oil and gas installations is only in pioneering stage in the North Sea, Australia and New Zealand where technological, environmental, regulatory, political and financial considerations are evolving. Decommissioning both steel and early concrete rigs are complex and costly operations.

3 Decommissioning Costs

An average of 120 offshore rigs are decommissioned annually worldwide. The task at hand is a total of 2000 aging offshore oil and gas platforms, subsea wells and related assets. Offshore oil drilling at Gulf of Mexico is estimated at $26 billion. Future closures in the UK continental shelf alone, is estimated at £50 billion. IHS Markit predicts that decommissioning expenses will increase from $2.4 billion in 2015 to $13 billion a year by 2040 [3]. Europe alone accounts for half of global decommissioning spending, with its North Sea installations.

The 1998 Convention for the Protection of the Marine Environment of the North-east Atlantic (OSPAR) prohibits the dumping of oil installations and requires operators to restore the healthy condition of the marine environment on vacating the site. In this respect, the total removal of 470 platforms, 5,000 wells, 10,000 km of pipelines and 40,000 concrete blocks will cost £17.6 billion between 2016 and 2025 [3].

Of concern is the financing of decommissioning costs. Whilst major oil companies may have previously set aside funds from the revenue of new projects to defray the decommissioning costs of older rigs, the low oil prices today do not allow much budget flexibility. At the same time, governments need to protect their budgets from increasing costs of decommissioning and it is necessary to structure alternative sources of funding and legislation to protect the environment and the North Sea economy [2].

4 Case for Repurposing

It is estimated that over 60% of the jack-up rigs available will be over 30 years old by 2025 [4]. The immense decommissioning costs pose complex challenges. The impetus for rig owners' to renew their fleets is affected by overall industry performance and projection. In an effort to reduce the gap between supply and demand rig owners are expected to scrap the excessive supply of rigs. Thus it would increase the day rates of operators relieving the economic pressure on rig owners running at a loss.

Rigs recently constructed but not deployed due to cancelled orders run the risks of mechanical failure when being "cold-stacked" over protracted periods of disuse. When rig demand is interrupted or orders cancelled due to oil pricing fluctuations, it becomes feasible to produce newbuilds with civilian applications and/or to refurbish rigs with serviceable structures for business opportunities outside the oil industry.

4.1 Physical Limits to Additional Structure

The extent to which an oil rig can support an additional number of floors affects the viability of a specialised vessel to accommodate its intended usage with usable floor area. For a jack-up rig the platform will be lowered to maximise the mast heights—100 m—to accommodate 25 storeys at 35 m depth of water, as illustrated in Fig. 1.

The following estimates were made with a Prefabricated Prefinished Construction (PPVC) system of hybrid steel-concrete cells each weighing 16.5 tons. If a 3-legged jack-up rig can carry up to 15,000 tons and the additional load is 81,770 tons then additional columns welded to the masts will be required, including additional foundations into the seabed. Additional structural reinforcements are necessary in cases where new loading exceeds structural limits of existing rig structure as illustrated in Fig. 2.

The use of lightweight material for modular construction—as shown in Fig. 3—of habitable units would reduce new deadweight and save construction time with the possibility of disassembly and relocation for deployability. This feature is crucial for cost recovery and flexibility in usage over the lifecycle of the repurposed structure.

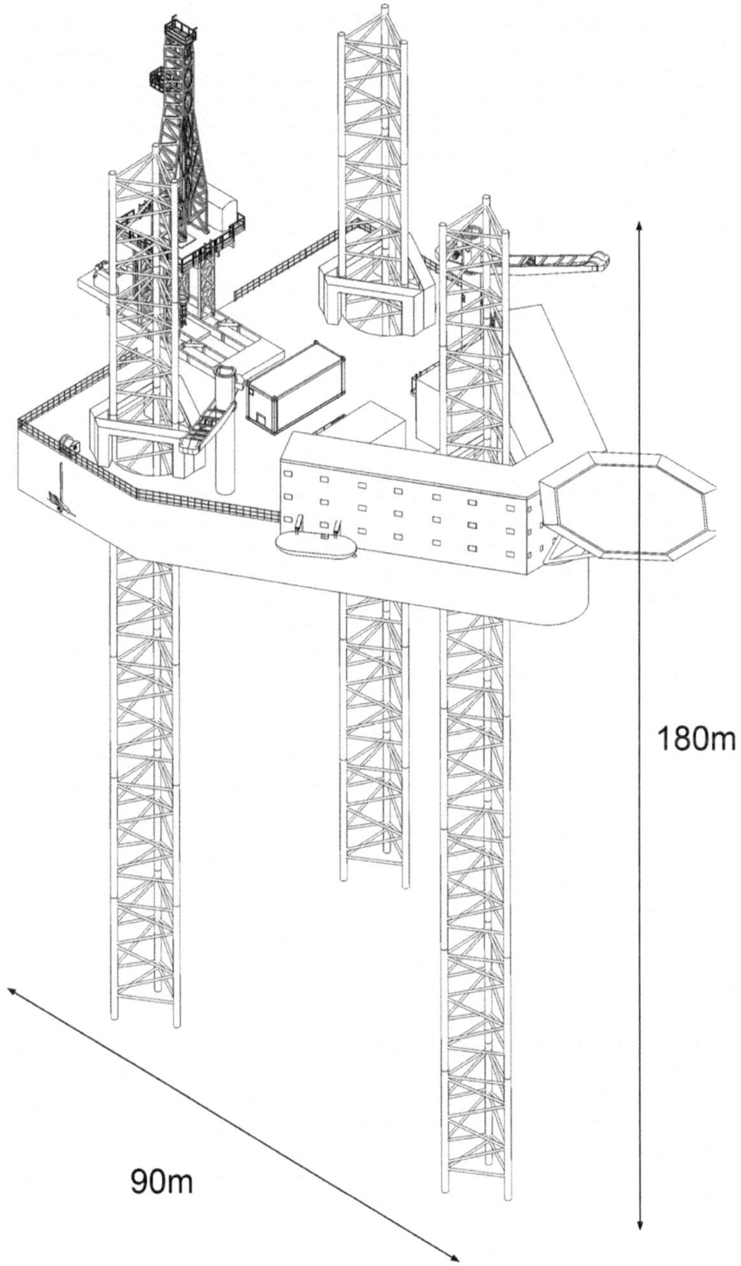

Fig. 1 Jack-up structure

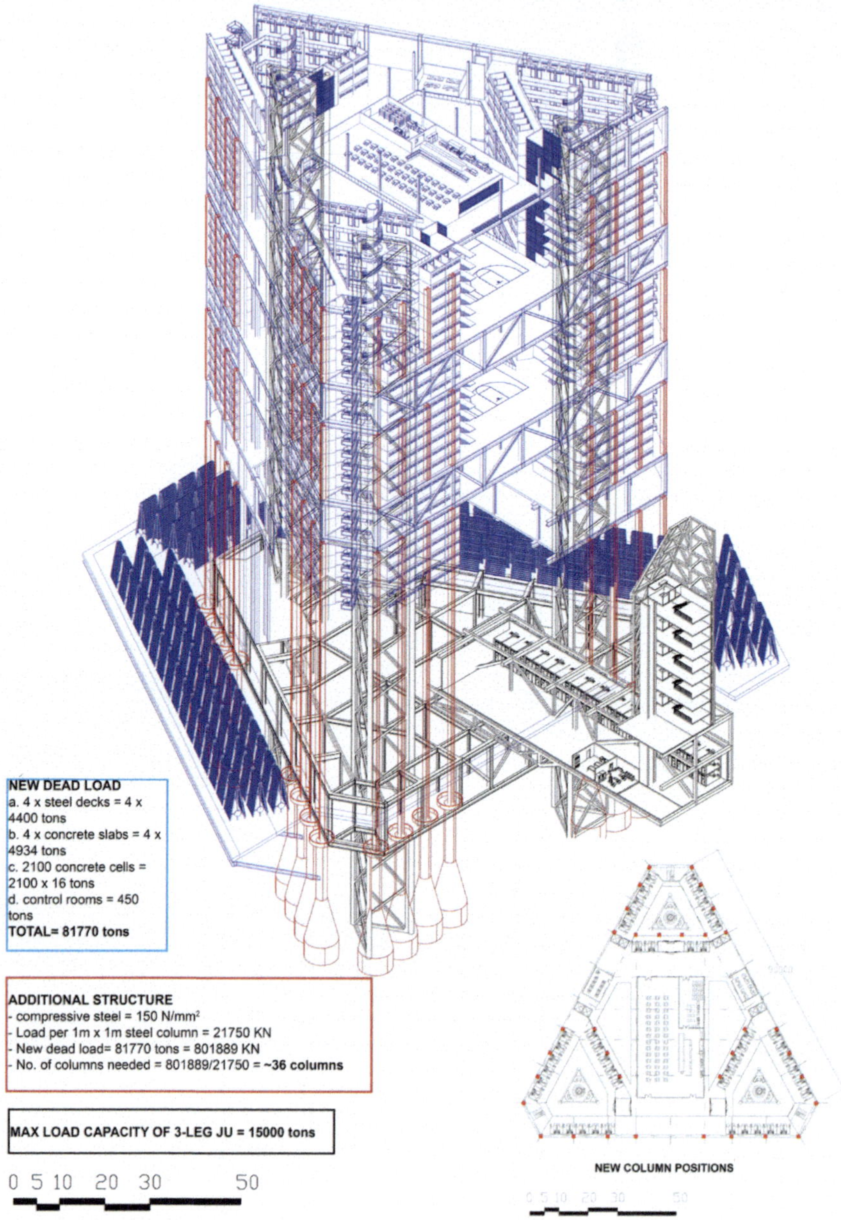

Fig. 2 Structural considerations

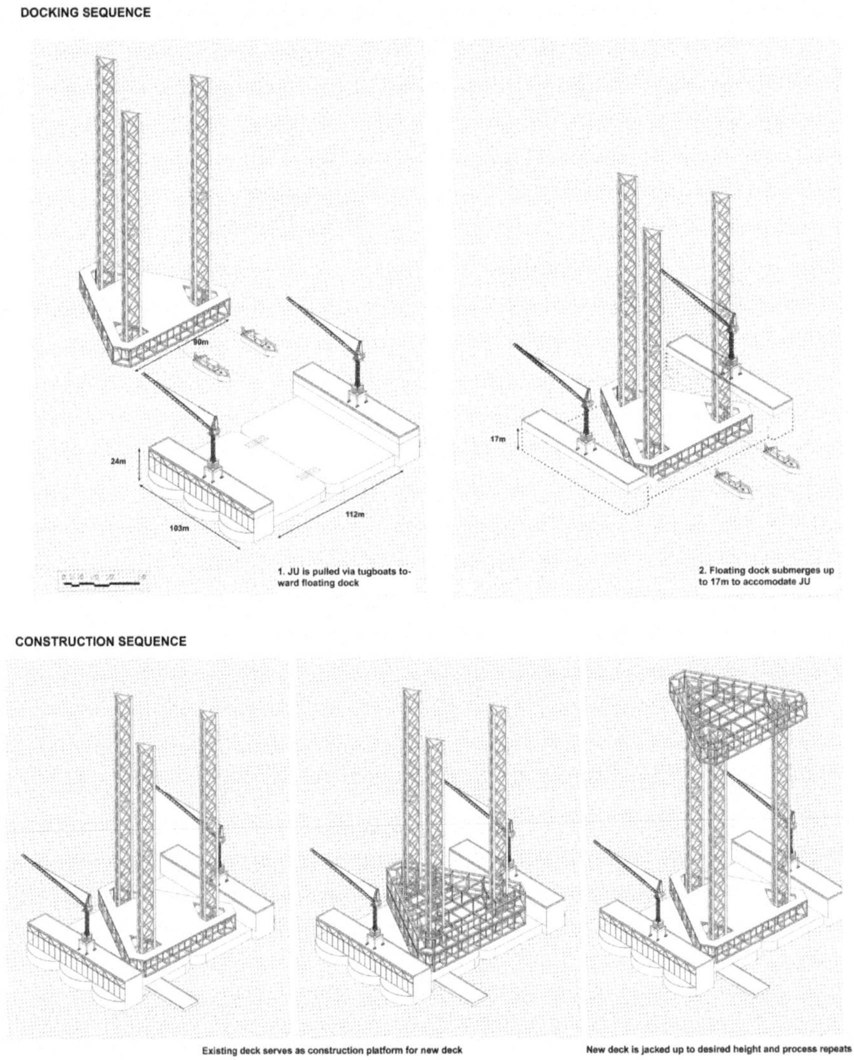

Fig. 3 Docking and construction sequence

4.2 Buoyancy and Toppling Estimates

For a semisubmersible, the limit to the number of storeys is constrained by flotation level and toppling in the transverse axis of the twin pontoons. Rigs designed for operating in harsh environments are heavier than those built for moderate environments. Harsh environment semisubmersibles have longer legs to maintain the air gap, and which increases the distance between legs and the pontoon size. Semi-submersibles

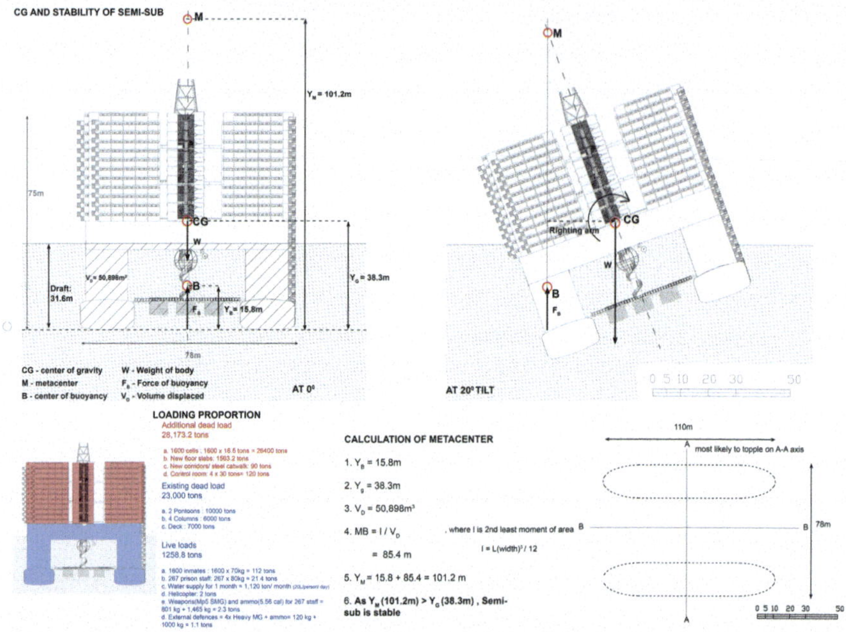

Fig. 4 Projected buoyancy and toppling estimates

built for harsh environments have longer and thicker columns than moderate environment units [6].

As illustrated in Figs. 4 and 5, when a semisubmersible structure that can carry 28,173.2 tonnes of LL and DL is subject to another 28,000 tons equivalent to twelve storeys of PPVC cells, the metacenter (M) is still higher than the center of gravity (CG), which means that the semi-submersible will not topple. However, approximately half the height of the semi-submersible will be underwater. Two countermeasures are possible, one is to significantly reduce the weight of the PPVC cell by using lightweight reinforced plastics or engineered wood including CLT LVL. The second countermeasure is to enlarge the pontoon dimensions of newbuilds to increase its displacement. Excluding the engine rooms of semisubmersible propulsion systems, which are retained, the tanks in pontoon hulls and legs may be freed up for use.

4.3 *Scale of Repurposing Oil Rigs*

Owing to costs involved in stabilizing additional loads of new structure, private investors are conservative in refurbishing individual oil rigs. For example, the Seaventures Centre is a 25-room hotel and diving school built onto the structure of an oil

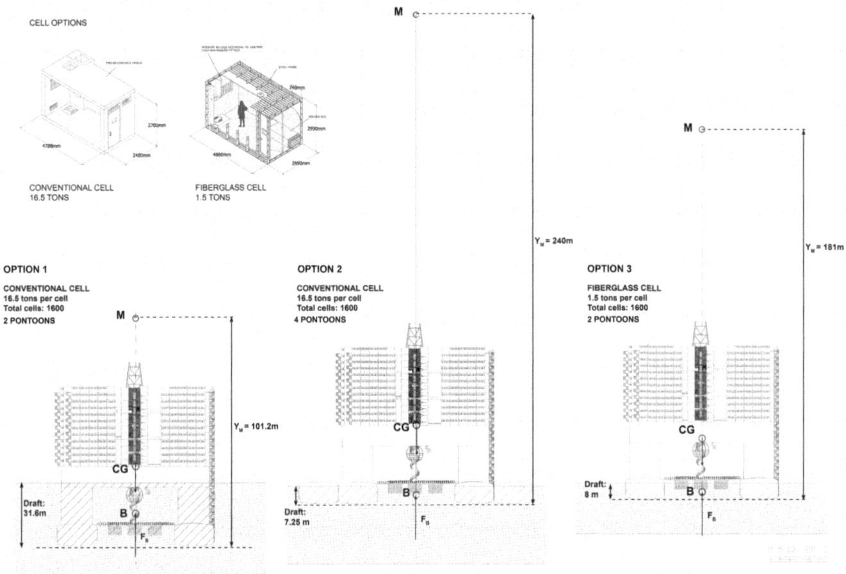

Fig. 5 Strategies to reduce draft in an example of a PPVC cell

rig near Sipadan, Borneo coast. Globally, there are plans of converting old rigs into platforms for other usages such as wind turbines, data storage centers and luxury homes. Most of these plans are scoped as individual single owner/operator entities with viable returns and operational costs where risks are manageable.

This study, however, argues that the mitigation of wastage and environmental risk in the oil rig construction industry cannot be sustained by individual installations and isolated investments. Instead, a wider agenda of accommodating global population growth on autonomous marine settlements demands a larger scale of investigation.

Thus design enquiries were based on strategies to replenish and regenerate marine biodiversity. The idea was for a settlement to be planned in ways that would not deplete food supply or natural resources at sea in pursuit of recreation and knowledge production with increasing population. Such settlements would take the form of a fleet of repurposed oil rigs forming entire marine settlements.

5 Global Challenges

It is expected that by 2050, more that 80 percent of 8.1 billion people on earth will live in cities, posing immense challenge to social and environmental sustainability. It is questionable that cities remain as models of sustainable development with unequal access to, and inefficient use of resources. It is critical to develop alternatives to megacities in order to avoid the depletion of natural resources and mitigate the damage to precious ecosystems remaining. Climate Central predicts that with a four degrees Celsius increase in temperature and a median global rise of 8.9 m by the end of this century, the dwellings of 627 million people will be affected.

Of concern is the vulnerability of coastal cities to sea level rise. This has prompted a rethink of coastal cities in the form of autonomous floating settlements, in terms of food, energy, housing and recreation in an environmentally sustainable way. The relentless extraction and commodification of natural resources will result in large ecological footprints.

World Health Organization cites fish as the primary source of protein for more than one billion people. Global fish catches are declining at rates which suggest overfishing. Such phenomenon occurs when fishes from the sea are caught at faster rates than they can be replenished by natural reproduction. Of particular concern, the extinction of fish food species is creating the impetus for fish farms. The required numbers for the global annual food fish consumption for a population of 7.4 billion people at present (148 million tonnes) and the projected population of 8.1 billion in the year 2025 (162 million tonnes) [7]. These two sources include sea catches and aquaculture production (the use of sustainable and innovative fish farming technologies in the current market to grow our fishes instead of harvesting them directly from the sea). The global fish catches have been on the decline due to overfishing, whilst aquaculture production is increasing over the years.

By 2025, aquaculture production (102 million tonnes) is projected to surpass catches from the sea at 78 million tonnes. The statistics illustrate the sole reliance on fishes caught from the sea is not sufficient to feed the growing world population, hence the shift towards an increasing aquaculture production is imperative to the global food security of the future. Moreover, the decrease in reliance on fish catches from the sea allows for the gradual regeneration of the heavily depleted marine wildlife.

For an offshore settlement to be sustainable, it must not only produce sufficient fish food for subsistence but also replenish the marine supply via a "fish battery" to recharge growth rates. Floating fish production facilities are thus an integral component of an offshore settlement.

6 Fish Farm Facility

For the offshore settlement to be sustainable [8], it has to lower its ecological footprint in energy and food production. Energy is produced by renewable sources of wave, wind or solar energy whilst part of energy may be recovered from waste treatment processes. Food production processes would be symbiotic (e.g. aquaponics) so that the provision of one necessity produces by-products for another with every precious

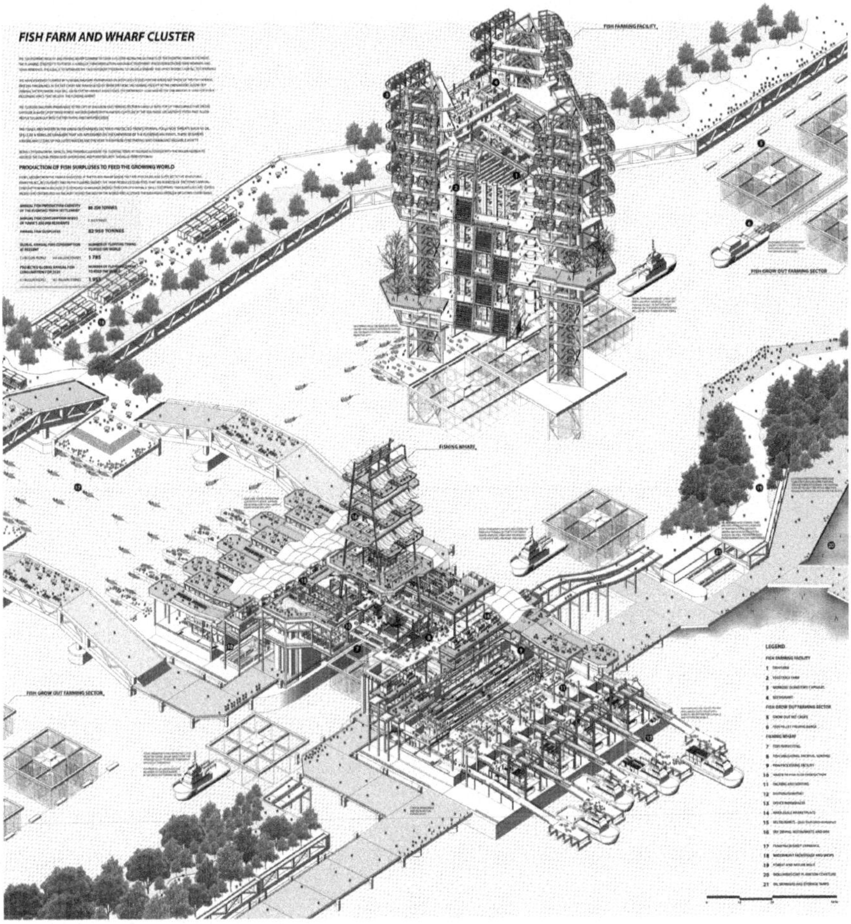

Fig. 6 Fish farm and wharf cluster oil rigs

unit of energy expended. The species farmed would be converted to many uses in addition to protein supply. Therefore, oil rigs can be strategically repurposed for fish production and distribution, wastewater treatment, energy production and housing. An offshore fish production facility as shown in Fig. 6 aims to intensify its oil rig footprint for annual fish production compared to coastal fish farms in Singapore. The architecture of the fish production facility and its wholesale markets comprise seafood, restaurants above the processing and distribution wharf. Attention is also paid to the quality of both the work spaces and user experiences.

The vertical fish farm using the structure of the oil rig offers a long term solution to fish food security in the light of increasing world population and to counter global overfishing which results in fishing wars. The main platform decks of both semi-submersibles and three-legged jack-ups accommodate nurseries and hatcheries

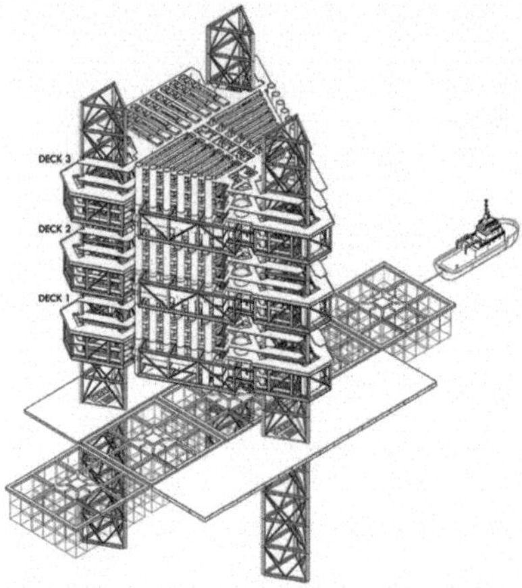

Yield per hatchery per year (1550sqm)	507 tonnes
# of hatcheries total (Deck 1-3)	12
total hatchery area	22,200 sqm
total yield per year	6,084 tonnes
total net cages needed	12
Total net cages from seven oil rigs	80
Yield per net cage	465
total yield from net cages	37,200 tonnes

Fig. 7 Projected fish production for three-legged jack-up oil rig. Fish production is limited by the number of hatcheries that can be accommodated on three platform decks of the jack-up and 12 growth cages per jack-up

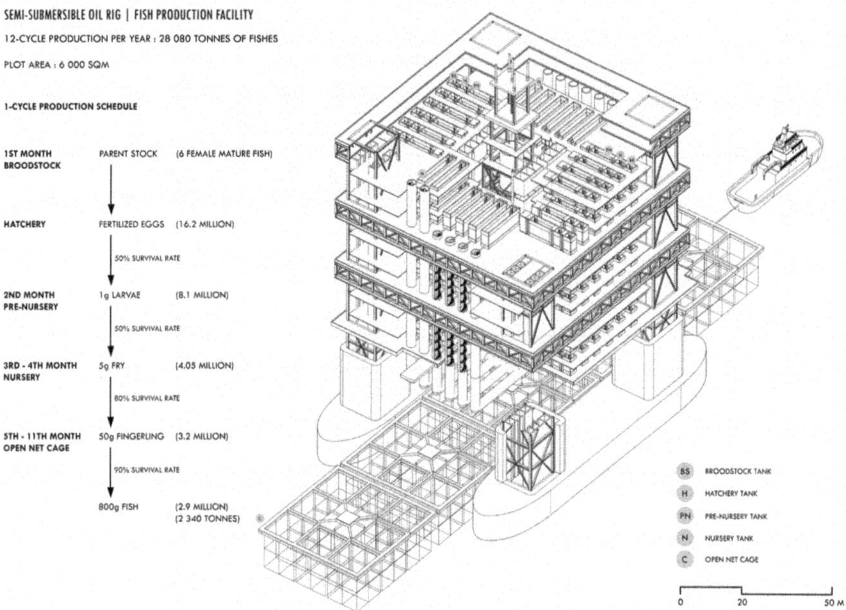

Fig. 8 Projected fish production for semi-submersible oil rig

whilst the grow-out sea cages and distribution centers are separately planned. In comparison, the three-legged jack-up oil rig has a greater annual production capacity of 37,000 tonnes per year as supposed to 28,080 tonnes of fish per year from the semi-submersibles as illustrated in Figs. 7 and 8.

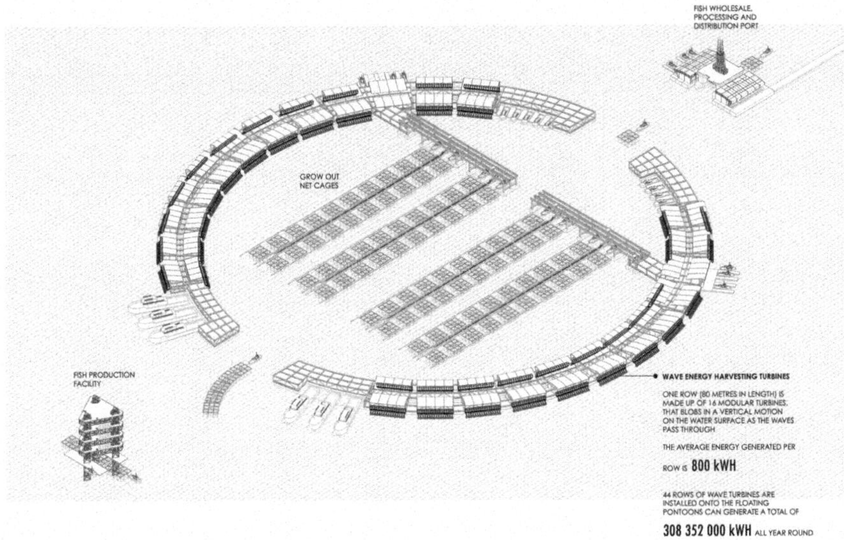

Fig. 9 Estimated energy output of wave turbines

The annual fish production by one jack-up fish farm facility exceeds Singapore's target of 15% as set by Agri-Food and Veterinary Authority of Singapore in the year 2050. Five of these jack-up fish farm facility can supply 100% of Singapore's fish consumption (168,000 tonnes in excess of 153,000 tonnes target in 2015). The global annual fish consumption of 162 million tonnes by 8.1 billion people in 2025, can be met by 4822 of such facility, occupying 642,290 ha of sea space.

The energy requirement of the offshore facility is met by renewable wave energy generated by wave turbines. Both waste and energy resources are looped to ensure that operations are non-polluting and self-sustained. 44 rows of waste turbines installed at floating pontoons circumventing the growth area for 80 net cages can generate an estimated 308 million kWh all year round as illustrated in Fig. 9. The total consumption for 33,600 tonnes of fish is only 167,400 kWh per year.

7 Wastewater Treatment Facility

The control of wastewater discharge quality for the protection of marine eco-systems is imperative for floating settlements. Residential, food production and consumption wastes are known types generated by settlement occupancy.

Wastewater treatment technology today enables compact systems to be up-scaled to serve 1.2 million population equivalent whilst the smallest ones can be in the hundreds of population equivalent. Secondary treatment comprises compact bio-

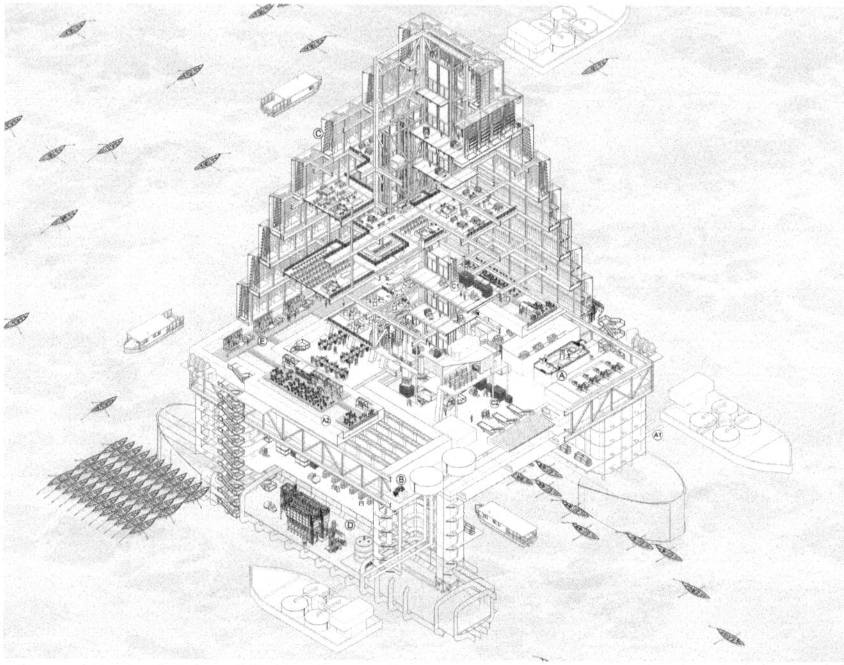

Fig. 10 Wastewater Treatment SSAU

processing components (aeSBR) and tertiary treatment comprises MF-RO membrane technologies. Such compact systems allow integration into plant rooms of oil rigs, and allow great flexibility in distributing population clusters within the settlement.

In selecting energy production systems, the most efficient systems using renewable sources were considered in relation to the energy loads and material flows required for food and water production. Using oil rig vessels as units of settlement plan, there are three scales: specialised vessel, flotilla and fleet. One such specialised vessel combines hydroponic production with wastewater treatment in a semi-submersible rig as shown in Fig. 10. The footprint of this specialised vessel is 6400 m^2. It treats 23 million cubic meters per year of wastewater for 250,000 people. Rainwater collection is estimated at 100,000 cubic meters a year. One SSAU provides 14 million cubic meters of water and 5309 tonnes of hydroponic vegetables a year.

8 Repurposing Superbarges for Displaced Populations

Superbarges by virtue of mobility and large displacement capacity are conceived as floating housing relief for post-disaster situations. The plight of refugee camps such as Kenya's Dadaab exist today since it was first established in 1992 for refugees of

Fig. 11 One module of superbarge housing

the Somalian war. The largest Dadaab camp, Hagadera has a capacity of 106,926 people within a space standard of 1.25 m^2 per person. By comparison, one module of the floating housing barge as illustrated in Fig. 11 accommodates 5184 people in a space standard of 2.5 m^2 per person. It does so with an ingenious tartan grid layout of container box housing. Food is grown on cantilevered farming decks with direct access to elevated housing. The cantilevered farm decks do not increase the footprint of the housing barges on water.

It takes 48 h to set up four housing barges, with on-board cranes. The housing units can be disassembled upon recovery of the disaster site, and be redeployed for successive disasters. The housing barges have the advantage of long term accommodation with minimal food and water aid from organisations when permanent housing for the displaced people is extensively delayed. A floating relief settlement of 107,712 people can be accommodated with 22 barge housing modules where one housing module of two superbarges holding 184 housing containers and 4896 people as illustrated in Fig. 12 and Table 1.

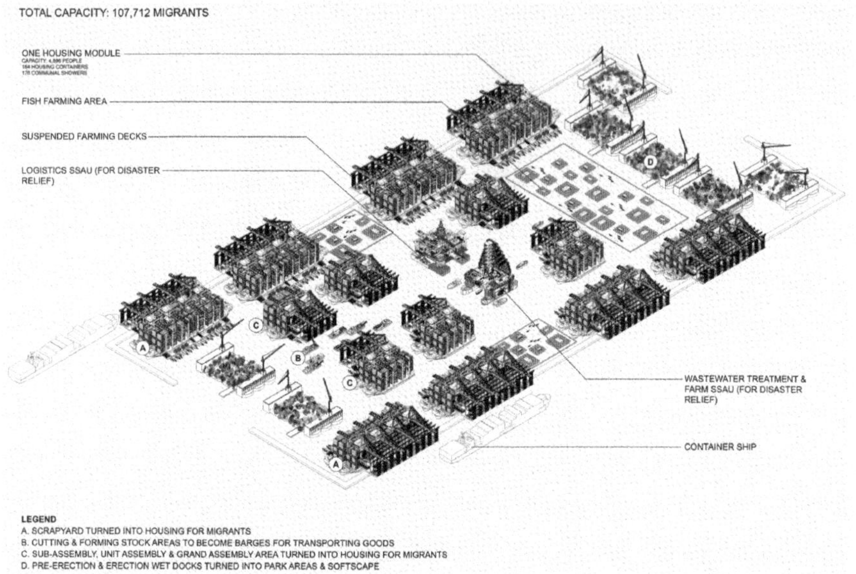

Fig. 12 Superbarge Settlement

9 Layout Determinants of High-Density Housing

Figure 13 illustrates housing variants generated on three types of oil rigs: 14 variants from three legged jack-up rigs, 10 variants from four legged jack-up rigs; and 9 variants from semi-submersible rigs.

The layout determinants of housing units are determined by several factors. Our research shows a four-room Singapore's Housing Development Board (HDB) of 90 m^2 flat forms the most number in public housing. So this was proposed in the oil rig housing with units having access to private terraces and to public sky gardens, made from additional platforms at intervals along the height of the oil rig housing as shown in Fig. 14.

In terms of plan and construction as illustrated in Fig. 15, units are configured around the masts/legs of the jack-ups to enable elevator access and escape stairs forming banks or wings of dwelling units enclosing central courtyards for ventilation and natural light. In jack-up housing oil rigs, each bank of apartments may be suspended from or supported in additional platform decks acting as transfer structures incorporating sky gardens and co-working spaces every 'n' floors. Lateral ties connecting all masts serve as corridor access with sea views. For semi-submersible housing oil rigs, symmetrical banks of apartments are configured to balance and distribute structural live loads to pontoons with limits of buoyancy, draft depths and toppling.

Table 1 Capacities of Superbarge Settlement

Superbarge Settlement

Vessel type	Two superbarge modules	*Area comparison*		
Structural additions	• Two panamax ship-to-shore gantry canes • Steel frame structure to support container stacks		Hagadera refugee camp (largest in the world)	Superbarge settlement
		Footprint (m^2/person)	81.3	1.45
		No. of people	106,926	549,504 [106 modules]
		Per housing unit	10 m^2 per 8 people	30 m^2 per 12 people
Total settlement area	12,000m^2	No. of superbarges needed for future projection?	106 [to accommodate 549,504 displaced people in a disaster of the scale as 2008 Aceh tsunami (4th largest in the world)]	
No. of people per superbarge module	5184			
No. of days to assemble each module	Cargo arrives at disaster site in 7 days and assembled within 48 hours			
No. of containers	624 x 40 ft containers [housing,bathroom] 648 DNV crash container frames [verandahs, corridors]	Energy source	LNG from cremation barges	
		Waste type and treatment strategies	<u>per superbarge module</u> 477,358m^3 wastewater per year treated at three WWT SSAUs [treating capacity of 69 million m^3 wastewater per year]	
No. of people per housing unit	12			
Total bathroom area	2880m^2			
Cooking and food preparation	432m^2			
Dining area	432m^2			
Farming yield capacity	575 tonnes/year [6,393m^2]	Water supply	42 million m^3 potable and non-potable water produced by three WWT SSAUs <u>per superbarge module</u> consumes 285,716 m^3 water per day	
Learning spaces	594m^2			
Play and recreational area	1296m^2			

Treatment plants of compact configurations with capacities matching load and wastewater volumes and potable consumption quantum are selected for use on the oil rigs. Wave energy power systems—specified in Table 2—are integrated into semi-submersibles hulls or anchored in adjacent locations. Derrick structures may be converted to chapel/shrine and observatories. These link to boat/water taxi jetties, market and food center facilities. The undercroft of jack-ups and semi-submersibles define the public gathering space for communal/recreational activities as illustrated in Fig. 16.

10 Settlement Plan

Settlement size is determined by the capacity of a specialised vessel incorporating elements of infrastructure. Decentralized water plants for 40,000 people at precincts or centralized water treatment plants for entire settlements of 250,000 are incorporated in rig types.

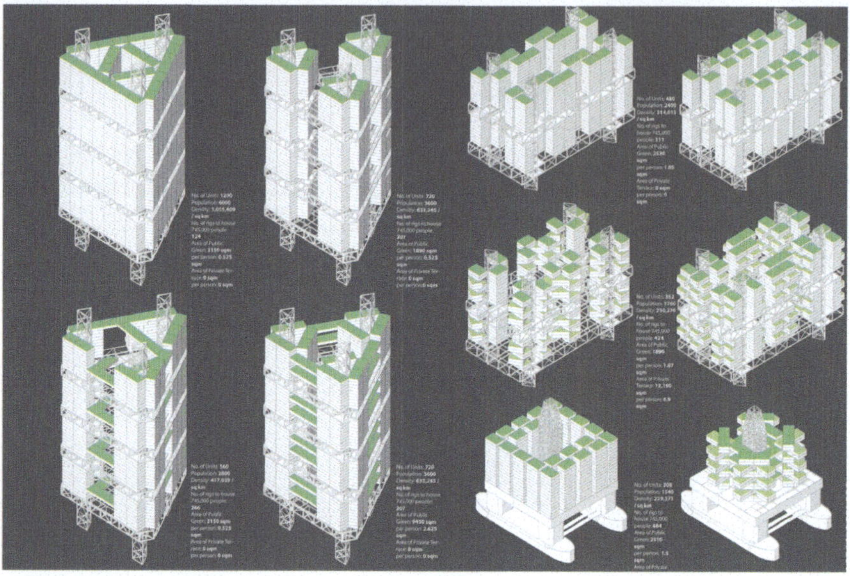

Fig. 13 Sample of Housing variants generated from 3 and 4 legged jack-up rigs and semi-submersibles rigs

Table 2 Capacities for Jack-up Housing

Jack-up Housing

Rig type	3-legged jackup	Total energy required	2,907,736 kWh/year [domestic usage] 1,109,600 kWh/year [machinery and service]
No. of apartments	672		
No. of residents	3,360		
Apartment sizes	64, 96, 128 m^2		
Total apartment area	64,512 m^2		
No. of private sky terraces	672 [21,504m^2]	Household foodwaste	467 kg/day average
No. of public sky gardens	12 [12,000m^2]		
		Biogas container	2 x 20 ft containers operates on 380w
No. of rigs to house population figure in 2050?	2,083,340 rigs to house 7 billion people or 28,000 settlements [38.5 km^2 per 250,000 people]	Electricity for battery storage (microgrid)	11,000 kWh/day
Potable water	650 tonnes [domestic usage] 800 tonnes [output]	Battery storage container	4 x 20 ft containers, one on each hull
Desalination containerized plant	2 x 40 ft containers on the bottom hull [consume 277,250 kWh/year]	Foodwaste to biogas energy	563,997 kWh/year
Greywater	130 tonnes [irrigation] 150 tonnes [output]	PV panels on roof (solar radiation x effective PV area)	6,896 MW/year [1,667kWh/m^2 x 4,137m^2]
Sewage treatment containerized plant	6 x 40 ft containers on the top hull [consume 831,750kWh/year]	Electricity generated from PV	1,531,000 kWh/year [22.2% of 6,896MW]
		Total energy generated (PV + biogas energy)	2,094,997 kWh/year [meets 50% of energy requirement]

Fig. 14 Jack-up housing

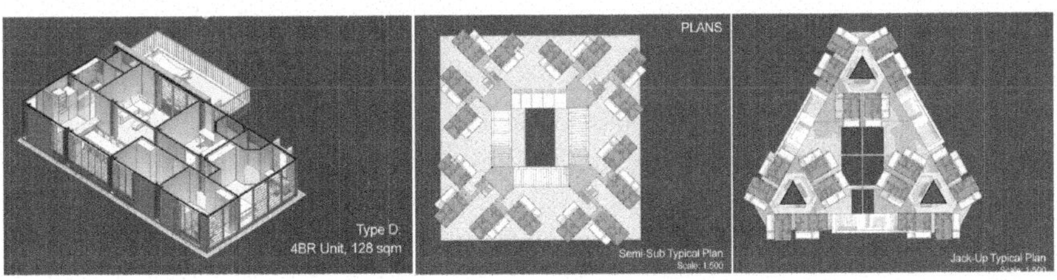

Fig. 15 Layout of 4-bedroom housing units on a semi-submersible and jack-up rigs

Fig. 16 Activities onboard semisubmersible oil rigs housing

Specialised vessels such as in Fig. 17 are arranged into flotillas illustrated in Figs. 18 and 19 to accommodate the number of inhabitants required to optimize the system in an efficient scale of operation. In this respect they form the building blocks of the settlement. Flotillas are also planned in configurations, which exploit promenade and water courtyard relationships with water-edge pathways enclosing waterside activities. The population size of a settlement fleet would be capped at an efficient maximum scale for water treatment with a compact footprint of one semi-submersible and the provision of one hospital for 250,000 people. Tertiary education is sized at one college per 100,000 population. Recreational spaces are planned as outcomes of communal activity in all forms of production and consumption. To reduce settlement footprint, key systems such as water treatment plants are integrated with orchards/crop production facilities and conceived as urban landscapes.

The following key considerations guided permutations of alternative settlements on water. They are broadly categorised into three branches: Landscape Integration, Zoning and Water Separation, and Transport Plan.

Landscape within settlement emphasises on proportion of open recreational spaces such as greenery and water courtyards to form urban vibrant nodes. Whereas landscape as defence mechanism uses floating wetlands along the periphery of

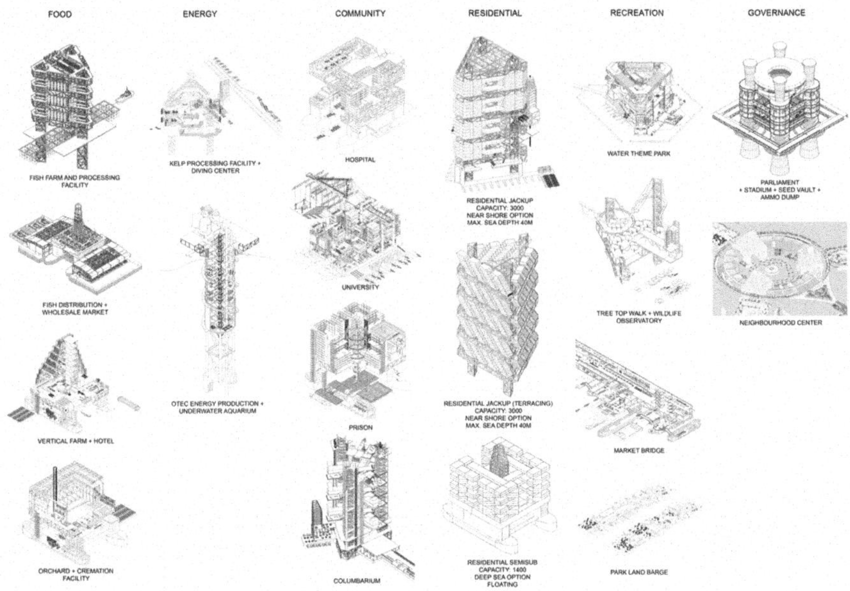

Fig. 17 Types of specialised vessels

Fig. 18 A flotilla of 30,000-40,000 population

Fig. 19 A settlement of seven flotillas accommodating 226,800 people

the Masterplan as a buffer from tidal waves and tsunamis which also acts as the outermost protective ring to the settlement.

This zoning and water separation accommodates daily recreational activities within the inner settlement of less turbulent water. The landscape buffer also acts as a primary filter to maintain water quality around and within the settlement. No wake zones strategy partitions the waterscape within the settlement so that the quality of water for farming is kept in good quality and is separated from those for recreational and circulation purposes. Outside no wake zones, wave energy would be harnessed at the outer boundaries of the settlement where the wave conditions would be optimal for harvesters. The wave energy harnessers also serve as wave attenuators. The extent to which the perimeter of the settlement may be deployed for wave energy harvesters with efficient transmission networks is a key consideration.

Luminescent plankton forms a landscape strategy for intrusion detection, while creating a unique ambience for the settlement during the night. A significant proportion of landscape in the settlement serves the greater ecology of marine wildlife and migratory species. This also creates a unique experience for the settlement in the form of opportunities for wildlife watching of migratory birds and marine wildlife.

In the transport plan, integrating promenades and waterways with nodes in settlement infrastructure and landscape elements intensifies the network of urban accessible space. Ambulatory networks and water taxis are coordinated with scenic vistas to arrive at urban destinations ranging from nodes of intense activity to wildlife locations of solitude. Service and waste disposal networks are to adopt efficient layouts to maximise space. Outside no wake zones, wave energy can be harnessed at the outer boundaries of the settlement where the wave conditions are optimal for harvesters. These wave energy harnessers also serve as wave attenuators. The extent to which the perimeter of the settlement may be deployed for wave energy harvesters with efficient transmission networks is a key consideration for urban occupancy and enjoyment. Programs and hybrid typologies of related waste loops can be in proximity to reduce service distance.

Settlement fleets may be distinguished by the quality and quantum of water integrated spaces provided for recreational activity, living space per person, and the accessibility to all parts of its boundaries with a promenade network. Settlement fleets like character, specialisation of knowledge and other forms of production, climate and migratory species may distinguish cities sighted from specific offshore locations. Living and recreational spaces have unique qualities when integrated with production activity. Specialised vessels are not just machines but they provide the spatial backdrop for settlement vibrancy as a counterpoint to the city. Whilst settlement fleets are kept to less than four kilometers radius in size, they have different population capacities, production outputs and proportion of recreation areas per person. The four kilometers radius factors in a pedestrian range in the planning of transportation forms, which do not have significant energy requirements.

Settlement fleets have a spacing based on the UNESCO minimum biosphere reserve area [9] of 9.7–21.9 km radius and a 22 km radius based on territorial waters for full sovereign rights. Biosphere Reserve areas of 150,000 ha are believed to enable a balanced relationship between man and nature. In proliferating settlements, the UNESCO guideline for Biosphere Reserve Area of 44 km was observed. As illustrated in Fig. 20, instead of repeating 200 settlements at 44 km apart within 500 by 500 km of water space, 30 settlements spaced at 240 km apart were used in the proposal. This was an attempt to lessen the impact of settlements on the marine ecology and differentiate settlements by distance beyond the UNESCO guideline.

Where each of the settlement had 245,00 population equivalent, 30 such settlements could be organised on 250 km^2 of sea, thus only 54.25 million square kilometers of sea may be occupied by 1.6 billion in 2050. This is an estimated 20% of the projected population in 2050. The location of these settlements depended on 'safe zones' as shown in Fig. 21 with the following characteristics: (a) shows areas with a sea air temperature range of 5–30 °C for human thermal comfort and food farming, (b) away from marine wildlife and migratory pathways, both airborne and water, (c) away from hurricane pathways and waves higher than six meters, and (d) away from pirate and commercial fishing zones, rubbish gyres and shipping routes.

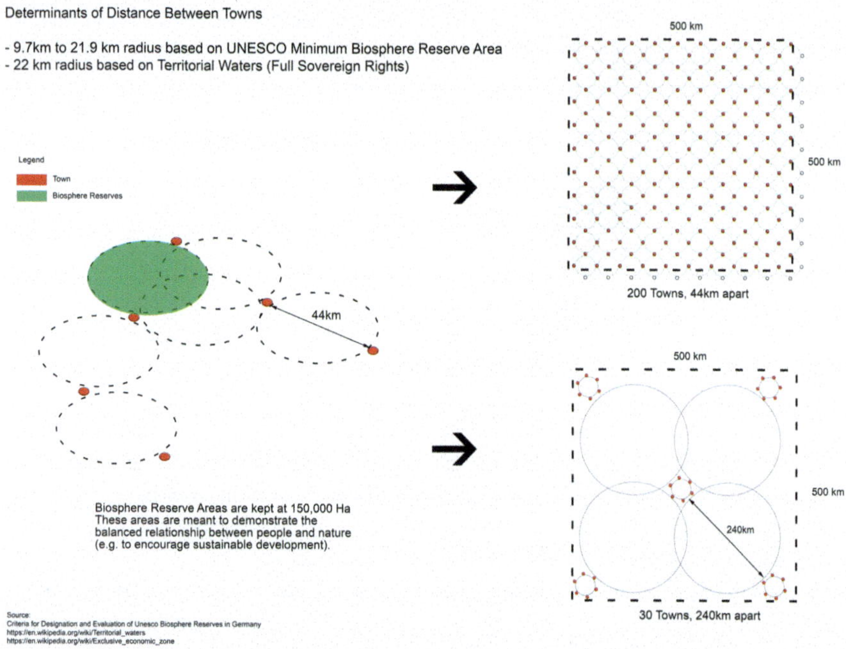

Fig. 20 Siting an offshore settlement

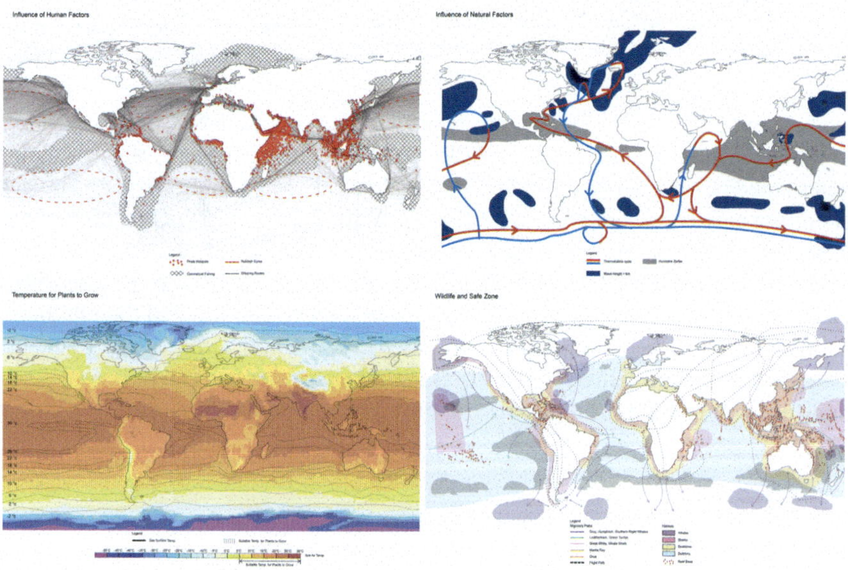

Fig. 21 'Safe zone' considerations for floating settlements

11 Conclusion

11.1 Comparing Settlement Variants

Figures 22 and 23 compare settlement outputs in respect of sustainability in fish farming, harnessing energy and population density. Across all the permutations, there are surpluses in fish production with the highest coming from the *Seastar* and Quadrant settlement at 130 kilotons per year. 1247 of each of these two settlements can produce enough fish to meet the world population in 2025. The settlements with the highest amount of fish production are those that had their fish production facilities within its periphery. This strategy does not confine the growth of the fish production facility and allows for easy expansion outwards. The planning is also suited for exporting the surplus fish at wharf/market nodes along its coastal border.

In contrast, settlements that place their fish production facilities in the middle are confined by the water area and thus limiting the capacity of their fish produce.

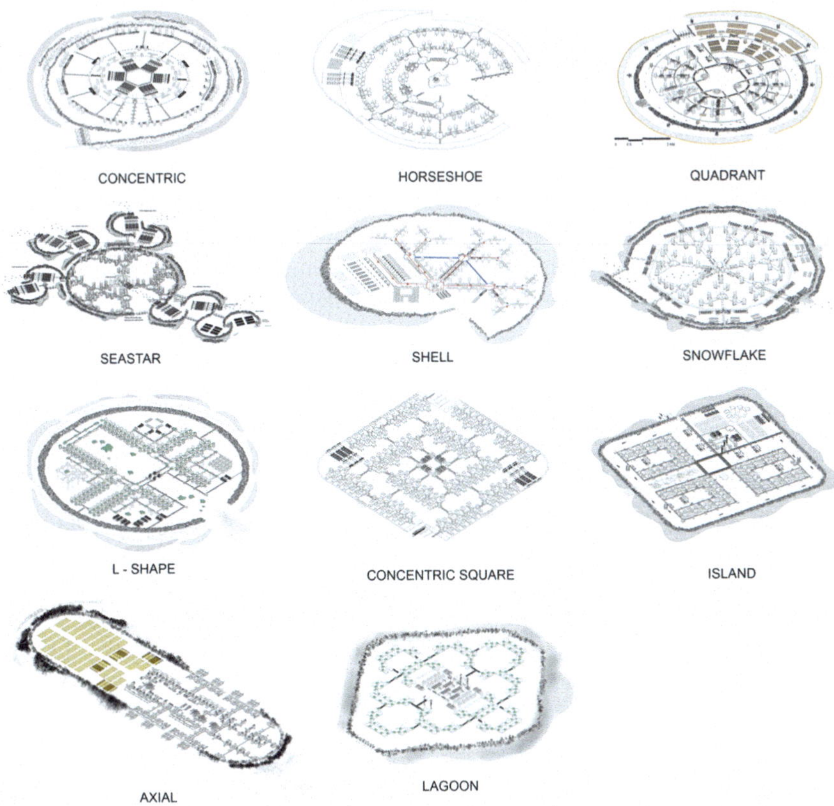

Fig. 22 Exploration of Masterplan permutations

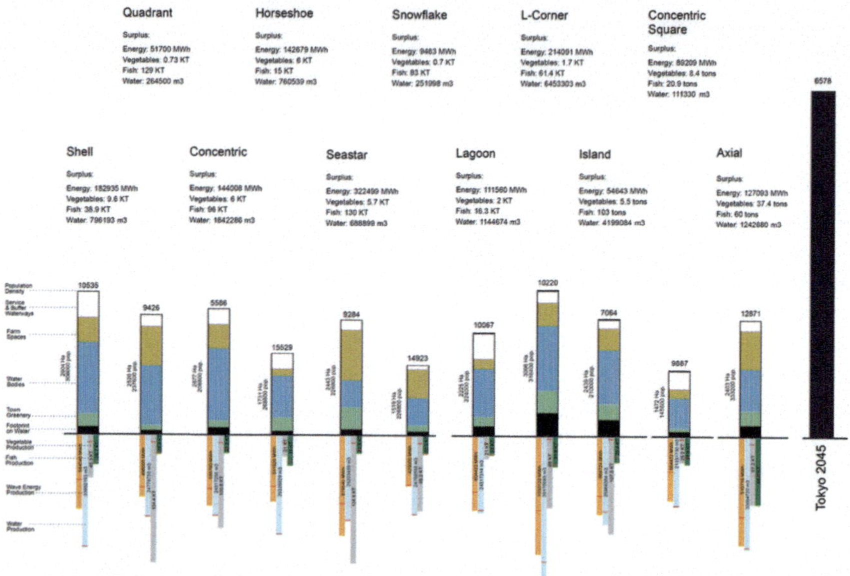

Fig. 23 Overall comparison of settlement variants

However the strategic benefit for such placement comes from the food security and protection from external or environmental threats. Therefore there exists a tradeoff between production capacity and food security when it comes to the planning of fish production facilities within a settlement.

The bulk of the energy supply of the settlement comes from wave energy harnessing devices that are placed along its security perimeters. These security perimeters also double up as green spaces for the enjoyment of residents as well as a habitat for migratory species. Recreational area and wave energy generation is created simultaneously.

A longer perimeter will allow for more wave energy to be harnessed. However an interesting observation arose from the permutations. The *Seastar* permutation, which does not have a single large boundary perimeter, is still able to achieve a high wave energy yield of 819,450 MWh from its multiple short circumferential borders enclosing ten different farming clusters. In deep sea locations, L-corner, Axial, Island, Lagoon and Concentric settlements have energy, which make ocean thermal energy conversion plants viable. This is due to the extent of length of settlement boundary that can be used for wave energy harnessing. All other settlements of 208,000–268,000 population equivalent have an energy yield ranging from 500,000 to 900,000 MWh.

Fig. 24 An example of a market bridge that connects various nodes in the floating settlement and maximising recreational activities

11.2 Comparing Floating Settlement and Town

The flotilla comprises nine jack-up oil rigs with different housing typologies, connected to three parklands repurposed from barges, where they converge to the central neighbourhood ring. Connected to the parkland barges are the semi-submersible oil rigs where vertical farm production spaces are integrated into the dwelling zones. This neighbourhood ring can be used to hold large events for the residents, capitalising on spectator panoramas of the waterscapes as illustrated in Fig. 24. Whilst jack-ups can accommodate 2000–4000 people equivalent to one HDB precinct, semi-submersibles range from 1000 to 2000 people. A flotilla can accommodate a population in excess of a HDB neighbourhood of 30,000 people. A settlement fleet has a capacity equivalent to a New Town at 300,000. The projected density of the oil rig housing types exceeds the HDB types on land. This is achieved in spite of providing larger recreational area, and private terraces to dwelling units of a floating settlement.

Comparing densities in persons per square kilometers, Ghim Moh has a moderate density of 163,765–242,100 persons per square kilometers whilst Pinnacle Duxton has a high density of 367,000 persons per square kilometer. In comparison, jack-ups housing ranges from 230,000 to 600,000 persons per square kilometer whilst semi-submersibles housing ranges from 100,000 to 280,000 persons per square kilometer.

A floating settlement has the challenge of not diminishing biodiversity and open space with a concentration of population enabling the advantages of specialisation in forms of economic production, pedestrian accessibility and efficient energy and waste treatment and systems. Future settlements have to look beyond infrastructural solutions to sustain their existence but also to sustain natural assets beyond their physical boundaries. The challenge is to put back more than what we take out from nature and one attempt with marine settlements is to replenish food supply in an attempt to mitigate fish wars. Physical configurations and settlement programs which generate a surplus of water and energy resources need to be tested in efforts to diminish water wars and the environmental impact of new technology on ecological footprint.

In closing, the physical advantage of the oil rig structures to planning settlements lies in the following aspects: (1) their minimal water footprints with semi-submersible pontoons and jack-up legs, (2) the ability to enclose large communal spaces under elevated platform decks and in between their legs, (3) the ability to define open water courtyards for production and recreational programs when planned in clusters viz. they form nodes of urban space and communal activity and for connection to urban networks, (4) a proportion of usable space within the platform deck structure and the semi-submersible hulls. They are in essence elevated tower and podium types in the urban form of cities; and (5) the substantial extent of airspace above platform decks to accommodate varying housing densities, hybrid farming and recreational spaces with infrastructure, the integration of transportation nodes with promenade networks, mixed usage including skyparks and workspaces in vertical configuration to increase green surface area per person.

These allow the floating settlement the advantage of being configured in different ways to integrate both polycentric and monocentric urban forms in order to flexibly change urban scale and hierarchy. The use of cluster flotillas specialising in forms of production or recreation also increase open water space. These facilitate the experiments with re-reefing and recharging marine ecologies. It also overcomes the limitations of landed cities with a hierarchy using only the vessel and the flotilla as units of a settlement plan to accommodate global population increase with minimal footprint on water.

References

1. Marks, P. (2018). What it takes to dismantle oil rig. *BBC Future*. Available at: http://www.bbc.com/future/story/20160804-what-it-takes-to-dismantle-an-oil-rig. Accessed December 2018.

2. Pearce, F. (2018). As North Sea oil wanes, removing abandoned rigs stirs controversy. *YaleEnvironment360, 26* June 2018. Available at: https://e360.yale.edu/features/as-north-sea-oil-wanes-removing-abandoned-rigs-stirs-controversy. Accessed December 2018.
3. Newman, N. (2018). The sharp rise of platforms decommissioning. *SPARKS*. Available at: https://www.eniday.com/en/sparks_en/platforms-decommissioning/. Accessed December 2018.
4. Offshore Jack-up Rigs. Daleel oil and gas supply chain portal. Available at https://www.scmdaleel.com/category/offshore-jack-up-rigs/213. Accessed December 2018.
5. Lamont, T. (2018). Where oil rigs go to die. *The Guardian*. Available at: https://www.theguardian.com/business/2017/may/02/where-oil-rigs-go-to-die. Accessed December 2018.
6. Kaiser, M., Snyder, B., & Pulsipher, A. (2013). Offshore drilling industry and rig construction market in the Gulf of Mexico. *Coastal Marine Institute Report*.
7. FAO. (2016). The state of the world fisheries and aquaculture 2016. Contributing to food security and nutrition for all, Rome.
8. Moles, R., Kelly, R., O'Regan, B., Ravetz, J. & McEvoy, D. (2002) *Methodologies for the estimation of sustainable settlement size*. Ireland: Environmental Protection Agency.
9. Technical Guidelines for Biosphere Reserves. Available at https://en.unesco.org/op-wnbr.

Floating Infrastructure Large Scale Public Spaces on Water

Gerard Ronzatti and Petar Lovric

Abstract The strain and pressure on the global city infrastructure can be efficiently relieved by allowing this infrastructure to step "onto" the water and thus break this ever existing barrier. Apart from the fact that by doing so we manage to create new and attractive public space in the environment which is usually scarcely utilized, floating infrastructure also provides extremely high resilience while minimizing the environmental impact. Important thing to keep in mind is that these spaces of water often exist in the very heart of our cities where availability of building sites tends to be extremely limited. We have provided practical demonstration on how large public spaces can successfully exist on water and how floating architecture has a potential to deliver new realities by physically mirroring the city on this recently empty water space. Our experience and projects successfully delivered to date remove any doubt that large floating infrastructure projects such as floating hotels, floating hospitals, floating sports activity centre with Olympic size swimming pool, floating museum , floating climbing centre, floating bars and restaurants and many others can successfully exist in the unpopulated environment of bodies of water that meet the city shores. Apart from this fact, floating architecture is able to deliver solutions that provide high resilience while minimising the environmental impact.

Keywords Floating architecture · Floating infrastructure · Public spaces on water · Resilient city infrastructure

1 Introduction

Society's interest in floating architecture was ever present throughout the ages. On some occasions as envisioned and sometimes even obtained means of basic survival, while on some others a pure expression of art, joy and wealth. In its simplest form, the description of this elementary challenge remained the same to current day; how does one allow this specific land based form of architecture to step onto the water in a

G. Ronzatti (✉) · P. Lovric
Seine Design, Paris, France
e-mail: seinedesign@ronzatti.com

C. M. Wang et al. (eds.), *WCFS2019*, Lecture Notes in Civil Engineering 41,
https://doi.org/10.1007/978-981-13-8743-2_3

Fig. 1 The Louvre under Louis XV Nautical performance of Servandoni for the wedding of Louise-Elisabeth with the future Duke of Parma, August 29, 1739

way which is functional, attractive, but above all resilient and safe. Figure 1 provides an insight in the role that one specific floating structure played in the creation of attraction and stunning impact worthy of peer admiration at the French court. Even if created exclusively to spark wonder and admiration it inevitably drew attention to other, more practical uses of this specific knowledge and technology.

As expected, nature and form of large scale floating architecture keeps on evolving and looking at the period of 1861 when the first edition of the book "Paris dans sa splendeur [1]" was first published, we can "discover" a new reality in collection of monuments, views, historical scenes, descriptions and history which serves as a living witness of strong presence of floating architecture on the river Seine. Here we can see fully defined floating structures, large in scale and scope and mostly intended and designed for practical and public use. The floating architecture has supported the city's step onto the water and has thus enabled the utilization of this large and attractive space. Hotels, markets, restaurants and even swimming pools are here portrayed as the centerpieces of the city life, fully integrated into the network of bridges, navigational routs and busy promenades (Fig. 2).

Surprisingly, future development of mega city floating infrastructure was not delivered with the same drive and enthusiasm which resulted in demise of its presence and inevitably, loss of perception, ideas and knowledge crucial to its implementation.

Fig. 2. "Paris dans sa splendeur", Monuments, vues, scènes historiques, descriptions et histoire by Henri Charpentier

Today, web searches on the term "Floating architecture" deliver results filled with countless renderings, sketches and illustrations which sometimes portray utopian solutions and suggest realities which have not passed the test of technical implementation. Nevertheless, some successful examples of large floating structures are of course present and are ranging from highly advanced offshore accommodation units, to 200-room 5-star hotels.

2 Can We Quantify the Field of Floating Architecture?

Our world is the world of oceans, "the total area of the Earth is approximately 510 million square kilometers and the oceans cover about 71 percent of the Earth's surface, which is about 360 million square kilometers [2]". In comparison, "land area covers only 150 million square kilometers [3]" and even that land surface is further divided into different types, parts of which are not permanently habitable: 20% is land covered by snow, 20% mountains, 20% dry land, 30% good land that can be farmed, 10% land that doesn't have topsoil. [4]"

Predictions show that "by year 2050, 75% of world's population will be inhabiting a 2 million km^2 areas of cities. An important fact to note here is that this mega city land area is supported by estimated 2 million square kilometers large areas of protected waters (estuaries, bays, rivers and lakes) [5]". So, if one of the obvious answers is to live on water, then the consequent relevant comment must be "**Yes, but how?**".

As per the elaboration shown below, we are looking at three different approaches to this topic and each comes with its specific type of implementation challenges and advantages. These three approaches can be defined as:

Fig. 3 Reclamation project Palm Island, Dubai, 2002/2008

Fig. 4 Palafite structure in Manila, Philippines

(i) **Land reclamation**—extremely large environmental impact and limited resilience. Examples of this approach can be found all over the world, from land reclamation on the coast of the mega cities such as New York, all the way to Dubai based "Palm Island" projects (see Fig. 3) and similar.

(ii) **Palafites/Piling**—medium impact and medium resilience. As per one of the piling techniques which is efficiently applied is an example from Manilla (see Fig. 4), we can observe that the approach even though limited in its environmental impact remains sensitive to increased changes in water levels which are becoming a significant risk factor even in the historically less volatile environments.

(iii) **Floating structures**—minimal environmental impact and strong resilience. Floating architecture allows us to utilize the large territory of bodies of water with minimal impact and high resilience to changing water levels while providing support to the diversity of projects and the ambitions of the community. As a fully utilized habitable space, floating architecture has a potential to provide a "mirror" to the city.

Figures 5, 6, 7, 8 and 9 provide only a portion of the multiple successful examples delivered by Seine Design:

Fig. 5 L'ADAMANT, Paris. Floating psychiatric hospital

Fig. 6 Rosa Bonheur, Paris. Floating community center

Fig. 7 LE OFF PARIS SEINE, Quai d'Austerlitz, Paris. Floating hotel

Fig. 8 FLUCTUART - CENTER OF FLOATING URBAN ART Paris. Fluctuart will be opening its doors to public in May 2019

3 Le OFF Paris Seine—A Paradigm of the Floating Architecture

In our view, project of design, construction and delivery of the floating hotel "LE OFF SEINE" can be well used to describe the paradigm of today's large scale floating infrastructure. The general specification of the structure is described below:

- Dimension: 75 m × 18 m (two hulls)
- Terrace: 400 m^2 quay, seasonal

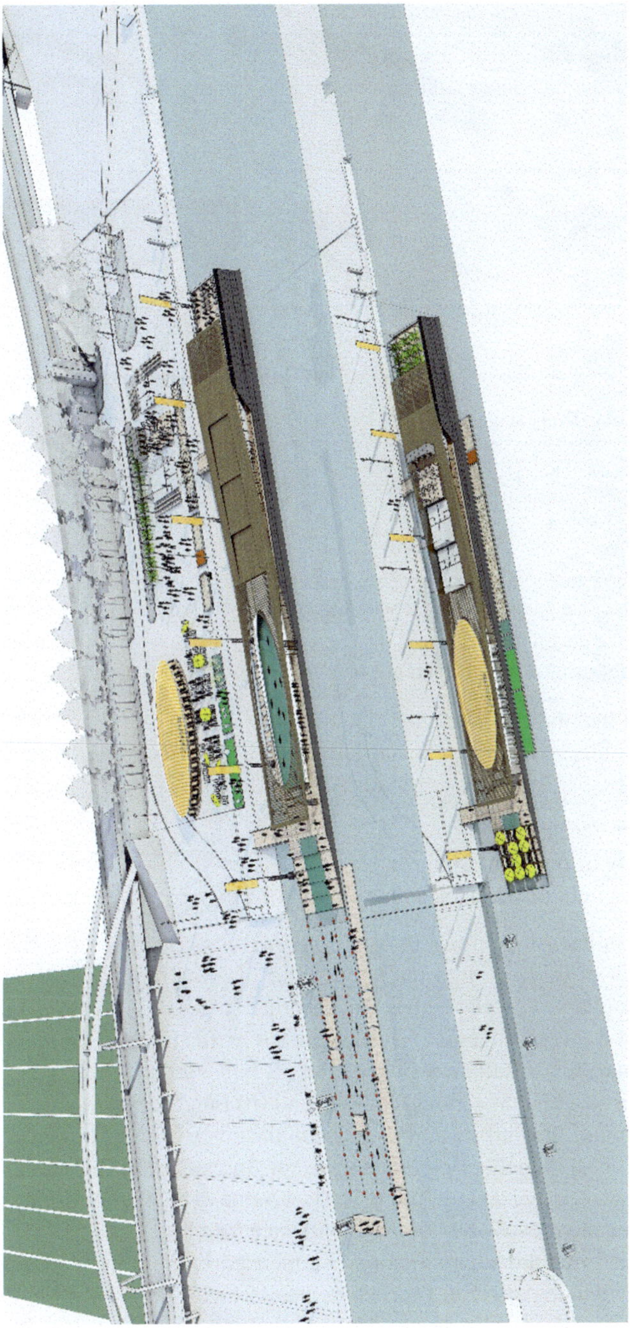

Fig. 9 L'ARCHE—PORT DE JAVEL, Paris. 150 m long sports and activity centre with an Olympic size swimming pool shown in two modes of its existence, with and without its summer terrace. L'ARCHE will be delivered to the owner in 2020

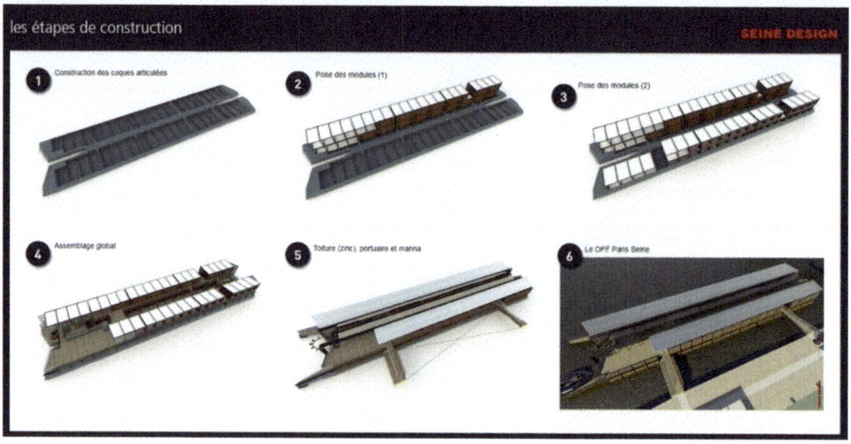

Fig. 10 Illustration of delivery stage

- Pool: 10 m × 2 m
- Weight: 700 tons
- 54 rooms (15 m^2)
- 4 suites (30 m^2)
- Capacity of the restaurant: 80 people
- Noble and durable materials: wood, copper, leather, glass and zinc.

Major limiting factors which heavily influenced the design of the structure itself, and consequently the process of project delivery are often found in many urban localities:

- Lack of suitable yard facility
- Limited capacity of transportation routs
- Installation sites in highly urbanized (work authorization).

Owing to the impact of limiting factors described above which prohibited the delivery of project of this size into the very center of a metropolis such as Paris, France, project delivery was planned in six separate stages, as shown in Fig. 10. The description of delivery stages: (1) Hull construction, (2) Installation of modules to hull number one, (3) Installation of modules to hull number two, (4) General outfitting, (5) Access and connections to shore, and (6) Final delivery.

The construction of the hull stage took place in the city of Dieppe, France which sits on the shores of La Manche. Both hulls were constructed in the shipyard, loaded onto the transportation barge and towed up the river Seine to the Port of Rouen where the second stage of the process was due to commence (see Fig. 11).

Assembly of the modular superstructure commenced in the Port of Rouen. The port facilities provided logistical infrastructure necessary for this operation. Prefabricated modules were lifted of the trucks and installed onto both hulls (see Fig. 12).

Fig. 11 Fabricated hulls are getting ready for transport

Fig. 12 Installation of modular superstructure

Finally, after last set of towing operations, the two twin buildings with an inner street in the center which served as metaphor of Paris, with its right and left banks, were assembled on site and connected to the shore (see Fig. 13).

Le OFF Paris Seine was a pilot project, an example of sustainability, both in its design and in its exploitation, a bridge between the Paris of yesterday and tomorrow. In the end, delivered structure presented an eco-responsible approach, which reduces the environmental impact and delivers a resilient building, which adapts to the flood.

Fig. 13 Final assembly of two hulls and connections to the shore

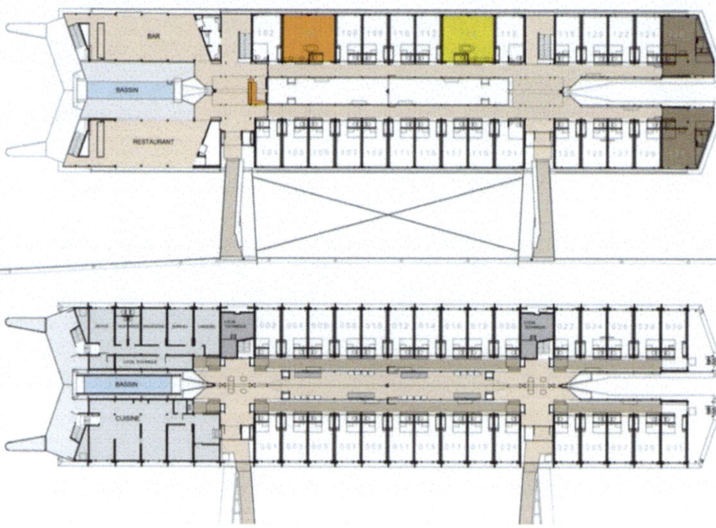

Fig. 14 General arrangement

General arrangement and the cross section view of the structure can be seen in Figs. 14 and 15.

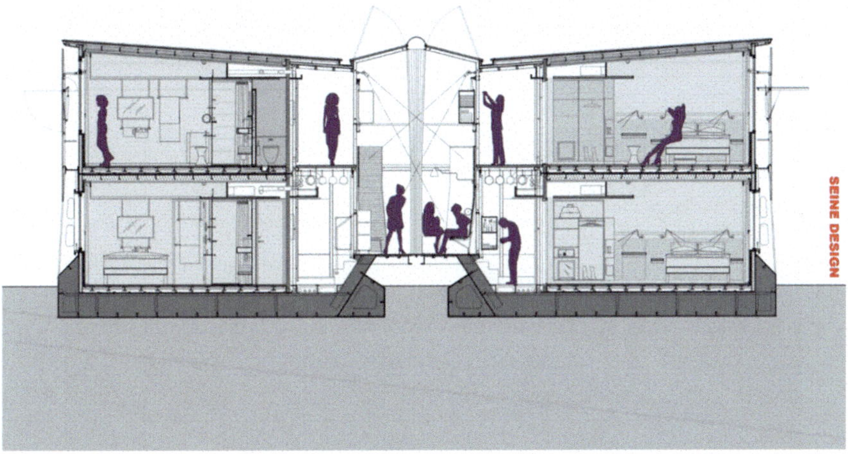

Fig. 15 Cross sectional view

4 Looking Towards the Future

In all our discussions about the future of implementation of the floating infrastructure, we can clearly see growth in both scale and scope. Growing population, increased concentration of people around the urban city areas and changing weather conditions are likely to further emphasize the need for resilient, smart and efficient large scale floating public infrastructure.

With this in mind, we are keeping our focus on development of solutions which will adequately address above stated challenges. One of such solutions is the MOOR, a multifunctional floating facility which is designed a new standard of living afloat (see Fig. 16). The MOOR is completely aligned with its immediate environment and is able to respond to its movement and changes (current, tides, wind, floods and so on). As the rest of the floating architecture, this structure provides high resilience and delivers minimal environmental impact. Solution itself is perfectly scalable and it can range from a single floating unit with up to a 100 rooms to a series of floating units that form a block, a neighborhood, perhaps the community of its own.

The MOOR is structured into three main elements, forming a stable triangular shape which can hold up to 100 rooms as shown in Fig. 17 or alternatively provide more than 8000 square meteres of multifunctional space. The design of the connections between these elements is improved in order to absorb energy from the waves and create comfort of living aboard. The two sides are the main constructions with three decks on top of the open hull. Independent floating structures are connected

Fig. 16 The MOOR

to these main buildings. The third side is the bridge which supports the semicircular gardens and is used for access, reception and other public spaces. Successful realization of this project heavily draws on physical data previously developed and implemented in delivery of multiple successful large scale floating infrastructure projects. This data comprises research and development studies from multiple technical fields relevant for delivery of floating infrastructure such stability studies, studies of water banks suitable for fixing/mooring, studies of construction and outfitting material and similar.

5 Concluding Remarks

The project examples discussed herein remove any doubt that large floating infrastructure projects such as floating hotels, floating hospitals, floating sports activity center with Olympic size swimming pool, floating museum, floating climbing center, floating bars and restaurants and many others can successfully exist in the unpopulated environment of bodies of water that meet the city shores. Apart from this

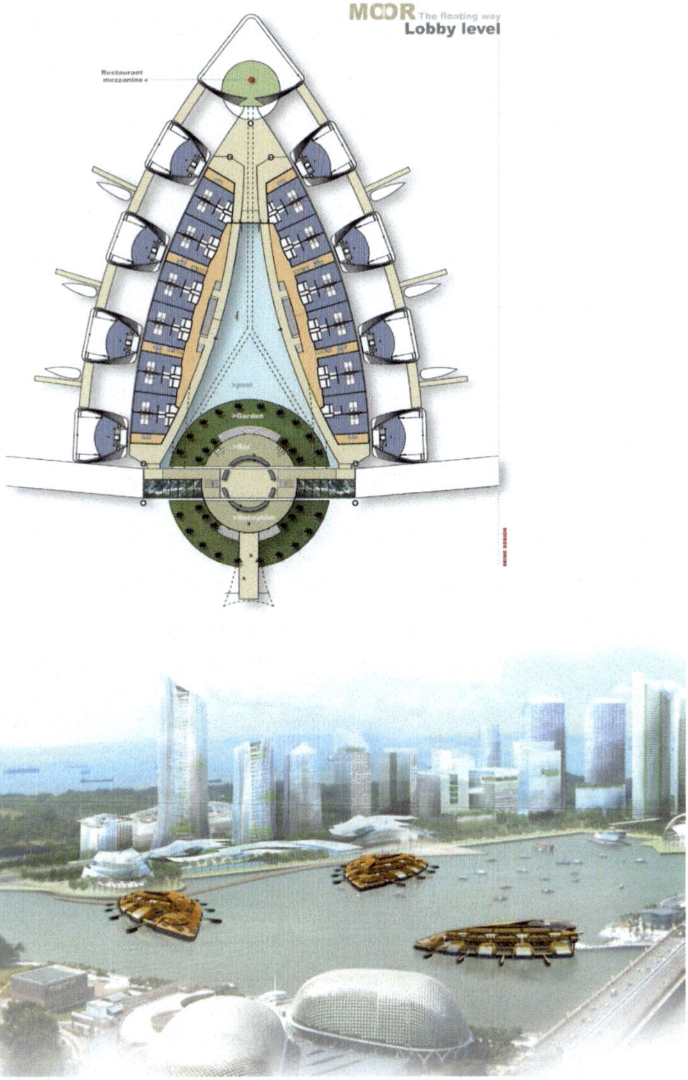

Fig. 17 The MOOR in Singapore

fact, floating architecture is able to deliver solutions that provide high resilience while minimizing the environmental impact and is thus a preferred solution when it comes to addressing the issues of growing population, increased concentration in and around metropolitan areas, changing weather conditions, site accessibility, cost of implementation and many others.

Appendix 1: The Moor—Flexible, Efficient and Easily Transportable

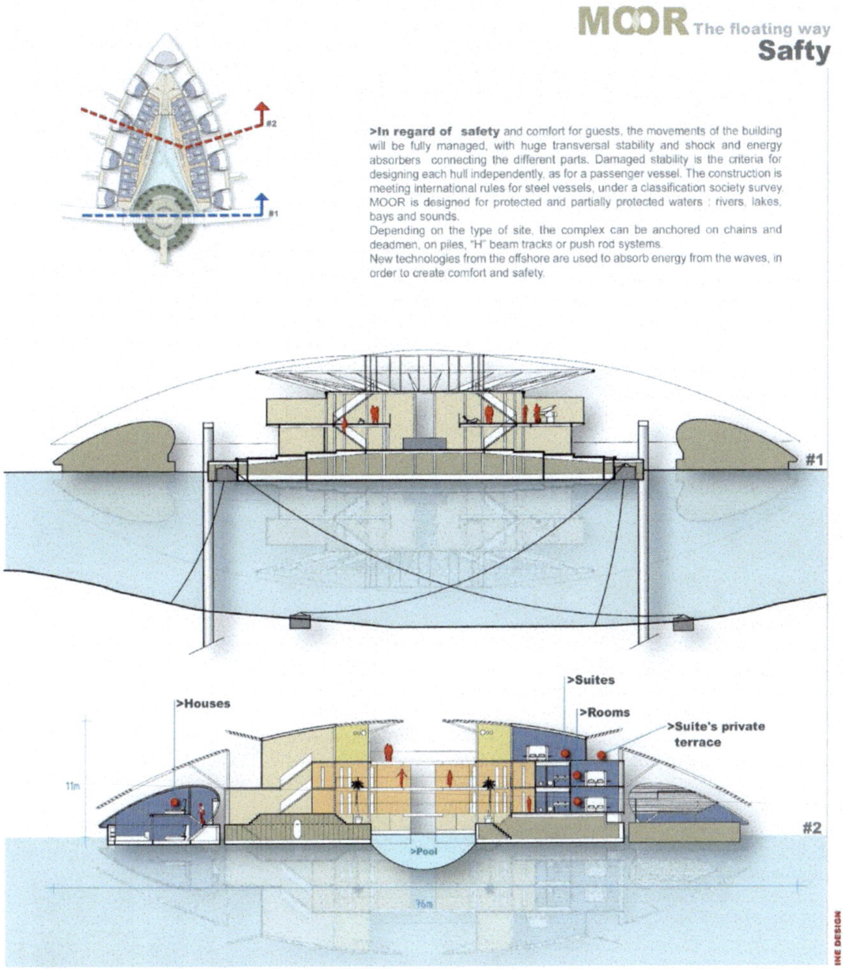
MOOR The floating way
Safty
#2
#1
>In regard of safety and comfort for guests, the movements of the building will be fully managed, with huge transversal stability and shock and energy absorbers connecting the different parts. Damaged stability is the criteria for designing each hull independently, as for a passenger vessel. The construction is meeting international rules for steel vessels, under a classification society survey. MOOR is designed for protected and partially protected waters : rivers, lakes, bays and sounds.
Depending on the type of site, the complex can be anchored on chains and deadmen, on piles, "H" beam tracks or push rod systems.
New technologies from the offshore are used to absorb energy from the waves, in order to create comfort and safety.
#1
>Suites
>Houses
>Rooms
>Suite's private terrace
11m
#2
>Pool
96m
INE DESIGN

MOOR The floating way
Friendly Environmental
>Friendly Environmental
Either connected to the shore, or totally autonomous, MOOR must be exemplary for low impact on the environment.
Energy consumption will be reduced by improving natural ventilation, thermal insulation, shading and installing heat pumps and efficient lighting.
Wind, waves and solar energy are easily accommodated and will support a large part of the needs. When shore power is not available, a power plant will be installed.
Water consumption will be reduced by using vacuum collection system.
Fresh water is treated on board by rain storage, filtration and desalination.
Grey and black water are perfectly treated before discharge overboard.
SEINE DESIGN

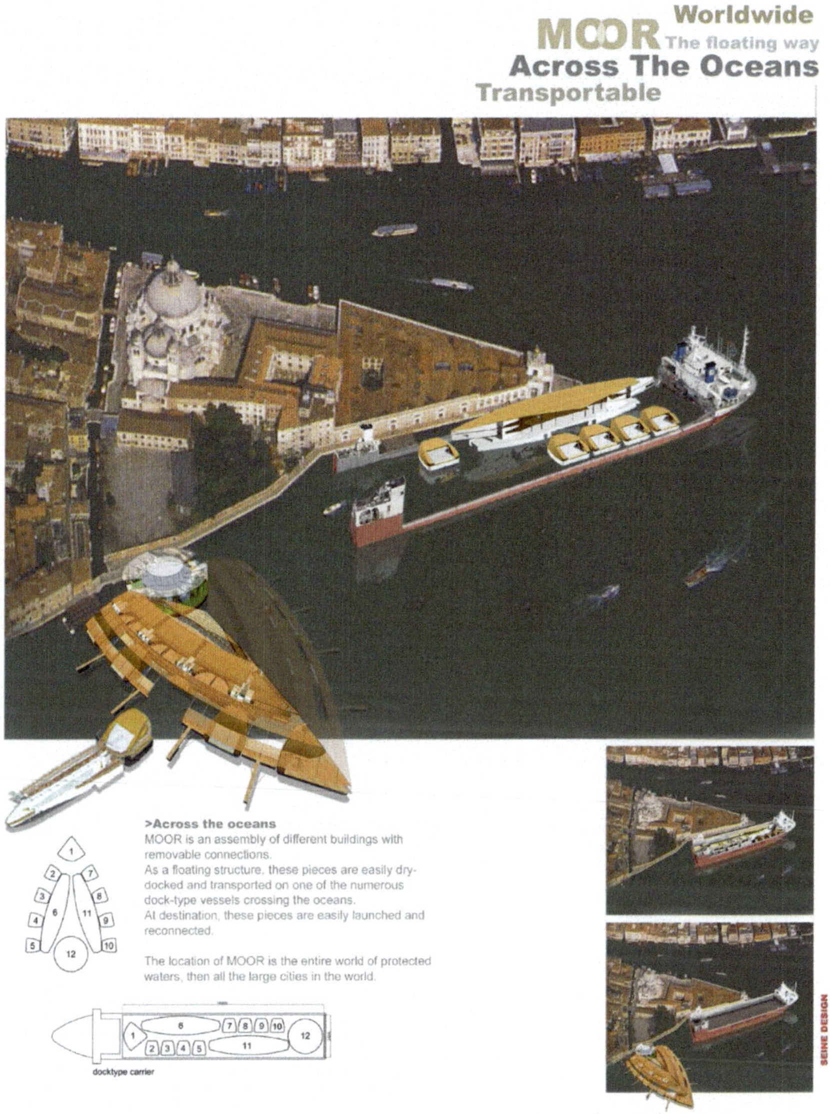

References

1. Charpentier, H. (1861). *Paris dans sa splendeur, monuments, vues, scènes historiques, descriptions et histoire*. Paris, France: Henri Charpentier.
2. Science Desk Reference. (1999). *American scientific*. New York: Wiley.
3. Lutgens, F. (1992). *Essentials of geology*. New York, USA: MacMillan.

4. Glenn, E., Chen, D. (2019). *Area of earth's land surface*. The Physics Factbook. Available at: https://hypertextbook.com/facts/2001/DanielChen.shtml [22nd January, 2019].
5. Euzen, A., Gaill, F., Lacroix, D., & Cury, P. *The ocean revealed.* CNRS Éditions, Paris, 2017 [data synthesis by the Author].

Floating Shipyard Design: Concept and Application

Cliff Ohl, Adrian Arnold, Hannes Uys and Miguel Andrade

Abstract Shipyards often incorporate floating structures as elements within their facilities, for example floating docks for drydocking vessels requiring repairs or maintenance and pontoons to provide berths for smaller support vessels, such as tugs and workboats. However, the concept of a shipyard comprised exclusively of floating structures is considered herein. Floating shipyard facilities are particularly suitable for locations with naturally deep water, where floating structures may be more economical due to lower capital construction costs relative to land based facilities (e.g. floating docks versus graving docks). Similarly, fabrication of floating structures can be conducted off site, often benefiting from modular construction and providing more economical solutions relative to in situ construction. In addition, modular approaches to aspects such as power production may be incorporated, e.g. using modular power plants on floating barges. For appropriate operational conditions within the yard facility and acceptable downtime, relatively tranquil metocean conditions are required at the site, i.e. limiting structural movements for the benefits of equipment (e.g. cranes) and personnel. As such, typical suitable locations will be either naturally sheltered or provided with adequate breakwaters. Benefits of such a floating shipyard include the flexibility to more readily adapt the facility when required and potentially re-locate all or part of the shipyard if required due to changes in market conditions. Floating shipyards are likely to be more suitable for ship repair and not for shipbuilding, which typically requires more extensive workshop and assembly areas. In addition to an overview of the floating shipyard concept, relevant examples are provided from an ongoing project under development in a remote location in Angola, West Africa, that is ideally suited to the concept.

Keywords Floating shipyard · Modular construction · Floating dock · Floating pier · Ship repair

C. Ohl (✉) · A. Arnold
Royal HaskoningDHV, Newcastle upon Tyne, UK
e-mail: cliff.ohl@rhdhv.com

H. Uys · M. Andrade
Bainave, Luanda, Angola

C. M. Wang et al. (eds.), *WCFS2019*, Lecture Notes in Civil Engineering 41,
https://doi.org/10.1007/978-981-13-8743-2_4

1 Introduction

Shipyards often incorporate floating structures as elements within their facilities, for example:

- Floating docks for drydocking vessels that require repairs or routine maintenance, e.g. hull structural inspection, marine growth removal or reinstatement of coatings;
- Pontoons to provide berths for smaller support vessels, such as tugs and workboats with relatively low freeboard;
- Barges and floating piers to serve as wet berths for ship repair and layby berths for vessels waiting to dock.

The concept of a shipyard comprised exclusively of floating structures is considered herein, including the principal opportunities, benefits and challenges. In addition to an overview of the floating shipyard concept, relevant examples are provided from an ongoing project, i.e. of a shipyard which is predominantly comprised of floating structures. This project is currently under development in a remote location in Angola, West Africa, at a site that is ideally suited to the floating shipyard concept.

2 Concept—Typical Aspects

2.1 Location

Floating marine structures are particularly suitable for locations with naturally deep water, where capital construction costs can be significantly lower relative to land based marine facilities, which typically require significant earthworks, retaining structures, piling, dredging and reclamation.

Water depth is particularly significant for shipyard facilities, whereby floating docks (i.e. floating dry docks) require adequate water depth for vessel docking (refer Fig. 1) [1], e.g. seabed level of the order of -15mCD for a dock with approximately 20,000 tonne maximum lift capacity. For many deep water locations, costs of mobilizing a floating dock (and associated floating support infrastructure) are likely to be significantly less than the civil and structural works associated with graving dock (i.e. land based dry dock) construction.

2.2 Fabrication

Floating structures (e.g. floating docks, barges, platforms, etc.) are typically constructed in shipyards that specialize in ship and/or offshore structure fabrication.

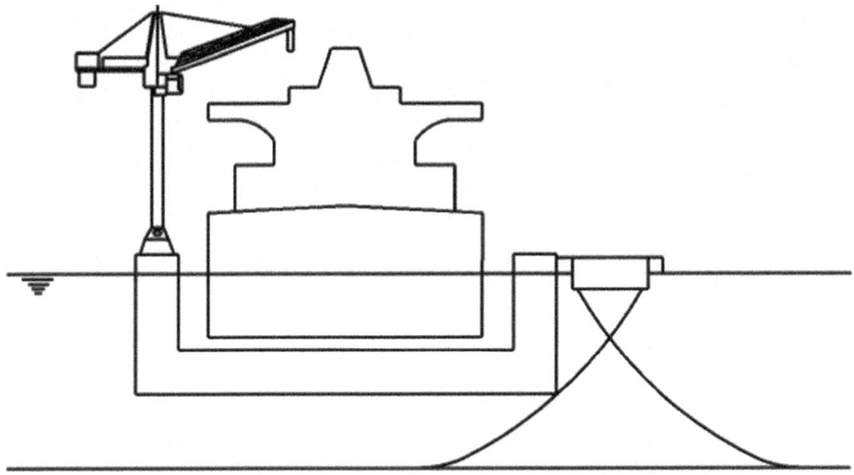

Fig. 1 Floating dock catenary mooring—operational docking and undocking of vessel (Note that the floating dock as shown is moored to a floating mooring dolphin, which is moored with a conventional catenary system to the seabed. While direct connection of mooring chains, anchors, etc. to floating docks is more common, connection via a floating mooring dolphin or similar will allow for faster disconnection from the mooring when required. As such, in advance of a significant storm event the dock can be rapidly de-mobilised, and faster reconnection and mobilization achieved after the event passes—i.e. resulting in less downtime for the facility.)

These facilities are generally off site, i.e. remote to the project location, and as such may benefit from access to stable work environments with skilled labour, established management and quality systems, etc. Such off site, shipyard fabrication can prove economically beneficial relative to typical in situ construction, e.g. construction of a reinforced concrete graving dock in a remote area.

2.3 Modular

Shipyard facilities require a range of infrastructure, including dry docks, piers (or wet berths), production workshops, canteens, administration buildings, accommodation, etc. Many of these facilities may be developed using a modular approach. For example, providing modular accommodation barges to house workers can provide flexibility to increase or decrease accommodation to suit demand. Similarly, modular power production may be readily incorporated in floating shipyard developments, e.g. using modular power plants on floating barges.

The possible use of very large floating structures (VLFS) as floating shipyards has been proposed, and may be monolithic or comprise multiple module with either rigid or flexible connections [2]. While rigid floating structures may be appropriate for some applications, use of smaller, modular structures with flexible connections to form a floating facility with adequate combined working area to support a shipyard may be a lower cost alternative. Such modular structures may be assembled with flexible joints or hinges to avoid the development of large bending moments/stresses and will generally allow the use of structures with lower structural depth, lighter sections and lower fabrication cost. However, such flexibility will result in relative movements between the modular sections and will require adequate ramps, transitions, etc. for movement of vehicles, equipment and personnel.

2.4 Environmental Conditions

Similar to many operational industrial facilities, shipyards require acceptable operational conditions to function with an acceptable amount of downtime. In addition to wind speed limitations for onshore facilities, floating shipyards require low wave heights and current speeds for efficient operations.

Relatively benign metocean conditions result in smaller structural movements (lateral, vertical and rotational velocities and accelerations) and relative movements/displacements. Such absolute and relative motions between floating structures are critical for operation of equipment (e.g. cranes and vehicles) as well as personnel.

As noted above, to reduce motions to accept operational thresholds, typical suitable locations will be naturally sheltered. Alternatively, floating shipyards in more exposed locations may be provided with:

- More robust restraint systems (e.g. piled dolphins, catenary moorings, seabed anchors, etc.) to limit the range of displacements;
- Equipment suitable for operation in a larger range of motion (e.g. cranes design for offshore/dynamic lifting);
- Fixed or floating breakwaters to reduce incident wave heights.

However, the above options result in additional capital cost, and will have limited application. For example, the catenary mooring system illustrated in Figs. 1 and 2 can be designed for a range of operational conditions, but the system weight requirements (mooring chain and anchor) increase significantly as environmental loads increase. In addition, the mooring system illustrated provides for relatively rapid demobilization of the floating structure, i.e. by decoupling from the mooring dolphin as shown. In addition, cranes and lifting equipment may be designed for a range of platform motions; however, shipyard operational activities such as welding and cutting, materials handling, etc. are also affected by such motions.

Similarly, while floating breakwaters may be provided, such structures require specialist numerical and/or physical modelling for adequate design and may only be effective for a relatively narrow range of incident wave conditions [3, 4].

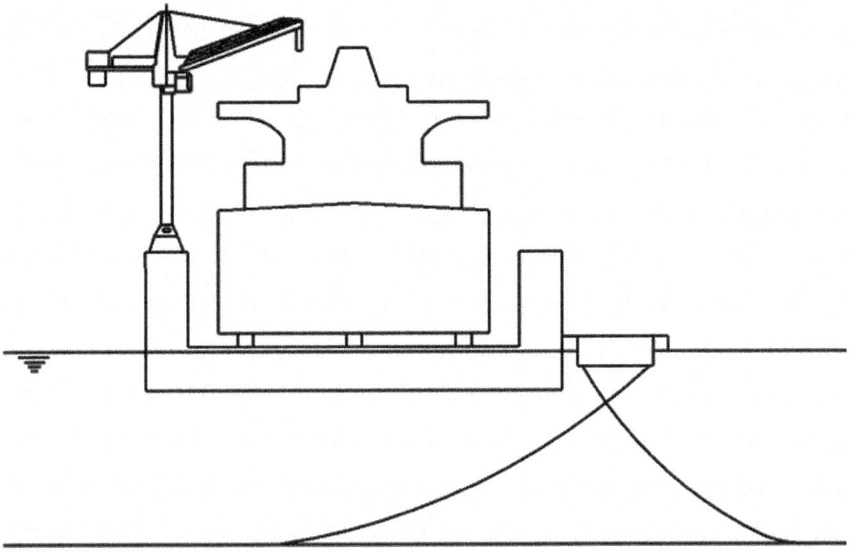

Fig. 2 Floating dock catenary mooring—mooring system design with lateral load (For the floating dock and mooring system in Fig. 1, the system displacement due to a significant lateral load, e.g. operational wind, vessel berthing, etc. is illustrated.)

Finally, as noted generally for floating structures [2], floating shipyards are not directly affected by earthquakes and, relative to onshore facilities, will not be significantly affected by tsunamis.

2.5 Adaptability

Relative to onshore facilities, floating shipyards may be more readily adapted when required due to changes in market conditions, for example:

- Relocation of the entire facility to another region to capitalize on a new, developing market;
- Reduction in facility size by removing items that are not required/no longer economically beneficial (e.g. removal of a floating dock that cannot accommodate the current vessel sizes);
- Introduction of new facilities to increase productivity (e.g. additional accommodation to support an increased workforce);
- Reconfiguration of existing facilities to improve operational efficiency, capacity or changing market conditions.

2.6 Application

Shipyards are typically characterized in function as either ship repair or shipbuilding (or a combination of these). Shipbuilding yards typically require more extensive workshop and assembly areas relative to ship repair, particularly for construction of some of the larger vessels. As such, floating shipyards (i.e. comprising exclusively floating structures) are likely to be more suitable for ship repair and not for shipbuilding. However, it is noted that floating structures such as floating docks and floating cranes are common features of some shipbuilding facilities to provide assembly platforms or for materials handling (e.g. as part of block transport). In addition, floating facilities for dry docking will be limited to the lifting capacities of floating dry docks, whereas onshore graving docks are more commonly used to construct larger vessels.

2.7 Summary

In sum, shipyards comprising of floating structures have the following principal characteristics:

- Location—suitable for naturally deep water
- Fabrication—typically constructed in shipyard facilities
- Modular—facilities that may be scaled to suit requirements
- Environmental conditions—requiring acceptable operational metocean conditions
- Adaptability—to suit market conditions
- Application—better suited to ship repair yards and constrained by floating dock capacity.

3 Application—Example Project

3.1 Background

Bainave is developing an international ship repair yard to service the offshore oil production needs of Angola and the west coast of Africa at Baía Farta, as shown in Fig. 3. Baia Farta is located on the west coast of Angola approximately 215 nm (400 km) south of the capital Luanda. The project site is on the eastern side of the peninsula (Ponta de São José) and faces east toward the relatively sheltered bay of Baia Farta.

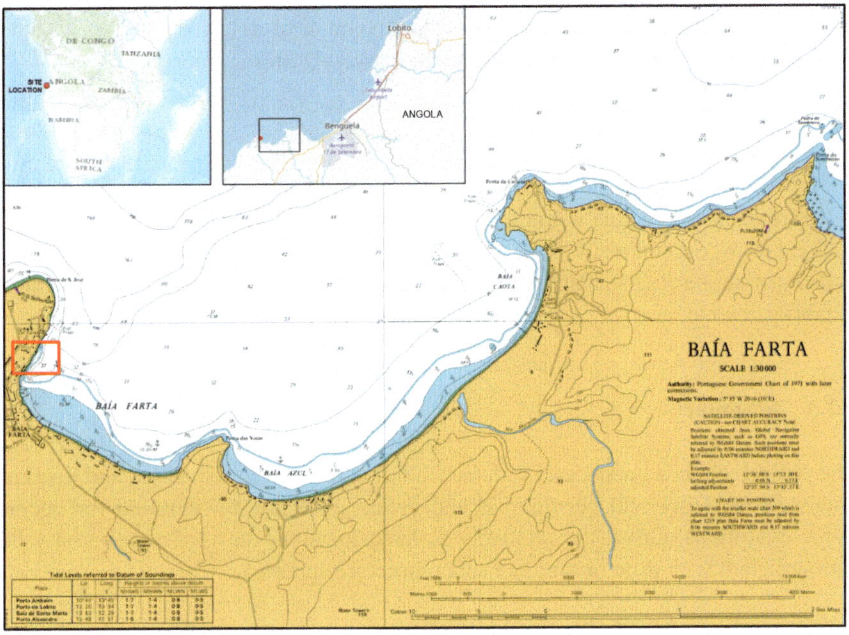

Fig. 3 Project location

3.2 *Site Overview*

The project site is characterized by the following principal aspects:

3.2.1 Maritime

- Natural deep water close to the shoreline, e.g. seabed level −20 mCD at ~100 m from the shore;
- Spring tidal range is 1.2 m and negligible tidal currents, i.e. less than 0.5 m/s at the project site;
- Offshore waves are typically from the southwest, and the project site is well protected for operational purposes with operational wave heights typically less than 0.5 m at the project site;
- Winds are typically from the southwest and generally low, e.g. maximum 34 kn (17 m/s) for a 100 year return period.

While the above conditions characterize the operational design parameters, previous metocean studies and associated modelling have noted the possibility of more extreme wave conditions at the site, e.g. due to offshore storm events. As such, the project is progressed with the assumption that meteorological forecasting systems must be in place as well as response plans to prepare in advance for potential adverse conditions, e.g. relocating significant floating assets to storm moorings.

3.2.2 Onshore

- Restricted onshore area with limited opportunity for storage
- Good transport connections, including good roads and airport;
- Lack of supporting industries, i.e. insufficient local subcontractors to outsource fabrication etc.

3.2.3 Summary

The project site is a good location for installation of floating structures and floating docks, as dredging will be minimal and breakwaters are not required for operational conditions. Given the relatively isolated location, the yard will initially be self-sufficient with little or no local subcontractor support.

3.3 Yard Requirements and Facilities

3.3.1 Marine

The ship repair yard will ultimately have two self-contained floating docks with approximate dimensions 200 m × 50 m and 300 m × 70 m (external dimensions), both with submerged draft of approximately 15 m (i.e. to be accommodated largely within existing bathymetry).

The floating docks will be fully self-contained with:

- Electrical power generation, mechanical and electrical services, etc.;
- Small workshops and accommodation which will prove useful at start-up;
- Dock wall travelling craneage on both dock walls;
- The ability to dock and undock vessels during operational environmental conditions and to demobilize to a dedicated offshore storm mooring prior to extreme storm events.

For access to the floating docks and to assist with mooring in adequate water depth, barges (typically 100 m × 30 m floating transition platforms) will be positioned and restrained. Where appropriate, barges will also be provided with fixed cranes for use at wet berths and to support materials handling generally.

In the later development phases of the project, a floating pier (of the order of 300 m long) will extend from the shoreline, with wet berths on both sides and craneage as described below. The pier will comprise multiple 50 m long × 30 m wide modular sections, connected with articulated hinges to allow vertical rotation and lower bending moments/stresses.

The barges and floating pier structures will be capable of supporting deck loads up to 10 t/m^2. Though load distribution will be dictated by stability requirements such as freeboard and list angle.

Given the relatively protected conditions at the site (i.e. low operational wave heights and wind speeds), mooring systems for the floating structures (floating docks and pier) will be as indicated in Figs. 1 and 2. That is, the principal floating infrastructure will be moored to:

- Barges at the landward end, i.e. in shallower water;
- Floating mooring dolphins at the seaward end, i.e. in deeper water depth, with more substantial catenary mooring systems.

As noted in Sect. 2, the mooring system adopted will allow for relatively rapid demobilization of the principal floating structures, e.g. for temporary positioning on a storm mooring in advance of a significant storm event (which may be forecast typically more than three days in advance using modern systems).

3.3.2 Onshore

Piled abutments will be constructed at the shoreline, with modular linkspans providing access to the floating structures (i.e. barges, floating docks and floating pier). Provided principally through a central services corridor, adequate mechanical and electrical services will be required for all facilities (buildings, wet berths, floating docks, etc.), including:

- Electric power (assumed by portable generators) and lighting (24 h);
- Potable water, seawater, grey water and production water;
- Production gases for welding, cutting, etc. (supplied in bottles);
- Compressed air for pneumatic tools, cleaning, etc.
- Telephone, broadband and marine radio (VHF).

Craneage is required for efficient, safe yard lifting operations throughout the shipyard, and will include provision of:

- Mobile gantry cranes, for use in laydown areas and, where applicable, production workshops;
- Mobile cranes (up to 500t maximum capacity), for use both onshore and on the barges and floating pier;
- Fixed tower cranes (approx. capacity 10t at 30 m) on barges and the floating pier, with ability to operate at list angles up to 2°.

Principally of steel portal frame and fabric construction, workshops on the shipyard site will be provided for engineering and fabrication. Management/administration offices and personnel facilities will be provided of masonry construction.

3.4 Development Phasing

The yard will be developed in phases as outlined below and in the figures overleaf, with low initial capital outlay and early cash flow from operations to generate funds for further development.

3.4.1 Phase 1

Figure 4 shows the project layout for Phase 1.

- One floating dock moored to the north of the site, perpendicular to the shore and connected via a linkspan to a fixed position barge (moored and connected via a linkspan).
- Minimal onshore development will be present, consisting mostly of temporary buildings.
- The barge and floating dock will allow four wet berths to be established.

3.4.2 Phase 2

Figure 5 shows the project layout for Phase 2.

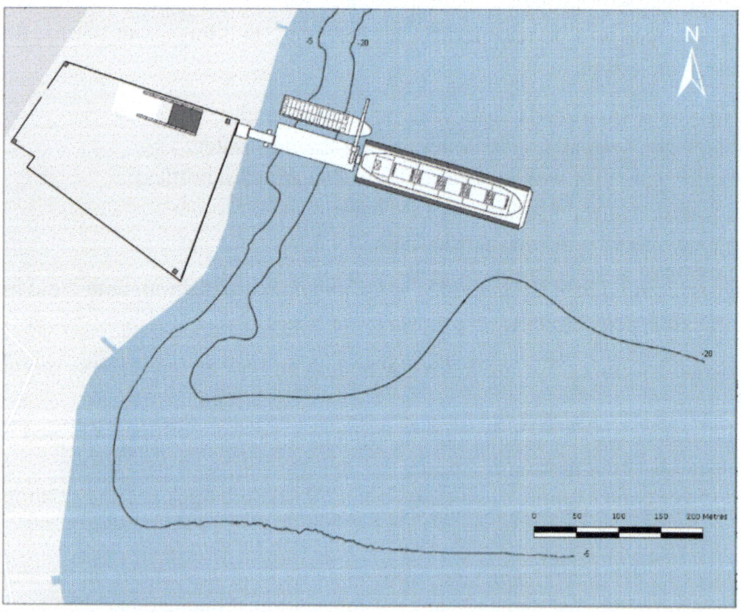

Fig. 4 Project layout—Phase 1

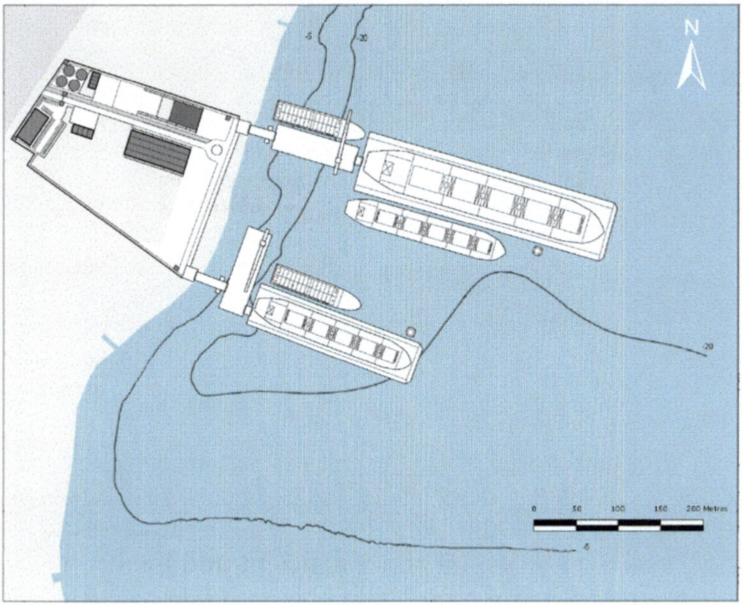

Fig. 5 Project layout—Phase 2

- A second floating dock will be provided, with both docks moored perpendicular to the shore at the south of the site and access provided via removable ramps to a barge moored parallel to the shore.
- The previously installed barge to the north of the site will remain, thus providing five wet berths in total.
- Further developments onshore to provide additional workshop space and administration will take place.

3.4.3 Phase 3

Figure 6 shows the project layout for Phase 3.

- A floating pier of nominal length 300 m will be provided, increasing the number of wet berths.
- All workshops and buildings will be completed and equipped, and services installed to suit.
- Both floating docks will be relocated to the north side of the project site.

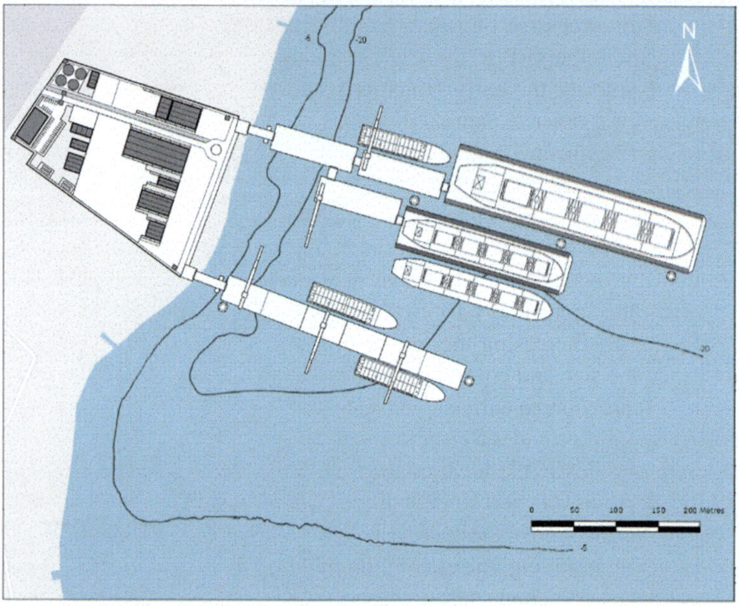

Fig. 6 Project layout—Phase 3

3.5 Summary

3.5.1 Facilities

Table 1 presents a summary of the facilities provided in each phase.

3.5.2 Manpower

Table 2 provides a summary of the approximate manpower estimated for each phase.

The manpower estimates are based on the following:

- Initial values in accordance with similar facilities in Europe;
- Increases to allow for operational efficiencies likely at the project site;
- Likely shortage of subcontractors at the project site.

Table 1 Facilities provided in each phase

	Phase 1	Phase 2	Phase 3
Marine			
Floating Docks	1No. 200 × 50 m	1No. 200 × 50 m 1No. 300 × 70 m	1No. 200 × 50 m 1No. 300 × 70 m
Pier	–	–	1No. 300 × 30 m
No. of Wet Berths	4	5	7
Total wet berth length	600 m	1250 m	1400 m
Onshore			
Total Land area	10,000 m^2	24,000 m^2	35,000 m^2
Buildings	1No. production + 1No. gatehouse + 4No. watchtowers	9No. production + 1No. gatehouse + 4No. watchtowers	12No. production + 1No. gatehouse + 4No. watchtowers
Services			
Electricity	Portable generators	Portable generators	11 kV 3 phase 50 Hz
Lighting	Portable	Portable	Full site
Potable water	Water tanks	Water Tanks	3 bar 150 mm ring main
Gasses	Supplied on pallets. Stored in 40 ft containers		
Compressed air	Mobile units		

Table 2 Manpower provided in each phase

	Phase 1	Phase 2	Phase 3
Shifts per 24 h	1	2	3
Direct Labour	150	400	500
Staff (support)	50	130	150
Total	200	530	650

4 Conclusion

The concept of a shipyard comprised exclusively of floating structures has been considered herein. Such facilities are more suitable for locations with naturally deep water and relatively tranquil metocean conditions, i.e. to accommodate floating docks and other floating structures. Benefits include flexibility to adapt the facility when required and potentially re-locate all or part of the shipyard if required due to changes in market conditions. Floating shipyards are likely to be more suitable for ship repair and not for shipbuilding.

A relevant example has been provided from an ongoing project under development in a remote location that is ideally suited to the concept, i.e. with relatively sheltered and natural deep water. As noted above, onshore areas are available for development at the site, and onshore facilities provided for the project comprise production and support infrastructure (e.g. fabrication, management, administration and accommodation), while the major ship repair assets are comprised of modular floating structures (floating docks, pier and barges).

A comparable ship repair facility comprised wholly of floating structures would require additional floating structures to support the following principal facilities:

- Material offloading/supply quay;
- Production workshops;
- Power generation;
- Accommodation for workers and staff.

As noted above, such additional facilities could readily be incorporated through additional modular floating structures, e.g. dedicated accommodation barges, floating supply piers, etc., if required in future phases.

For projects where onshore areas are either unavailable or inaccessible, the floating shipyard facility would have the aforementioned benefits of flexibility and adaptability with respect to facilities and location.

References

1. BS 6349–3. (2013). *British Standards, maritime works—Part 3: Code of practice for the design of shipyards and sea locks.*
2. Lim, S. H. (2016) Applying VLFS for shipyard operations. In *The 14th East Asia-Pacific Conference on Structural Engineering and Construction*, Ho Chi Minh City, Vietnam, January 6th, 2016.
3. BS 6349-6. (1989). *British standards, code of practice for maritime structures—Part 6: Design of inshore moorings and floating structures.*
4. PIANC. (1994). *Floating breakwaters, practical guide for design and construction, report of working group 13 of Permanent Technical Committee 2.*

Floating Bridges and Submerged Tunnels in Norway—The History and Future Outlook

Torgeir Moan and Mathias Egeland Eidem

Abstract To improve the efficiency of land transport, bridges, submerged tunnels and subsea tunnels are introduced to replace ferries to cross straits. For wide and especially straits with a large depth or very soft bottom, floating bridges or submerged tunnels are attractive. Modern floating bridges can be traced back to the pontoon bridge design implemented in the 1940s, and the notable Hood Canal bridge in 1961. More recent floating bridges include the two Norwegian floating bridges: the 845-m long Bergsøysund and the 1246-m long Nordhordland floating bridges built in 1992 and 1994, respectively. Submerged floating tunnels have been considered as an option for strait crossings, especially wide crossing such as the Gibraltar and Messina straits and the Høgsfjord in Norway. So far submerged floating tunnels have not been built, while immersed tunnels have been used in many places, essentially in relatively shallow water. Currently, the Norwegian Public Road Administration (NPRA) is assessing replacing ferries across 8 fjords by providing bridges or submerged tunnels on the Coastal Highway Route E39 Project. The width of the strait crossings is up to 5 km and the water depth is up to 1300 m. The NPRA is currently considering three alternative floating bridge concepts: curved, end anchored floating bridge or straight, side anchored floating bridge with mooring system and floating suspension bridge with pylons supported by TLP or spar floating bodies; as well as submerged tunnel type concepts. This paper presents an overview of relevant concepts, their characteristic behaviour and design criteria for serviceability and safety, especially dynamic response due to environmental and accidental loads, with a highlight on development trends.

Keywords Buoyant bridge concepts · Design · Serviceability · Safety · Dynamic response

---

T. Moan (✉)
Department of Marine Technology, Centre for Autonomous Marine Operations and Systems, The Norwegian University of Science and Technology, Otto Nielsens Veg 10, 7491 Trondheim, Norway
e-mail: torgeir.moan@ntnu.no

M. E. Eidem
Fjord Crossing Project, Western Region, Norwegian Public Roads Administration, Askedalen 4, 6863 Leikanger, Norway

C. M. Wang et al. (eds.), *WCFS2019*, Lecture Notes in Civil Engineering 41,
https://doi.org/10.1007/978-981-13-8743-2_5

1 Introduction

Modern floating bridges can be traced back to the pontoon bridge design that was implemented in the Hobart Bridge, Australia in 1943, and was a first of its kind in the world. Other early designs were the floating bridges built across the Lake Washington in the Seattle area in 1940 and 1963. Other examples of modern floating bridges include the 1988 m long Hood Canal Bridge built in 1961 [1]. A main reason for choosing floating bridges was the long span and the depth of water and thickness of a layer of soft soil to solid ground. Currently four floating bridges are in operation in the state of Washington [2]. Two floating bridges are in operation in Norway, namely the 845 m long Bergsøysund Floating Bridge built in 1992 over a fjord depth of 320 m and the 1246 m long Nordhordland Floating Bridge built in 1994 over a 500 m deep fjord (see Fig. 1) [3–5].

So far submerged floating tunnels (SFTs) have not been built while immersed tunnels (resting/buried in the seabed) have been used in many places, essentially when the water depth is relatively small [6]. Submerged floating tunnels would be competitive when the water depth is large and wave and wind conditions are severe and there is a heavy ship traffic. There have been projects considering submerged floating tunnels for strait crossing, especially wide crossing such as the Gibraltar and Messina straits and Qiandao lake in China [7].

Since the 1960s, the Norwegian Public Roads Administration (NPRA) has been heavily involved in developing SFTs. For instance, the Høgsfjord SFT project aimed at developing the technology needed for a submerged floating tunnel across the Høgsfjord south of Stavanger in Rogaland County [3, 7–9]. Four different SFT concepts were considered; three of which used a concrete tunnel and the fourth a steel tunnel. Three were connected to pontoons on the sea surface and the fourth used vertical tethers to the seabed. All of the concepts consisted of a 1400 m long tunnel between the abutments, an inner diameter of 4.75 m, allowing two traffic lanes and a narrow footpath on one side. The minimum depth for ship passage was 20 m. Significant research and engineering effort were devoted to this SFT. Even if the technology was

(a) Nordhordland bridge (b) Bergsøysund bridge

Fig. 1 Existing floating bridges in Norway

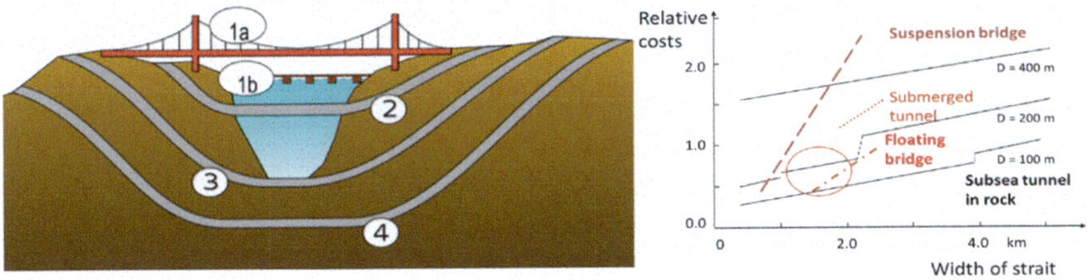

Fig. 2 Indicative relative costs of strait crossing based on tunnel in the ground (rock), submerged floating tunnel and floating bridge or free span suspension bridge for a fjord depending on its width and a depth D. Adapted after [7]

found to be feasible the Høgsfjord SFT was never built, due to a political re-decision of the road system in the region.

Currently, NPRA is assessing replacing ferry transport across 8 fjords by providing bridges, submerged tunnels or subsea rock tunnels on highway E-39 from Trondheim to Kristiansand, through the Coastal Highway Route E39 Project [10–12]. The width of the strait crossings is up to 5 km and the water depth is up to 1300 m. Owing to the maximum allowable slope of roadways, a subsea tunnel in rock will be very long and expensive (Fig. 2). Due to the width and depth of the fjords, the free span for bridges with towers on land, will in most cases get an excessive span. The largest free span of existing bridges is about 2 km. Some deliberations on using a 3700 m single span suspension bridge have been made by NPRA [13]. NPRA is considering three possible floating bridge concepts and a submerged tunnel type of concept for strait crossing (Fig. 3). Floating bridge concepts in general include [4, 11]:

A: Curved, end-anchored floating bridge on discrete pontoons combined with a cable stayed high bridge
B: Straight, floating bridge on discrete pontoons anchored by a mooring system combined with a cable stayed high bridge
C: Floating suspension bridge with two pylons supported by TLPs or spar floating bodies.

Concept A will be an extension of the existing two floating bridges in Norway (Fig. 1) [4]. However, increasing the length implies additional challenges as discussed below. Feasibility studies have been carried out or are underway, e.g. for the Sognefjord, Bjørnafjord and Sulafjord. as an example, the crossing site of Bjørnafjord is approximately 5 km wide and 550–600 m deep. Concepts A and B are the finalists for this site [4, 11].

Another issue is that the bridges need to be designed for passage of ships. This would normally imply that part of the bridge needs to be high as indicated for the Nordhordland bridge in Figs. 1a and 3a, b. This feature will be directly accounted for by a floating suspension bridge (Fig. 3c) and a submerged floating tunnel (Fig. 4).

The development of technology for floating bridges has benefitted from transfer of technology from other engineering sectors and implemented by consideration of the

(a) Curved, end-anchored bridge

(b) Side-anchored straight bridge

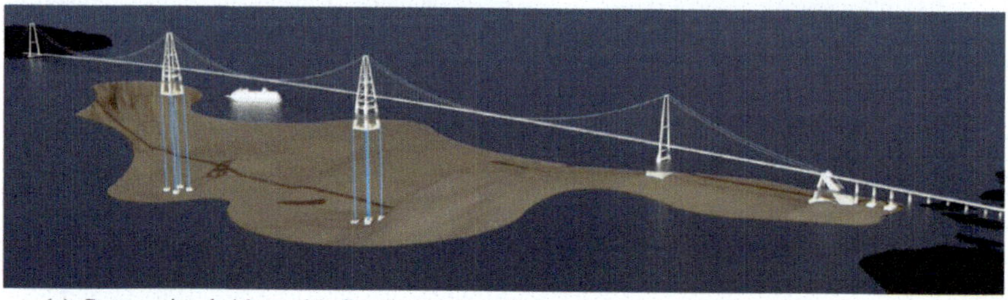

(c) Suspension bridge with floating pylons (towers)

Fig. 3 Floating surface bridge concepts. *Courtesy* NPRA

Fig. 4 Submerged floating twin tube tunnel: **a** aerial view of pontoon based support; **b** pontoon support; **c** tension-leg support. *Courtesy* NPRA

unique serviceability requirements and the target safety level relevant for bridges. In particular, the principles and methods developed for oil and gas platforms and ships as well as for floating airports in the Mega-Float project in Japan during 1995–2000 and the MOB project in USA, especially in the period 1997–2000 [14–16], have provided a basis for design and analysis of floating bridges and submerged tunnels.

In this paper, design and analyses of floating surface bridges and submerged tunnels are addressed; by describing the evolution of typical concepts, briefly summarizing design criteria and characteristic features of the behaviour of floating bridges and submerged floating tunnels, with an emphasis on the response to environmental and accidental loads.

2 General Considerations

2.1 *Societal Issues for Design*

Bridges make up a part of the road infrastructure in order to improve the efficiency of the transport system. At the same time, bridges should in general fulfill sociopolitical criteria that address the aesthetics, environmental sustainability, budgetary and legal constraints. Moreover, they should not represent a hindrance to ship traffic or other activities in the fjord.

The bridge should satisfy serviceability and safety requirements that should be maintained during the operational life, of say, 100 years, of the bridge, with a minimum of life cycle costs. In the design of floating bridges and submerged tunnels, the following loads should be considered: dead load, hydrostatic pressure (including buoyancy), live load, wind-, wave-, seismic-, temperature change and ocean current loads; effects of tidal change, seabed movement, movements of bearings; snow load, effects of tsunamis, effects of storm surges, ship waves, seaquake, brake load, possible effects of drift ice and ice pressure, and effects of marine growths, accidental loads—such as impact loads due to collision of ships with the floating structure—fires and explosions, and be durable under corrosion and other chemical attacks. Typical criteria for floating bridges are given in e.g. [17–20].

2.2 *Serviceability*

The purpose of serviceability criteria is to ensure that the structure fulfils its function as specified by the owner; i.e. the comfort and safety of both drivers and pedestrians. Besides requirements to the width of the roadway, separation of traffic and pedestrians that use the bridge, the bridge should satisfy serviceability requirements to deflections and motions; e.g. [17–19], such that the down-time due to environmental conditions (strong winds, wave-induced motions) is limited.

For instance, serviceability criteria could limit the maximum deflection and rotation of the bridge girder to be less than 1.5 m and 1°, respectively for 70% of the characteristic traffic load. Moreover, the rotation about the bridge axis under 1-year static wind load should be less than 0.5° while the root mean square (rms) value in 1-year storm should be less than 1.5°. The rms value of the accelerations in a 1-year storm in any lane should be less than 0.3–0.5 m/s^2, depending on the speed limit. It should be noted that the design specification for the Hood Canal floating bridge had a requirement to the jerk (d^3u/dt^3) [17].

Since floating bridges need to accommodate passing road traffic, the end connections need provide a smooth transition from the roadway to the bridge under the action of vertical change of position of the bridge due to tidal water level variations and wave actions.

The downtime associated with non-serviceability due to possible weather restrictions in the use of the bridge, accidents/failures; repair/replacement of components, is an important societal measure of the bridge performance.

2.3 *Safety*

2.3.1 General

Safety means absence of accidents and faults that can lead to fatalities, or environmental damage or significant economic losses, including that caused by in-serviceability. Global structural failure, capsizing, sinking, and drifting off-station are failure modes which are taken care of under the safety criteria.

Floating bridges and submerged tunnel in Norway should in general be designed for the highest consequence and reliability class (CC3, RC3) [18, 21]. However, particular components might be designed for a lower consequence class (CC2). In addition to fatalities caused by damage or failure of the bridge structure, injury or loss of human life directly caused by water filling, fire, explosions, toxic liquids/gases especially for SFT, and directly by wind on FB, should be considered. In connection with accident scenarios on FBs or in SFTs monitoring to activate an alarm and stop traffic and initiate evacuation are essential.

2.3.2 Design Criteria

Floating bridges (FBs) and submerged tunnels (SFTs) are relatively novel structures with limited service experiences and not completely covered by existing civil engineering standards such as the Eurocodes [20]. The design basis also needs to be based on standards for offshore structures [22, 23]. Moreover, standards need to evolve through the design basis in projects in an interaction between the engineering consultants, advisors and the organization, NPRA, responsible for the road and bridge infrastructure, e.g. [19, 20], by accounting for service experiences and eventually be adopted in the NPRA bridge standard [18].

Safety is ensured by fulfilling limit state design criteria in terms of ultimate-, fatigue- and accidental collapse criteria (ULS, FLS, ALS). ULS design check of environmental load effects are based on characteristic values with reference to an annual exceedance probability of 10^{-2} and characteristic resistances and partial load and resistance factors, generally based on the suite of Eurocodes. The load effect analysis represents a particular challenge and might involve significant uncertainties. Hence, the partial load factor for environmental loads is (tentatively) taken to be 1.6, slightly larger than 1.5 in EN 1990 [21].

The layout is also chosen to minimize the effect of hazards; e.g. by providing passage for ships by using a high or submerged bridge. The layout also needs to be planned to facilitate evacuation in case of fires, explosion, water inflow (especially for submerged tunnels) as well as ship impacts. Robustness with respect to progressive failure, must be provided, taking into account the cause, failure mode and consequences by fulfilling ALS criteria (Sect. 5.1). This is done ensuring that the structure can "survive or be serviceable" after a damage due to accidental loads or other hazards at an annual probability level of 10^{-4}. Damages, determined by risk assessment [24], might include flooding of 2 compartments of a pontoon, loss of a pontoon (due to ship impact), failure of ballast system, loss of 1 or 2 mooring lines. The damaged structure should "survive" self-weight, environmental loads (at to an annual exceedance of 10^{-2}). Robustness is also required to fully utilize an inspection and repair strategy to ensure safety.

In addition, robustness is ensured by conservative assumptions of parameters affecting load effects and resistances, which are sensitive to the parameters.

The axial (compressive) force in curved end-anchored bridges is significant and affects the geometric stiffness of the bridge, especially when the bridge is slender, and could lead to parametric resonances. Criteria have been established to avoid this phenomenon and hence make the response more well-behaved, allowing conventional dynamic response analyses to be applied for this concept [25].

Water-tightness (and limited permeability of chlorides) of concrete structures, is ensured by requiring a minimum compression zone in the cross-section at the service load level. By using pre-stressing e.g. of pontoons, the ship impact resistance is also increased. In addition, no penetrations to sea, are allowed. Control of cracking during design is found to be more important than strength of pontoons for the floating bridges in the State of Washington [2]. Some of the early floating bridges have already been decommissioned and replaced by new ones [2].

Water-tightness of joints in SFTs that are fabricated by a modular approach, needs to be particularly addressed.

Owing to the corrosive sea environment, steel structures have to be protected by a coating or sacrificial anode or, as regards the bridge girder, be protected by ensuring low humidity in airtight compartments.

Possible degradation due to corrosion or crack growth (fatigue) requires a proper system for inspection, monitoring, maintenance and repair during use to ensure durability.

Regarding floating tunnels, there are no direct service experiences. Design specifications in various projects have been based on experiences with floating bridges, immersed tunnels and offshore structures, e.g. [19, 20].

2.3.3 In-Service Failure Experiences and Measures of Safety

The design requirements for various limit states in codes are established to imply a certain probability of failure. Structural reliability analysis (SRA) is for instance commonly used to calibrate the characteristic loads and resistances and safety factors in ULS requirements to correspond to a target failure probability level of structural components. However, only normal uncertainties due to fundamental variability and lack of data and not human errors and omissions, are accounted for in SRA.

In principle, the true (actuarial) annual failure rate, p can be determined by $p = n_f/N$, where n_f and N are the number of collapsed and existing structures in a year. Based on an estimated 15,000 bridges in Norway in 2011, Rimehaug [26] estimated the annual collapse rate of bridges to be about 10^{-4}. The statistical uncertainty of this estimate is significant due to the limited sample. Cook et al. [27] used a subset of 94,800 among the 604,415 bridges in USA and estimated the collapse rate to be 2×10^{-4}, i.e. quite close to that in [26]. It should be noted that the bridges in the inventory have different sizes, traffic volume (ADT) and CapEx. These estimated failure rates are one or two orders of magnitude larger that target levels mentioned in international standards [21, 22]. A reason for that may be that the mentioned standards refer to the failure probability estimated by structural reliability analysis, which represents a subset of the actuarial probability. Another aspect is that most design codes refer to component failures while the above estimates refer to "total collapse".

Cook et al. [27] categorized the failures according to the apparent physical-technical "causes" of failures as follows: Hydraulic (due to rainfall flooding, dam failure, etc.); Collision (by car, train or ship); Geotechnical; Fire and Explosion; Deterioration; Overload; Nature (environmental events) and Other (Fatigue, Design Error, Construction, Miscellaneous, or Unknown). Only in the category "Other" he touches upon the root causes of failures, namely human and organizational factors (HOF), probably because of lack of such information. As demonstrated e.g. for civil engineering structures, e.g. in [28], and oil and gas platforms e.g. in [29, 30], virtually all accidents are influenced by HOF, in terms of deficient design codes in the engineering community at large, errors or omissions by those involved in design, fabrication or operation, including QA/QC. Most likely the various categories of failure causes listed in [27] involve elements of HOF. For instance: fires, explosions and ship impacts are clearly due to errors. Deterioration is listed as a sole cause as well as contributing to more than 50% of the overload failures. Then HOF issues relate to how deterioration is treated in relevant standards and practiced in the different lifecycle phases.

There are limited experiences with floating bridges; with an accumulated bridge years of operation of the order of 200 and 50 for floating bridges in the State of Washington, USA and Norway, respectively. Yet "total" collapses have occurred for floating bridges—resulting in changed operational practice [2].

Besides defining safety by the failure probability of the bridge, it could be expressed directly by fatality rate of individuals that use the road/bridge. The fatality rate depends on the probability of structural failure and the conditional probability of fatality considering the traffic on the bridge and needs to be assessed by a risk analysis. Considering the rate of fatalities per km in the traffic as a reference, the target might be related to a limited acceptable increase in the fatality rate due to failure of the bridge. For instance, the fatality rate in road traffic per km is 3×10^{-9} for standard roads and 0.6×10^{-9} for highway in Norway, corresponding to a FAR of 25 and 5, respectively.

In societal terms fatalities and downtime can be expressed by a frequency (F) versus number (N) diagram such that F (N > n) e.g. represents the probability of number of deaths greater than n, as determined by risk analysis. The acceptance limit is defined by curves F (N > n) = C_i/n, typically defining three regions in the FN-diagram—unacceptable F, region where mitigation actions according to the ALARP principle should be applied and acceptable F, as practiced in the oil and gas industry [24]. The coefficient C_i is determined such that the expected number of fatalities corresponds to that implied by the individual fatality rate and the average daily traffic volume. Similar considerations can be made regarding downtime.

2.3.4 Lifecycle Structural Integrity Management

The desired safety is ensured by structural design, including the design against accidental conditions (ALS design check), QA/QC of analyses and design, inspection, monitoring, maintenance, repair and replacement of components. The role of QA/QC is emphasized in view of the fact that human and organizational factors (HOF) are the main cause of failures of load-carrying structures and especially in view of the novelty of concepts such as floating bridges and submerged tunnels.

An important issue is to balance the design with inspection/monitoring during operation and through design provide the necessary access for inspection, repair and replacement of components; i.e. according to the lifecycle perspective (Fig. 5).

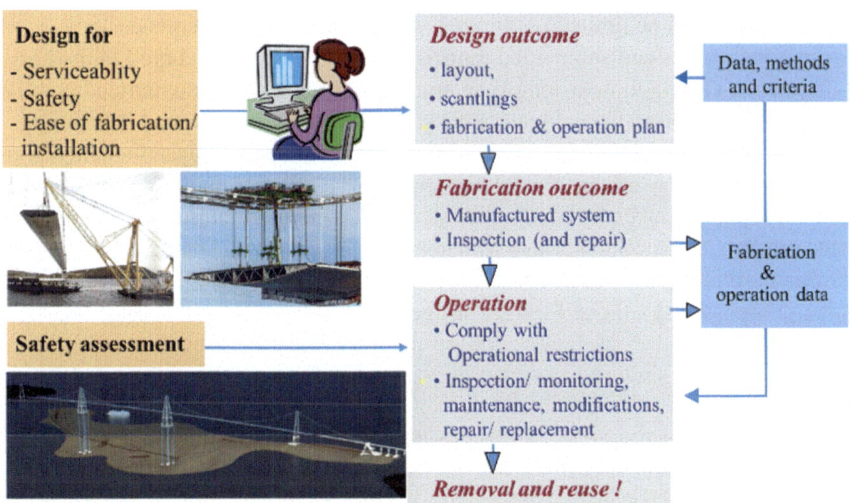

Fig. 5 Life cycle assessment of floating bridges and submerged tunnels

3 Bridge Concepts and Their Characteristic Features

3.1 General

As shown in Table 1, NPRA considers floating bridges and submerged tunnels as relevant concepts for several fjord crossings in the E39 project.

The main report of the Sognefjord engineering studies was published in [19] and in conference papers [31, 32]. Comprehensive studies have initially been made for four concepts for the Bjørnafjord during conceptual towards engineering design phases. Now, in phase 4 only end- and side-anchored low bridges with a high bridge for ship passage, remain.

A particular feature of floating bridges and tunnels is that the static self-weight and pay-loads are carried by buoyancy forces typically by pontoons. Such bridges are subjected to wave and wind actions that exhibit a time variation with a frequency that can cause dynamic effects in these bridges. Floating bridges have many natural periods that can be excited by ocean wave- and possible wind or seismic loading.

The floating bridges might be built typically with a box- or truss girder in steel while towers and pontoons could be in steel or concrete. Low pontoon bridges would normally be combined with a high bridge for ship passage (Fig. 6) while suspension bridges will have the necessary height. The impact risk on submerged tunnels due to surface vessels can be limited or eliminated by a sufficient submergence.

Floating bridges and submerged tunnels might be straight or curved in the horizontal plane. Owing to their long span straight ones need to be supported laterally by clusters of mooring lines. Floating bridges connected to the ground at their ends, experience water level variations that need to be accommodated by sufficiently flexible abutments. The effect of tidal variation on submerged floating tunnels with pontoon support is taken by bending moments in the tubes monolithically connected to the ground. A suspension bridge with tension-leg support will have very small displacements due to water level variations, since it will only cause elastic deformations in the pre-tensioned tethers. Straight bridges need to have an end bearing structure

Table 1 Relevant technologies for different fjord crossings in Highway E39 project, excluding sites where subsea tunnels and conventional suspension bridges have already been decided [11]

Fjord crossing site	Potential solutions				Year for technological resolution
	Classical suspension bridge	SFT	Floating bridge	Suspension bridge with floating pylons (TLP)	
Bjørnafjorden			x		2018
Sognefjorden	x	x		x	2022?
Vartdalsfjorden	x	x	x	x	2020
Sulafjorden	x	x		x	2020
Halsafjorden	x	x	x	x	2019
Familiar technology, to be further developed				New technology required	

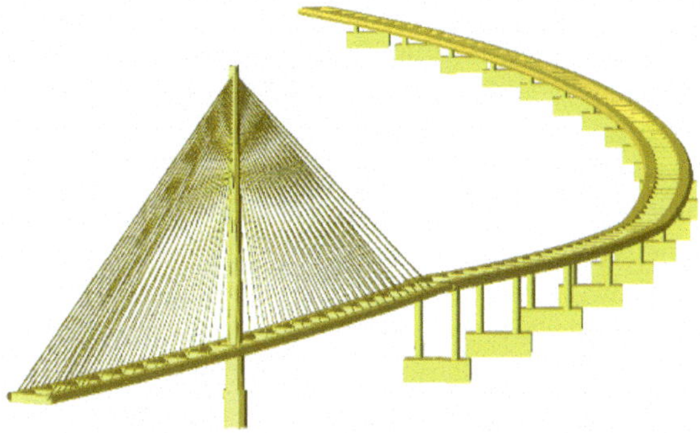

Fig. 6 An early curved, end-anchored floating bridge concept for the Bjørnafjord

Fig. 7 Flexible plates, 3.5 m long and 126 mm thick, connecting the bridge girder of the Nordhordland FB to its abutment [5]

that allow axial movements due to temperature effect on the whole bridge. Bridges with a curved shape need to be properly supported at the ends to carry especially the axial loads, yet flexible enough to accommodate imposed deformations due to water level variation (Fig. 7).

While the layout of the main girder of floating bridges is similar to that of conventional bridges, submerged tunnels need to be made in view of arrangement of the lanes for cars, bikes and pedestrians, a ventilation and light system that also work in emergency situations, a ballast system to compensate variation in weight, separate "tunnels" with escape doors to allow escape and evacuation and rescue, as well as a traffic and condition monitoring and alarm system. Alternatively, vertical escape tunnels to the sea surface has also been proposed [33].

3.2 Floating Bridges

In the vertical direction low bridges are supported by continuous or preferably discrete pontoons spaced at 100–200 m, while a suspension bridge over e.g. a 5 km span might have two pylons with free spans of a length of the order of 1.5 km. One of the concepts investigated for the Sognefjord site, with a width of 3700 m and depth of 1250 m, was a floating suspension bridge with three spans and onshore viaducts. The two floating towers are supported on concrete pontoons, moored to the seabed, while the towers themselves are made of steel. The bridge deck is an orthotropic steel box girder [31]. In the Bjørnafjord conceptual study, the pylons were assumed to be supported by a floater with tension-leg mooring [34], similar to the concept in Fig. 4c. In the study [35] the dynamic behaviour of suspension bridges with alternative floaters, was investigated.

To carry the relatively large horizontal forces due to waves and wind on low pontoons bridges two designs are envisaged. Straight bridges hence need a support by a cluster of catenary mooring lines anchored on the seabed, every 1–1.5 km. A circular or s-shaped curved bridge that carry the horizontal forces by axial forces due to the "arch effect" and with anchoring of the bridge ends.

It is noted that the stiffness relating to the water plane area of the pontoons of low bridges is important to limit the vertical displacement during quasi-static load conditions. Where the horizontal size of the structure is larger than the wave length, the resultant horizontal forces will be reduced given that different phases (direction and size) of the wave force will act on various parts of the structure.

For the floating bridges, the main hazards, besides functional-, wave-, current- and wind loads, are ship impacts and land- or underwater slides that can cause a severe transient wave condition or threaten the mooring system of side-anchored or tension-leg bridges.

3.3 Submerged Floating Tunnels

Further studies of SFTs have continued in the E39 project, notably for the Sognefjord and Bjørnafjord crossings, e.g. [9, 32, 36, 37]. In the Bjørnafjord studies, the SFT was competing with floating bridges. In the final assessment only the low bridges remain as alternatives, but the SFT is considered as an alternative for 5 other sites (Table 1).

The basic idea of the SFT is that by locating it at say 25 m water depth, it integrates well into the landscape and allows reduction of the acoustic noise pollution in addition to be less exposed to the uncertain environmental loads. The submergence also eliminates ship impacts if a tension-leg mooring is applied. If pontoon support is applied, it represents a potential target for ship impacts. Moreover, the first order wave loads are limited due to the submergence. Hence, the motions are limited and down-time due to severe environmental conditions will be negligible.

SFTs are designed by balancing weight and buoyancy to limit the costs of the tunnel. The weight, G includes that of the structure, runway, equipment, permanent and variable ballast, marine growth, water absorption while water "in-flow" is removed by pumping, road wear and accumulation of dust etc. and varies to some extent over time—which makes it important to monitor the weight. The buoyancy, B is given by the submerged volume and water density. While G and B may be of the order of 1500–2500 kN/m, their difference, in the absence of seismic loads, might be as low as 10–30 kN/m [19, 32]. This fact implies that a small percentage uncertainty in B and G has a big impact on G-B. Rather than using safety factors this uncertainty is accounted for by adding a variable fraction of G, Δ G as a certain percentage of G as a free positive or negative load, depending on the control of weight and buoyancy during fabrication and operation [19].

Regarding the layout, it can be curved with end support; i.e. carry lateral loads by the arch effect or straight with mooring system, generally considered to be a tension-leg system. The cross-section can be a rectangular or circular one or consist of two cylinders connected by horizontal cross-bracing or trusswork, for global strength and evacuation purposes. In the case of a single tube diaphragms need to be provided to ensure two separate parts so that in case of emergency evacuation can take place in the other part. In the Sognefjord conceptual study [19, 32], two circular concrete tunnels with a diameter of 12.6 m and a wall thickness of 0.8 m was chosen to accommodate two car lanes and a pedestrian/bike lane in each direction and ballast.

As mentioned above, there are two main concepts to provide vertical support; i.e. either pontoons or excessive buoyancy combined with a tension-leg mooring—to avoid slack in the tethers (Fig. 5b, c). The main geometry of the bridge tube is practically the same. In the early Høgsfjord studies anchoring to the seabed was preferred [9]. But in other situations pontoons might be chosen, depending on the ship traffic, water depth, environmental and seabed soil conditions. In the Sognefjord study [19], the focus was on pontoon support.

Various tension-leg systems have been proposed for submerged tunnels, the main difference is whether the tethers are vertical (like in Fig. 3b) or inclined. Clearly, inclined tethers provide both a large vertical and horizontal stiffness. However, inclined tethers are more likely to go slack and cause snap loading. While slack can be prevented by increasing the pre-tension, it will increase the need for buoyancy and hence the size and costs. For instance, in the oil and gas industry inclined tethers were discarded at an early stage of tension-leg platform developments.

The tether-support provides a large vertical stiffness, limiting the vertical motions and accelerations. This is the ideal concept with respect to dynamic response, but the anchoring depends on good soil conditions. Moreover, installation of tethers and foundations and inspection of the deep-sea parts of the bridge with tension-leg system is more complex than for e.g. the curved floating bridges.

The pontoons can be connected to the bridge by a solid column or a trusswork. As mentioned in Sect. 5.3 a so-called weak-link design might be considered for this connection to limit the response of the tunnel due to a ship impact load on the pontoon.

For submerged floating tunnels, the main hazards, besides functional-, wave-, current- and wind loads, are ship impacts (on pontoons), fires/explosions, flooding and land- or underwater slides that can generate severe wave conditions or directly threaten the mooring system of side-anchored bridges.

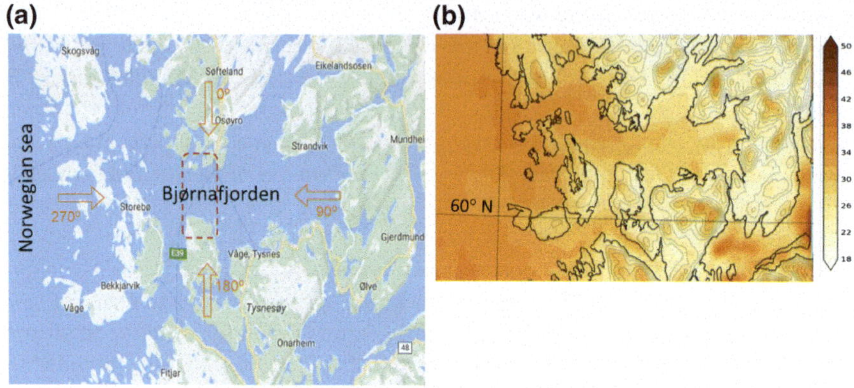

Fig. 8 **a** Topography of Bjørnafjorden, **b** Fifty year wind speed (m/s) with 10 min average at fjord at 10 m height. *Courtesy* Kjeller Vindteknikk [38]

4 Wave and Wind Induced Load Effects

4.1 Environmental Conditions

The relevant natural environment for floating bridges and submerged tunnels in Norway is wind, wave and current conditions. Seismicity might affect the response of floating submerged tunnels. The complex topography and hydrography of fjords imply complex wind and wave conditions as indicated for the Bjørnafjord in Fig. 8a, b. Wind conditions are predicted by a nested numerical approach starting with meteorological data and gradually accounting for the local topography by a refined mesh. The accuracy of predictions is validated by comparison with in situ observations made by NPRA [19, 20, 38]. Waves in the Bjørnafjord are generated by wind from North-West and East, generally by a fetch less than 20 km. Wave conditions are predicted by numerical hindcasting approaches and supported by field measurements. During a short period of time (typically 3 h) the sea surface elevation, η is assumed to be a Gaussian process, e.g. [39], specified in terms of an analytical wave spectral density, $S(\omega)$ for long-crested wave amplitudes with a given mean direction, θ_m and parameters, such as the significant wave height, H_s and peak period T_p; and a frequency independent directional distribution $D(\theta) \propto \cos^n\theta$

$$S_\eta(\omega,\theta) = S(\omega)D(\theta) \tag{1}$$

Owing to the topography, the local wind-generated waves have a limited height and period. The H_s and T_p corresponding to an annual exceedance probability of 10^{-2} at the Sognefjord and Bjørnafjord sites was initially estimated to be 2.2–2.8 and 4.5–7 s, respectively, depending on direction. Moreover, they will vary across the fjord. Very limited swell occurs at the bridge site Bjørnafjord. Tidal current and storm surge affect water level and current.

In the time domain the wave elevation may be described by a sum of long-crested waves specified by linear theory, with different amplitudes a_i, frequencies ω_i and phase angles ε_i which are uniformly distributed over $(-\pi, \pi)$.

It is noted that the sea state across the strait—along the bridge may be inhomogeneous; i.e. correspond to different mean direction, θ_m, spectra (H_s, T_p) and phase angles between the harmonic wave components [40, 41].

The current velocity is considered constant in time. However, its spatial variation is important, especially for submerged tunnels, for which vortex induced vibrations excited by an asymmetrical current across the fjord may be of interest.

The wind velocity field is described by three components; with U in the main—along wind direction. Moreover, U_w is split in a mean and turbulent component, $\overline{U}_w$ and u_w, respectively. The mean wind velocity is assumed to typically follow and exponential function over the height, h

$$\overline{U}_w(h)/\overline{U}_{w-ref} = \left(h/h_{ref}\right)^{\alpha} \tag{2}$$

Turbulent velocities u_w, v_w and w_w at two points i and j are described by wind spectra, such as that by Kaimal accounting for the coherence depending on the frequency of the wind component and the distance between the points. Further details are given in e.g. [18, 42].

4.2 *Wave and Current Loads*

The hydrodynamic modeling of wave loads for the floating pontoon and suspension bridges has been comprehensively addressed e.g. in [41, 44–46], based on the state-of-art in [39] and the review [43]. The following wave load categories might be relevant:

- first order, wave frequency loads induced by surface waves
- second order slow drift loads—at a difference and sum frequency. Such forces are an order of magnitude less than the first order loads but can impose significant response if their frequencies coincide with natural frequencies. Either a full second order theory or Newman's approximation may be used
- loads generated by internal gravity waves (especially for SFT) occurring in the sea due to layers with different density associated with influx of fresh water in a fjord, causing a layer with different density than the salt water in the fjord
- loads generated by a current inducing vortex induced vibrations (especially for SFT).

The pontoons are regarded as large volume structures, their hydrodynamic coefficients, such as added masses, radiation damping, and transfer functions of wave excitation forces, etc., are first estimated based on the potential flow theory. Possible hydrodynamic interaction between adjacent pontoons should be considered if the spacing between adjacent pontoons is of the same magnitude as the wave length. The added masses and radiation damping are then applied as radiation forces in time domain using the convolution technique [43]. The viscous drag forces on the pontoons are incorporated through the Morison's equation by considering only the quadratic viscous drag term.

The submerged tubular bridge might be subjected to vortex induced vibrations if the reduced velocity: $V_R = U_\alpha/(f_n\ D)$—where U_α is the current velocity, f_n is the natural frequency and D is the diameter of the tube, is in a critical range and the generalized "excitation force" (Eq. 5 below, with the "load" being the current velocity) is large [47, 48].

4.3 Wind Loads

Wind forces may stem from the mean pressure and the fluctuations in the incoming air flow (buffeting forces), vortices shed (vortex shedding) and oscillations of the structure itself (motion induced forces). The corresponding response to the three types of loads are treated separately, because they occur in well separately wind velocity regions, however, without clear borders. At the critical velocities, the response increases rapidly and it is assumed that the relevant forced response is determined for conditions outside the critical values [42].

Aerodynamic loads on the column, tower and cables are normally modelled by viscous drag forces. The wind load acting on the bridge girder is more complicated. In principle, it consists of three parts: the mean force due to mean wind velocity, the buffeting force due to fluctuating wind velocity in the two transverse directions, and the frequency-dependent force induced by girder motion [42]. For low bridges, the mean force and buffeting force are considered using quasi-steady theory, based on instantaneous cross sectional drag and lift forces and moment per unit length [42] while the frequency-dependent motion-induced aerodynamic forces are neglected.

Motion-induced aerodynamic loads need to be considered in determining the critical velocity of suspension bridge with respect to flutter and other instabilities [42, 49]. These phenomena especially relate to across wind direction and torsion. Xu et al. [50] estimated the critical velocity for a suspension bridge with two floating pylons, both by using the stability criterion and direct time domain analysis with very good agreement.

4.4 Structural Modelling

The global model of a long bridge will normally be based on flexible finite elements for the bridge girder, tower and cables; e.g. a beam model accommodating axial and shear forces, bending and torsional moments for the girder. Pontoons are modelled as simple elements with high stiffness or as rigid elements—eccentrically attached to the bridge girder. The pontoons are supported by buoyancy (relatively soft springs).

Given the global response of substructures (such as a section of the girder) local stresses; e.g. for fatigue analysis, is obtained by a detailed shell model. Owing to shear lag effects, the bending stresses might deviate from that of the simple beam theory. By providing additional longitudinal diaphragms in the girder, a more uniform stress distribution will be achieved, and the cost-benefit of this alternative design should be considered.

4.5 Dynamic Features of Response

The dynamic features of the bridge under harmonic loading can be conveniently judged based on the natural frequencies and mode shapes as well as damping in view of the distribution of excitation force over the bridge for different load frequencies, direction and space (phase angle for the load on different parts of the structure, e.g. [51, 52].

Figure 9 shows the mode shapes of an early 4600 m long floating bridge concept for crossing of the Bjørnafjord. The first 5 modes with natural periods are 56.7, 31.7 and down to 14.3 s are horizontal modes with some contribution from torsion. The natural periods for primarily torsional modes are 11.9 and 11.5 s, respectively. There are 20 vertical modes, some with contribution from torsion, with natural period in the range of 11.5–7.5 s, governed by the heave of the pontoons. Modes 17 and 18 have natural periods of 10.6 and 10.5 s, respectively. There are many modes, essentially torsional, with natural period in the range of 7–3.5 s [53]. Other floating bridges and submerged tunnels have similar dynamic properties.

The dynamic amplification factor, DAF of mode i with a natural frequency of ω_i for an harmonic excitation with frequency, $\bar{\omega}$

$$DAF = \left[\left(1 - \Omega^2\right)^2 + (2\varsigma\,\Omega)^2\right]^{-1/2} \tag{3}$$

where $\Omega = \bar{\omega}/\omega_i$ and ζ is the damping ratio, $\zeta = c/c_{cr}$, where c_{cr} is the critical damping. Moreover, the phase angle between the excitation and response, ϕ is

$$\varphi = arctg\left[2\varsigma\Omega/\left(1 - \Omega^2\right)\right] \tag{4}$$

It is convenient to express the effect of the spatial distribution of dynamic loading on the system by the generalized dynamic modal loading associated with a given excitation frequency.

$$q_i = \int_0^L q_{ext}(x)\phi_i(x)dx \tag{5}$$

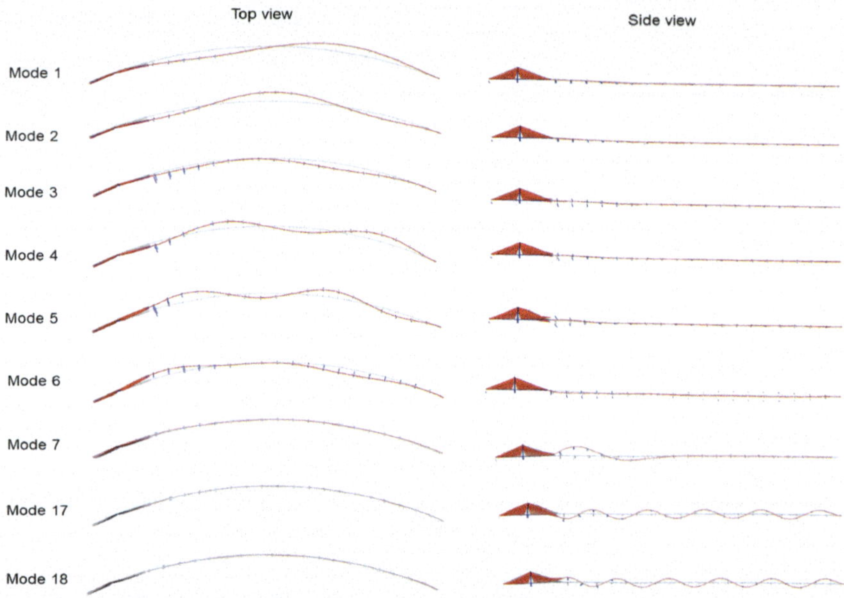

Fig. 9 Normal modes of the bridge of end-anchored bridge [53]

where q_{ext} and q_i are the spatial distribution of loading over a structure of length L at a given excitation frequency and phase angle. Analogous considerations can be made for the effect of turbulent wind.

In general, it is difficult to avoid that one of the natural frequencies of the bridge coincide with excitation frequencies. The resonant response that then occurs is very sensitive to damping. In general, the structural damping is small, the potential damping of floating bodies is limited to a certain frequency range. Wind-induced damping (at low frequencies) might therefore also be important.

4.6 Load Effect Analysis

The equation of motion may be written as [43, 45, 46]:

$$\mathbf{M}_s\ddot{\mathbf{u}}(t) + \mathbf{C}_s\dot{\mathbf{u}}(t) + (\mathbf{K}_s + \mathbf{K}_h)\mathbf{u}(t) = \underbrace{\mathbf{F}_{mean} + \mathbf{F}_{Buff}(t) + \mathbf{F}_{se}(t)}_{\mathbf{F}_{Aero}} + \underbrace{\mathbf{F}^{(1)}_{WA}(t) + \mathbf{F}^{(2\pm)}_{WA}(t) - \mathbf{F}_{Rad}(t)}_{\mathbf{F}_{Hydro}} \quad (6)$$

where M_s, C_s and K_s symbolize the still-air mass, damping and stiffness matrix, respectively, u represents the degrees of freedom of the finite element model. F_{Aero} represents the wind loads that consist of a time invariant part F_{mean} due to the mean wind velocity and a dynamic part due to turbulence in the wind field and the self-excited forces F_{se} generated by the motion of the structure F_{Hydro} represents the wave-induced loads, which consists of the radiation forces F_{Rad} induced by the motion of the submerged part of the structure, the hydrostatic restoring stiffness, and the wave excitation forces. Both first and second order difference and sum frequency wave forces, $F_{WA}^{(1)}$ and $F_{WA}^{(2\pm)}$, are considered.

The hydrodynamic radiation forces depend on the motion history and are expressed by convolution terms taking the fluid memory effect into account. It is however very time-consuming to solve the convolution integrals during a dynamic analysis, e.g. [39, 43].

A similar approach can be applied for the motion induced aerodynamic forces; i.e. starting with a frequency domain model of the self-excited force using flutter derivatives as proposed by Scanlan and coworkers and transformed into a time domain model [42, 49, 54].

A more efficient approach is to replace the convolution integral with a state-space model, e.g. [45], with the same accuracy as the initial formulation but with an order of magnitude faster computation.

Analyses can be implemented in standard Finite Element software like ABAQUS and ANSYS or software made for offshore floating structures. To account for the stochastic variation in time and space of the aerodynamic and hydrodynamic loading, some additional features need to be facilitated by additional subroutines.

4.7 Stochastic Modelling and Analysis

Limit state design checks are based on characteristic values of load effects given by annual exceedance probability, like 10^{-2} and 10^{-4} for ULS and ALS, respectively, in view of the fundamental short- and long-term variability of wave, current and wind conditions. The long-term probability distribution of load effects is obtained by superimposing the distributions for all short-term conditions multiplied by their probability of occurrence, e.g. [52]. While wave and wind processes are assumed to be stationary in 3 h and 10 min periods, respectively, a practical approach to account for the interaction between the response due to waves and wind, would be to use a 60 min period for both, with appropriate adjustment of the long-term data to refer to such a characteristic duration.

The short-term stochastic dynamic response under wave and wind-induced loads can be determined in the frequency or time domain [42, 43, 51, 52].

Since some natural and excitation frequencies may be as small as 0.01 Hz a long sample may be needed in a time domain analysis to capture the load effects, especially their extreme values. On the other hand, the time step needs to be small to capture all phenomena—including high frequency features. Proper use of “extrapolation” methods to determine extreme values, such as Weibull tail, global maxima or the ACER method is then crucial, see e.g. [52, 55].

Since it is time consuming to account for all short-term conditions in the full long-term analysis (FLTA), simplified methods such as the environmental contour method (ECM) [52, 56, 57] and simplified full long-term analysis (SLTA) [58] have been proposed. Xu et al. [46] found that the SLTA can predict the long-term extreme load effect as accurately as the FLA with about 10% of the effort. The ECM and SLTA methods are approximate and need to be validated by the FLTA, as e.g. illustrated for wind- and wave-induced response on a floating suspension bridge in [46].

The ULS design check is based on checking the nominal stress (load effects) as compared with the strength, considering yielding and possible buckling effects, in representative locations in the bridge girder cross section. Fatigue design checks are based on local "hot spot" stresses, including a stress concentration factor. The stresses used for ULS and FLS checks need to be obtained by directly relating them to the section forces and moments (normally referred to in the global analysis) based on a finite element shell analysis.

4.8 Physical Testing

Physical testing might be conducted to

- demonstrate feasibility or acceptance of a product
- provide data to support design
- provide basis for assessing the uncertainty in numerical models.

Small-scale experiments in controlled laboratory environments, e.g. [59], and field measurements in natural environments, are commonly used to assess uncertainties of predictions of the global behaviour, while full-scale laboratory tests of components are commonly carried out to validate the strength or durability of the components.

However, tests with combined wave and wind (and -current) on very large floating bridges with dynamic behaviour, is challenging due to scaling and the limited size of laboratories. The solution can then be to use a no-proper scaling, but with a proper scaling of the most important features, and use numerical analyses to generalize the model test results. An alternative to address the challenges would be to use a hybrid testing by combining a physical model, e.g. of wave loads, and a numerical model of wind loads, which is run in real-time with feedback from measurements of the physical model, e.g. as used for wind turbines [60, 61].

Kvåle [62] compared field measurements with numerical predictions of the response of the Bergsøysund bridge, with a focus on determination of natural frequencies and modes based on ambient response.

4.9 Example of Load Effects under Combined Wind, Wave and Current Loads

The effect of combined wave, wind and current on floating bridges is reviewed in [43]. An example of results for dynamic response is shown in Fig. 10 [44]. The plots display the standard deviation in each section and does not necessarily say anything about the simultaneous occurrence of the response. It is observed that the horizontal response is mainly induced by wind loads while wave excitation dominates the vertical response and torsion (not shown). It is noted that the turbulent wind can cause significantly larger low-frequency responses than second order wave excitation. A closer look at the results show that the current and a mean wind reduces the dynamic response due to their damping effects especially on the horizontal motion, axial force and strong axis bending.

5 Accidental Load Effects and the Collapse Limit State (ALS)

5.1 General

As outlined in [30] and briefly discussed in Sect. 2.3, structural failure can be attributed to

- Use of too small safety factors in ULS and FLS design to account for the inherent normal uncertainties
- Human and organizational factors (errors and omissions) in the life cycle
- Unknown phenomena to the engineering community at large.

While errors during operation typically materialize as accidental loads, errors in design and fabrication might lead to abnormal resistance, as estimated by risk analysis methodology [24]. For instance, a typical example of accidental situation relating to abnormal strength is failure of mooring lines. Hence, an ALS check of mooring systems is based on failure of 1 or 2 lines. In general, QA/QC is implemented to reduce the effect of errors and omissions. However, it turns out that this approach is not 100% effective. For this reason ALS criteria were introduced for offshore structures in 1984 to ensure more robust structures by a design check in two steps [63]:

- Estimate *accidental damage* (fabrication defects, deterioration or due to accidental loads) at an annual probability of 10^{-4} (and expected value of other loads).
- Check that the *damaged structure* survives gravity/payloads and environmental loads at a certain annual probability of occurrence (annual, 10 year or 100 year max, depending on the correlation between the damage situation and the environmental loads) and use of load factors equal to 1.0.

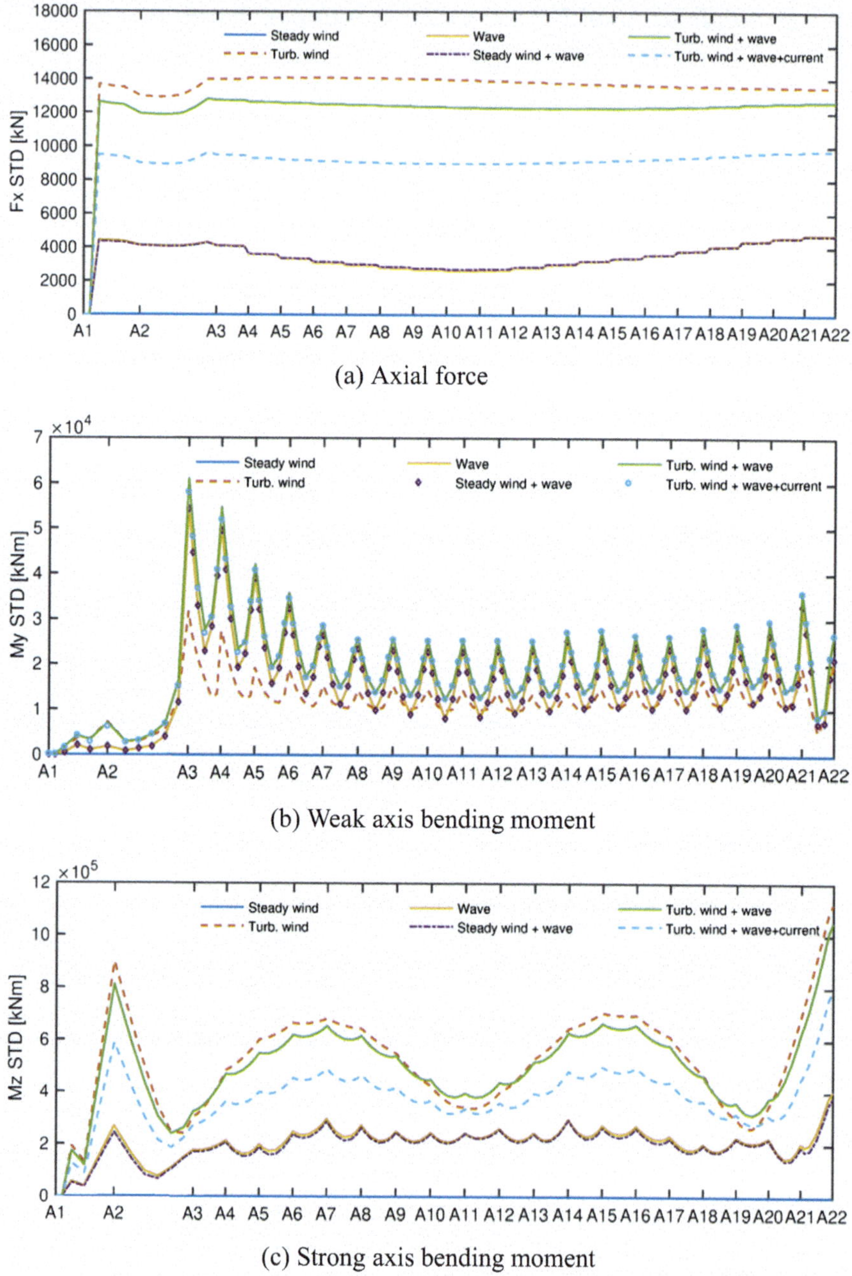

Fig. 10 Bending moments in bridge girder under wave and wind conditions. When waves are included: Hs = 2.4 m and Tp = 5.9 s. Principal wave direction = 270°. Short-crestedness parameter: n = 4. When wind is included: Uw = 31 m/s. Turbulence intensity, TI = 14%

The ALS criterion implies that events with an annual probability of 10^{-4} are allowed to cause damage. This means that a further risk assessment is needed to check the implication of the ALS and other design criteria for the bridge structure, on fatalities and traffic disturbance (downtime), as mentioned in Sect. 2.3.3.

5.2 Accidental Loads

For floating bridges and submerged tunnels the following accidental loads might be relevant:

- ship (and car) impacts
- accidental flooding of buoyancy chambers
- fire/explosions, including underwater explosions [33] (especially for SFT)
- rock or landslide—above or underwater, causing severe transient surface waves or failure of mooring anchors.

It is also noted that by introducing the ALS criteria with 10^{-4} accidental loads it was found natural to introduce ALS checks with environmental loads (wave-, wind-, current and seismic loads) at the same probability level to achieve a comparable safety level for such loads [63]. The reason is as follows: If the hazard curve: load intensity vs—log (prob of exceedance) is steep, such as for seismic loads, 10^{-4} load with safety factors equal to 1.0, may be more critical than 10^{-2} loads and safety factors of the order 1.3–1.6.

The treatment of ship impacts in design is briefly reviewed in Sect. 5.3. Accidental flooding is due to leaks due to ship impacts or ballast errors or as in the Hood Canal case [2]: an open hatch cover in a severe storm. Flooding affects people in the tunnel directly as well as the global strength. Fires and explosions occur in SFTs in similar manner as in conventional and immersed sea tunnels [6]. However, the consequences of fires and explosions for an SFT, besides the direct effect on people in the tunnel, might involve damage to the wall and hence potential flooding and global failure due to reduced bending/shear capacity and increased weight.

5.3 Ship Collisions

Bridges might be hit by passing vessels, see e.g. Fig. 11. To determine ship impact loads corresponding to an annual occurrence rate of 10^{-4}, risk analyses are performed [24]. Various scenarios are envisaged—e.g. involving various surface ship types (size, speed), impacting pontoons in the surface, the bridge girder of low bridges or submarines hitting the SFT or tension-leg mooring system. Various measures to reduce the collision hazard by specifying ship lanes (vertical and horizontal as well as keel clearance at the bridge), speed limits, safety zones, traffic control and surveillance, should be taken into account. The traffic pattern is based on long term

Fig. 11 Ships have impacted bridges. Ships might hit pontoons or superstructures of floating bridges (Courtesy: Dr. Yanyan Sha, NTNU)

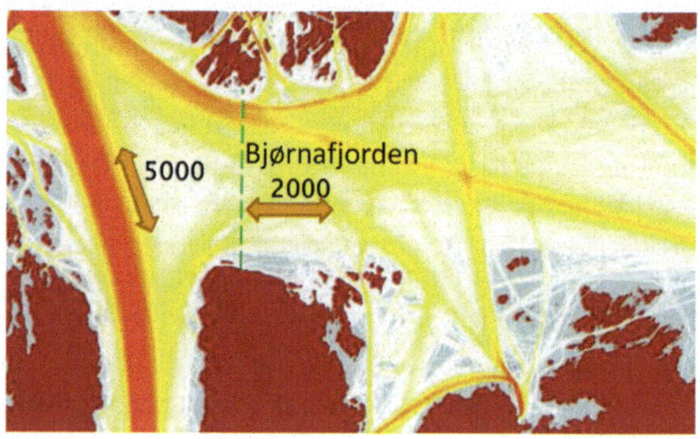

Fig. 12 Shipping in the Bjørnafjord based on the years 2015–2016. Numbers are annual passages of ships. Dashed line indicates one possible bridge site [64]

AIS recordings from the area (Fig. 12) and predicted route alterations resulting from the future presence of the fjord crossing and especially its navigation passage. Prognosis of the future change of ship traffic should also be considered. Empirically based mathematical models on expected frequencies of operational human errors and technical failures are used in the simulations. The resulting impact energy at a recurrence probability of 10^{-4} might be of the order of several hundred MNm, depending on risk reduction measures.

The consequences of various scenarios of ship impacts are first evaluated by determining the potential impact damage on the bridge. While the damage on conventional fixed bridge pillars is based on the assumption that the impact energy is absorbed

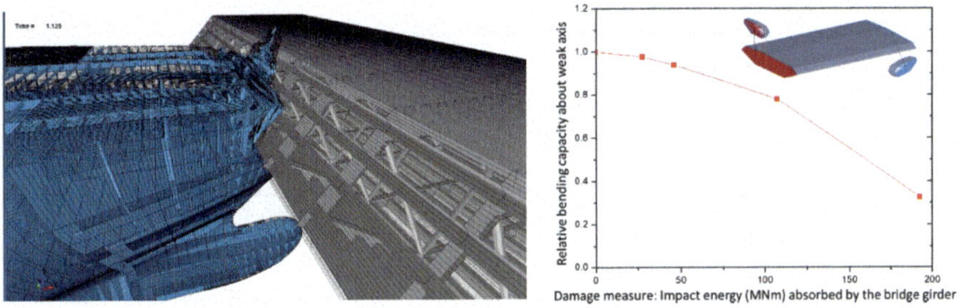

(a) Illustration of damage in a ship forecastle impact on bridge girder at moderate impact energy.

(b) Relative residual bending capacityof the bridge girder about weak axis.

Fig. 13 Ship impact on bridge girder of low bridge and residual strength after impact [67, 68]

by the ship, which facilitates an expression for the impact force as a function of the vessel displacement and velocity, FBs and SFTs are elastic structures and the kinetic energy in the impacting vessel is partly absorbed in elastic energy in the bridge and local dissipation of energy in the bridge and ship [65]. However, the local energy absorption in concrete elements of the bridge will normally be small [66]. The impact analysis is normally made by first calculating the energy absorption and force deformation relationship for the local impact, followed by a global impact analysis using the local impact features (force—displacement) with a global dynamic model. In the case study with an impact at the midpoint of the low bridge in Fig. 13 about 60% of the initial kinetic energy is dissipated in local damage, the remaining 40% is elastic energy in the bridge girder [65]. It is hence conservative to assume that the energy is absorbed locally. For moderate impact energy, say less than 100 MNm, the absorption (and damage) occurs in the ship (Fig. 13). For larger energies the bridge will have to contribute and get damage. The result of the impact is a reduced cross sectional capacity in the damaged area (Fig. 13), transient horizontal accelerations, and possible significant load effect in the bridge girder due to global response.

The consequences depend upon the ship properties (displacement, velocity, geometry, material properties, impact location, bridge geometry and material properties. In particular, it is noted that the pontoons and the SFT might be reinforced concrete. Consequences like loss of a pontoon or flooding of one or more compartments may be unavoidable. Then it needs to be shown that the bridge survives this kind of damage.

In the conceptual studies of the Sognefjord crossing the characteristic impact energy on the pontoon of the SFT at a probability level of 10^{-4} was estimated to be of the order of 1500 MNm. It was considered impossible to design for such an event. Instead, the idea of using a weak link design was introduced, e.g. [19, 32]. This implies that the pontoon-tube connection is designed to fail under large ship impact loads, thereby limiting the forces imposed on the bridge tube while for most load scenarios this pontoon-bridge tube connection is designed to be strong enough—with the necessary safety margins; i.e. also for ship impacts at a probability level of 10^{-2}, with a magnitude of the order of 50 MNm.

6 Fabrication and Installation

A modular fabrication and assembly of the full bridge takes place in a benign area. Experiences with marine operations for the installation of floating bridges are available [4, 5]. The bridge is towed to the site and hooked up. The submerged floating tunnel (SFT) could also be fabricated and installed in a similar manner [19, 32]. However, for the SFT with tension-leg mooring the tube tunnel needs to be equipped with temporary towers to provide buoyancy, and ballast, to control the lowering of the bridge at the site to connect the tethers. However, "simultaneously" connecting many tethers will be a challenging operation, but can be built upon experiences with TLPs in the oil and gas industry. The installation of the suspension bridge starts with the tension-leg or spar pylons. With the pylons in place, the installation methods used for conventional suspension bridges could in principle be used for the cables and bridge girder, even though the compliant support of the pylons cause some challenges.

7 Operation, Inspection, Monitoring, Maintenance, Repair and Replacement

Inspection or monitoring is necessary to validate design assumptions and for assessing the condition or health of the structure; catenary mooring/tether systems and their foundations and roadway; and cathodic corrosion protection system, water level sensors/pumps/ballasting systems as a basis for deciding on maintenance, repair or replacements. Weight monitoring is particularly important for SFTs due to the fine balance between weight and buoyancy that the design is based on. With a planned service life of, say, 100 years there might be parts of the bridge that has to be designed to be replaced during the operation. Even if the planned service life for bridge infrastructure is 100 years, it might be necessary to carry out repairs or replacements of structural components. It is important that inspection as well as repair and replacement of relevant components are facilitated during design. Moreover, these activities should be planned to be executed with limited road traffic disturbance.

For instance, AASHTO service life for bridges in the State of Washington in USA is 75 years. Mooring cables have a galvanized coating on outer wires and impressed current cathodic protection and are inspected by WSDOT's Underwater Inspection team. The upper 4–6 m are inspected every 2 years, while a full inspection by ROVs is carried out every 6 years. Anchor cables have a useful life of 25–30 years and replacements is a major preservation item [2]. In the first 20 years of operations of the Nordhordland bridge the main concern has been the corrosion protection of the girder bottom plates and movable parts (expansion joints, bearings). Moreover, an incident with a small cargo ship contact, caused a high stress level in stud bolts connecting the box girder to the flexible plate of some concern [5].

The inspection/monitoring, maintenance/repair procedures can be based on those for conventional bridges and tunnels (for the inside structure of SFTs), but also offshore structures, especially for underwater parts. Inspection of mooring lines and anchoring in deep fjords is challenging due to the limitations of divers capability and observation accuracy through surveillance by AUVs/ROVs. Hence, it is important that design and inspection - monitoring are balanced for an optimal cost.

8 Conclusions

To improve the efficiency of transport systems across straits, floating bridges or submerged tunnels are attractive to replace ferries, especially with reference to the Coastal Highway Route E39 project in Norway. Designing reliable and cost-effective floating bridges or submerged tunnels for a wide and deep fjord is very challenging because of the complex environmental conditions and the multiple natural frequencies of such bridges, and the potential effect of "man-induced" hazards such as ship impacts and fire/explosions. Relevant serviceability and safety criteria as well as floating bridge and submerged tunnel concepts are briefly reviewed. It is emphasized that serviceability and safety should be achieved by a lifecycle approach, recognizing proper design criteria, including accidental collapse limit states to ensure robustness, adequate competence for the innovative tasks, and QA/QC, and balanced efforts by inspection-monitoring, maintenance, repair and replacement. Significant uncertainty is associated with demonstrating durability in a planned service life of 100 years. The characteristic features of various concepts are discussed. While currently floating pontoon bridges have been built, suspension bridges with floating pylons and submerged floating tunnels also have a potential applicability. The latter SFT concept probably offers particular advantages in areas with deep water, severe wave conditions and surface ship traffic. The novelty of this concept suggests that it is first built for a strait crossing with a limited span.

Dynamic response analysis of floating bridges subjected to wave- and wind loads and ship impact loads in view of ULS and ALS criteria, respectively, is briefly reviewed. It is highlighted that the dynamic response to environmental loads needs to reflect the long-term variability in time and space of the environmental conditions and proper account of the multiple eigen-modes, which might be excited by the environmental loads. In an example, it is shown that the horizontal loads and response is governed by wind loads, while the vertical response is mainly induced by wave loads, significantly influenced by short-crestedness and inhomogeneity in the wave conditions, (as well as gravity loads due to self-weight and traffic loads.

Further work remains regarding collection of environmental data and modelling of the complex environmental conditions in fjord areas and developing simplified ULS design checks¸ including global instability of curved floating bridges, considering the variation in time and space of the combined wind-, wave- and current load effects, and the inherent uncertainty and its implication on the partial safety factors used in the design of such structures. This includes model tests of hydrodynamic and aerodynamic loads and the wave-wind interaction effect on aerodynamic loads on low bridges. In addition further assessment of the probability of ship impacts and their consequences as well as the associated risk reduction measures, is needed. Moreover, the overall actuarial risk should be estimated and compared with target frequency-consequence curves relating to fatalities and downtime. Finally, the efforts in creating innovative engineering solutions should be made with due consideration of the life cycle expenditure.

Acknowledgements The first author would especially like to acknowledge the support and cooperation with the Norwegian Public Road Administration over several years and the cooperation with colleagues in research and development projects at NTNU as well as standardization efforts for offshore structures and floating bridges.

References

1. Wang, C., & Wang, B. (2015). *Large Floating structures. Technological advances*. Singapore: Springer.
2. Rodda, N. T. (2018). *Washington state's floating bridges. Lessons learned.* Presentation (only) at the NPRA technology days, Trondheim, Nov. 01, 2018.
3. Skorpa, L., Jakobsen, B., & Østlid, H. (2017). Prof. Torgeir Moan and the record-breaking fjord crossings in Norway. Paper OMAE 2917-62659. In *Proceedings of the 36th International Conference on Ocean, Offshore and Arctic Engineering, 25–30 June 2017*. Trondheim, Norway: ASME.
4. Eidem, M. E (2017). Overview of floating bridge projects in Norway. Paper OMAE2017-62714. In *Proceedings of the 36th International Conference on Ocean, Offshore and Arctic Engineering, 25–30 June 2017*. Trondheim, Norway: ASME.
5. Skogland, J. O. (2013). The Nordhordland Bridge—twenty years in service. In *Proceedings of the 6th Symposium on Strait Crossings, 16–19 July 2013, Bergen, Norway* (pp. 330–339).
6. International Tunnelling and Underground Space Association. In *Working Group II: Immersed and Floating Tunnels*. http://www.itaaites.org/index.php?id=75&tx_ttnews%5Btt_news%5D=33&cHash=19d6b88689d67b82fd9d1a04c3946f90.
7. https://en.wikipedia.org/wiki/Submerged_floating_tunnel.
8. Skorpa, L. (2009). The Høgsfjord SFT project; what happened, why and what have been the consequences? In *Proceedings of the 5th Symposium on Strait Crossings, Trondheim, Norway*.
9. Jakobsen, B., Larssen, R. M., Egseth, G., Karlsrud, K., & Bjerkeli, L. (2013). Various SFT concepts for crossing wide and deep fjords. In *Proceedings of the 6th Symposium on Strait Crossings, Bergen, Norway, 16–19 July 2013* (pp. 430–440).
10. E-39 Coastal Highway Project. https://www.vegvesen.no/en/roads/Roads+and+bridges/Road%20+projects/e39coastalhighwayroute.
11. Choice of technologies for fjord crossings: The Coastal Highway Route E39 [10]. Accessed October, 2018.
12. Hove, K. J., & Strand, K. (2018). Improved and ferry free E39. Progress and development strategy, NPRA Technology days 2018. Trondheim, Norway. https://www.vegvesen.no/_attachment/2487050/binary/1294859?fast_title=Progress+and+development+strategy.pdf.
13. Isaksen, B et al. (2013). A 3700 m single span suspension bridge. In *Strait Crossing Conference 13–19 June 2013, Bergen*. Oslo, Norway: Proceedings Published by the Norwegian Public Road Administration.
14. TRAM (Technical Research Association of Mega-Float). (2002). Summary of Practical Research on Mega-Float Airport in 2002 (in Japanese).
15. Suzuki, H. (2005). Overview of megafloat: Concept, design criteria, analysis and design. *Marine Structures, 18*, 111–132.

16. *Mobile offshore base science and technology program* (Final Report, Technical Report TR-2125-OCN). Port Hueneme CA: Naval Facilities Engineering and Expeditionary Warfare Center, December 2000.
17. Washington State Department of Transportation. (1980). *Design criteria for the Hood Canal floating bridge* (for the reconstructed bridge of 1982), 15 Sept 1980. Seattle.
18. Norwegian Public Road Administration. (2017). *N400 handbook for design of bridges, ferry quays and other load carrying structures* (in Norwegian).
19. Norwegian Public Road Administration. *Feasibility study Sognefjorden: floating (suspension) bridge and submerged floating tunnel* (Reinertsen Olav Olsen Group Report 11744-ROO-R-002 to NPRA 31.05.2013). Subproject: Fjord Crossings/Downloads [10].
20. Norwegian Public Road Administration. (2017). *Design basis for the Bjørnafjord: Side- and end anchored floating bridges* (NPRA, 2017). *Submerged floating tube bridge* (Reinertsen—O Olsen—Norconsult, 16.04.2015). Floating suspension bridge (NPRA, 30.06.2016).
21. Norwegian Standard. (2016). NS-EN 1990:2002 + A1: 2005 + A: 2016 *Basis of design* (Norwegian version of EN 1990).
22. ISO 19900. (2013). *Petroleum and natural gas industries—offshore structures—Part 1: General Requirements and parts for different types of structures*. London: International Standardization Organization.
23. NORSOK N-001. (2012). *Structural design (—and associated suite of standards)*. Oslo: Norwegian Technology Standards.
24. Vinnem, J. E. (2014). *Offshore risk assessment*. London: Springer.
25. Øiseth, O. et al. (2018). *Dynamic stability of elastic nonlinear systems subjected to random excitation* (NTNU Report to NPRA).
26. Rimehaug, E. (2001). Norwegian bridges—reliability levels. In K. Krokeborg (Ed.), *Proceedings of Conference on Strait Crossing*. Swets and Zeitlinger Publishers Lisse.
27. Cook, W., Barr, P. J., & Halling, M. W. (2015). Bridge failure rate. *Journal of Performance of Constructed Facilities, 29*(3), 04014080.
28. Pugsley, A. G. (1966). *The safety of structures*. London: Edward Arnold.
29. Moan, T. (1983). Safety of offshore structures. In *Proceedings of 4th International Conference on Applications of Statistics and Probability in Soil and Structural Engineering*. Firenze: Pitagora Editrice.
30. Moan, T. (2018). Life cycle structural integrity management of offshore structures. *Structure and Infrastructure Engineering., 14*(7), 911–927.
31. Jakobsen, S. E., Norum Larsen, P., Sundet, E., Karlsrud, K., & Bjerkeli, L. (2013). Concept development of a Sognefjord floating bridge crossing. In *Proceedings of the 6th Symposium on Strait Crossings, Bergen, Norway, 16–19 July 2013* (pp. 136–141).
32. Fjeld, A., Haugerud, S. A., Einstabland, T., Saur Brandtsegg, A., Søreide, T., & Hillestad Sekse, J. (2013). Development of a submerged floating tunnel concept for crossing the Sognefjord. In *Proceedings of the 6th Symposium on Strait Crossings, Bergen, Norway, 16–19 July 2013* (pp. 593–602).
33. Seo, S. I., Sagong, M., & Son, S. W. (2013). Global response of submerged floating tunnel against underwater explosions. In *Proceedings of the 6th Symposium on Strait Crossings, Bergen, Norway, 16–19 July 2013* (pp. 825–835).
34. Veie, J., & Holtberget, S. H. (2015). Three span floating suspension bridge crossing the Bjørnafjord. In *Conference on Multi-Span Large Bridges, 1–3 July 2015, Porto, Portugal* (pp. 373–380).
35. Fredriksen, A. G, Bonnemaire, B., Lie, H., Munkeby, J., Nesteby, A., & Buckholm, P., et al. (2016). Comparison of global response of a 3-span floating suspension bridge with different floater concepts. Paper OMAE2016-54892, In *Proceedings of the 35th International Conference on Ocean, Offshore and Arctic Engineering*. American Society of Mechanical Engineers.
36. Minoretti, A., Myhr, A., Haugerud, S. A., Sekse, J., & Fjell Egeberg, T. (2016). The submerged floating tube bridge: The invisible bridge crossing the Bjørnafjord. In *19th Congress of IABSE*. Stockholm.

37. Eidem, M., Minoretti, A., Xu, X., Aasland, T. (2018). The submerged floating tube bridge as an alternative for a crossing: pros and cons. In *Proceedings of the 40th IABSE Symposium on Tomorrow's Mega-structures, 19–21 September 2018*. Nantes, France.
38. Ágústsson, H. (2018). E39—measurements, simulations and analysis of wind conditions for long bridges. In *Met-ocean Workshop, 12 March 2018*. Trondheim: NTNU/SINTEF.
39. Faltinsen, O. M. (1993). *Sea loads on ships and offshore structures*. Cambridge: Cambridge University Press.
40. Cheng, Z., Svangstu, E., Gao, Z., & Moan, T. (2018). Field measurements of inhomogeneous wave conditions in Bjørnafjorden. *Journal of Waterway, Port, Coastal, and Ocean Engineering., 145*(1), 05018008.
41. Cheng, Z., Gao, Z., & Moan, T. (2018). Wave load effect analysis of a floating bridge in a fjord considering inhomogeneous wave conditions. *Engineering Structures, 163,* 197–214.
42. Strømmen, E. (2010). *Theory of bridge aerodynamics*. Springer.
43. Moan, T. (2018). Dynamic action effect analysis for structural integrity management of very large floating bridges. In *9th GRACM International Congress on Computational Mechanics Chania, Greece 4–6 June 2018*. Technical University of Crete Publishing House.
44. Cheng, Z., Gao, Z., & Moan, T. (2018). Numerical modeling and dynamic analysis of a floating bridge subjected to wind, wave, and current loads. *Journal of Offshore Mechanics and Arctic Engineering, 141*(1), 011601.
45. Xu, Y., Øiseth, O., & Moan, T. (2018). Time domain simulations of wind- and wave-induced load effects on a three-span suspension bridge with two floating pylons. *Marine Structures, 58,* 434–452.
46. Xu, Y., Øiseth, O., Moan, T., & Næss, A. (2017). Prediction of long-term extreme load effects due to wave and wind for cable supported bridge with floating pylons. *Engineering Structures, 172,* 321–333.
47. DNV GL. (2014). *Environmental conditions and environmental loads* (DNV-RP-C205). Oslo, Norway: Det norske Veritas as.
48. Deng, S., Ren, H., Fu, S., Moan, T., & Gao, Z. (2019). Experimental study of vortex-induced vibrations for submerged floating tunnel using twin rigid cylinders on elastic support (submitted for publication).
49. Scanlan, R. H., & Tomko, J. (1971). Airfoil and bridge deck flutter derivatives. *Journal of the Engineering Mechanics Division, 97*(6), 1717–1737.
50. Xu, Y., Øiseth, O., Næss, A., & Moan, T. (2017). Prediction of long-term extreme load effects due to wind for cable supported bridges using time domain simulations. *Engineering Structures, 148,* 239–253.
51. Clough, R. W., & Penzien, J. (1975). *Dynamics of structures*. New York: McGrawHill.
52. Næss, A., & Moan, T. (2012). *Stochastic dynamic analysis of marine structures*. New York: Cambridge University Press.
53. Cheng, Z., Gao, Z., & Moan, T. (2017). Hydrodynamic and load modeling and analysis of a floating bridge in homogeneous wave conditions. *Marine Structures, 59,* 122–141.
54. Chen, X., Matsumoto, M., & Kareem, A. (2000). Time domain flutter and buffeting response analysis of bridges. *Journal of Engineering Mechanics, 126,* 7–16.
55. Cheng, Z., Gao, Z., & Moan, T. (2019). Extreme response analysis of an end-anchored floating bridge. OMAE2019. In *Proceedings of the 38th International Conference on Ocean, Offshore and Arctic Engineering, Glasgow, Scotland, UK*.
56. Giske, F. I. G., Leira, B. J., & Øiseth, O. (2017). Full long-term extreme response analysis of marine structures using inverse. *Probabilistic Engineering Mechanics, 50,* 1–8.
57. Haver, S., & Winterstein, S. (2009). Environmental contour lines: A method for estimating long term extremes by a short term analysis. *Transactions, Society of Naval Architects and Marine Engineers, 116,* 116–127.
58. Videiro, P. M., & Moan, T. (1999). Efficient evaluation of long-term distributions. In *Proceedings of the 18th International Conference on Offshore Mechanics and Arctic Engineering*. ASME.

59. Løken, A. E., Oftedal, R. A., Aarsnes, J. V. (1990). Aspects of hydrodynamic loading and responses in design of floating bridges. In *Proceedings of the 2nd Symposium on Strait Crossings*. Trondheim, Norway: The Norwegian Public Road Administration.
60. Sauder, T., Chabaud, V., Thys, M., Bachynski, E. E., & Sæther, L. O. (2016). Real-time hybrid model testing of a braceless semi-submersible wind turbine: Part I: The hybrid approach. In *Proceedings of the 35th International Conference on Ocean, Offshore and Arctic Engineering*.
61. Bayati, I., Belloli, M., Facchinetti, A., & Giappino, S. (2013). Wind tunnel tests on floating offshore wind turbines: A proposal for hardware-in-the-loop approach to validate numerical codes. *Wind Engineering, 37*(6), 557–568.
62. Kvåle, K. A. (2017). *Dynamic behaviour of floating bridges exposed to wave excitation—a numerical and experimental investigation* (Ph.D. thesis). Faculty of Engineering, NTNU, Trondheim.
63. Moan, T. (2009). Development of accidental collapse limit state criteria for offshore structures. *Structural Safety, 31*(2), 124–135.
64. SSPA Sweden AB. *Risk assessment for the planned crossing of Bjørnafjorden* (Report First Draft 26.10.2014).
65. Sha, Y., Amdahl, J., & Dørum, C. (2018). Local and global responses of a floating bridge under ship collisions. *Journal of Offshore Mechanics and Arctic Engineering* (Accepted).
66. Sha, Y., & Amdahl, J. (2019). Numerical investigations of a prestressed pontoon wall subjected to ship collision loads. *Ocean Engineering, 172*, 234–244.
67. Sha, Y., & Amdahl, J (2018). Design of steel bridge girders against ship forecastle collisions. *Engineering Structures* (Under review).
68. Sha, Y., & Amdahl, J. (2018). Numerical investigations of the collision damage and residual strength of a floating bridge girder. In *Proceedings of the 37th International Conference on Ocean, Offshore and Arctic Engineering, Madrid, Spain*.

Dynamics of Super-Scale Modularized Floating Airport

Haicheng Zhang, Daolin Xu, Shuyan Xia, Qijia Shi, Guangyu Yang and Rui Ding

Abstract This paper reviews a study on the nonlinear dynamics of a super-scale floating airport that consists of multiple floating modules with a flexible connection system. A novel network structure dynamics method is proposed for the dynamics prediction of the floating structure. A network modeling method is developed for the super-scale floating airport with arbitrary topological configuration and connection and the experimental validation is conducted in a wave basin. Nonlinear dynamics and network synergetic effect of the floating airport are elaborated, especially for the physical phenomenon of "amplitude death" that plays a key role in the system stability. The mechanism for the occurrence of amplitude death (AD) in non-autonomous systems is revealed and further the mathematical criterion is derived. The stability analysis based on amplitude death mechanism is carried out. Some applications of the network structure dynamics method in ocean engineering is illustrated. Finally, the prospective of the methodology is addressed, potentially extendable to many engineering problems with network structure alike.

Keywords Floating airport · Network · Flexible connector · Amplitude death · Control

1 Introduction

In order to ease the pressure on existing heavily-used land space and to develop the marine economy, the governments of the island and coastal countries have resorted to expand into the ocean space. In response to both the aforementioned needs and problems, engineers and scholars have proposed the construction of very large floating structures (VLFSs) for airports, industrial space, entertainment facilities and even military bases [1]. In comparison with the traditional land reclamation solution, VLFSs have some advantages, such as cost effective in large depth water;

H. Zhang · D. Xu (✉) · S. Xia · Q. Shi · G. Yang · R. Ding
Centre for Marine Technology, College of Mechanical and Vehicle Engineering, Hunan University, Changsha, China
e-mail: dlxu@hnu.edu.cn

C. M. Wang et al. (eds.), *WCFS2019*, Lecture Notes in Civil Engineering 41,
https://doi.org/10.1007/978-981-13-8743-2_6

environmentally friendly for marine eco-system; easily and rapid construction and expansion [1].

The VLFS concept may be traced back to 1924 when Edward Armstrong [2] proposed the concept of a seadrome as 'aeroplane' supply and navigating stations for aircrafts flying across the oceans due to the planes could not travel long distances at that time. Japan has made great progress by constructing the world's first sizeable floating runway which is one-kilometer long in 1998 in the Tokyo bay [3]. This so-called Mega-Float test model was built with the view to conduct a series of studies on the effects of wave motion on instruments for landing and taking off tests on a floating runway as well as the effects of having a very large floating structure on the sea environment and water quality. Unlike the mat-type VLFS, The Office of Naval Research of US has conducted studies on the technical feasibility and costs of building a semisubmersible mobile offshore base (MOB) [4]. A MOB is a self-propelled and modular floating system that could be assembled into lengths of the order of one mile to provide logistic support of US military operations where fixed bases are not available. Apart from the contributions to the research field of VLFSs in Japan and USA, some other countries, such as Singapore, South Korea, Norway, The Netherlands and China have also proposed or constructed several types of VLFS.

For these super-scale floating structures, the safety and stability design of VLFSs based on hydrodynamic response analysis is a significant field [5]. In the last two decades, many scholars had proposed response prediction methods and studied the dynamic characteristics of VLFSs under wave action and impact loadings. The modeling methods mentioned above are almost based on the integral modeling idea which cannot be suited for different topology configurations. Considering the massive size of structures, fluid-structure interaction and flexible connection among modules, the multi-modular floating structure operating in violent environment is typically a complex network system. Our group pioneered the application of the network science method, which comes from the humanities and social science fields, to the ocean engineering [6]. By introducing the most advanced network theory, the nonlinear dynamic characteristics of the floating airport is investigated in the past few years. The study covers the methodology development of network modeling [7], the nonlinear dynamic characteristics of the floating airport [8], the global stability of the floating airport based on the mechanism of "amplitude death" [9] and critical conditions [6], its application for configuration design of the flexible connectors [10] and stability control [11], network catastrophe predication [12] and modular configuration of an artificial floating island. This paper provides a new perspective for the engineering safety design of the floating airport and reviews the research progress and future research work for the network modeling and analysis methods.

2 Network Structure Dynamic Method

2.1 Network Science

Network science is a discipline which studies the evolutive law of complex networks such as communication networks, computer networks, biological networks and social networks, by considering distinct elements represented by nodes (or vertices) and the connections between the elements as links (or edges). The earliest work in this field can be traced back to the famous problem of Euler Seven Bridges of Königsberg in 1736. More recently, the works that Watts and Strogatz [13] reported the dynamics of small world in *Nature* and Barabasi and Albert [14] revealed the scale-free feature of network in *Science* triggered the vigorous development of network science.

Actually, engineering system also has a lot of network structures such as graphene network structure, photonic crystal network architecture. Network method has unique advantage in dealing with complex modular system because of its methodological features such as modularized modeling method and new concept to analyzed the natural characteristics of the network systems. Therefore, we propose the use of the network method to the dynamic response prediction of very large modular floating structures (see Fig. 1).

Fig. 1 Huge network structure in ocean engineering—modular floating city

2.2 Modeling Method of Network Structure Dynamics for Modular Floating Airport

In a view point of network science, a modular floating structure linked by connections in a particular topology is formed as a typical network structure where each module can be viewed as an network node and a connector is viewed as a network coupling between the nodes. The excitation forces of waves can be viewed as network action and the mooring system is viewed as network constraint. Thus, we can use the network science method to build the governing equation of motions for the modularized floating structure [7],

$$\mathbf{M}_i\ddot{\mathbf{X}}_i + \sum_{j=1}^{N}\left(\mathbf{A}_{ij}\ddot{\mathbf{X}}_i + \mathbf{B}_{ij}\dot{\mathbf{X}}_i\right) + [\mathbf{S}_i + \delta(i, i_0)\mathbf{K}_i]\mathbf{X}_i = \mathbf{F}_{i,w}e^{-\mathrm{i}\omega t} + \varepsilon\sum_{j=1}^{N}\Phi_{ij}G(\mathbf{X}_i, \mathbf{X}_j),\ i = 1, \ldots, N \quad (1)$$

where the special term $\varepsilon\sum_{j=1}^{N}\Phi_{ij}G(\mathbf{X}_i, \mathbf{X}_j)$ is the coupling term, ε indicates the coupling strength and its physical meaning is stiffness of the flexible connector. The coupling function $G(\mathbf{X}_i, \mathbf{X}_j)$ denotes the material and geometric properties of the flexible connector which may be a nonlinear function. The symbol Φ is the coupling topology matrix where its element Φ_{ij} is set to 1 if the model i is connected with the module j, otherwise Φ_{ij} is equal to 0. By introducing the topology matrix, the dynamic model enables to deal with the arbitrary topology configurations of VLFSs by only changing the assignment of the matrix element. The basic idea is shown in Fig. 2

Network structure dynamic method has three advantageous features: (1) network modeling method is a modular modeling approach which enable rapid construction

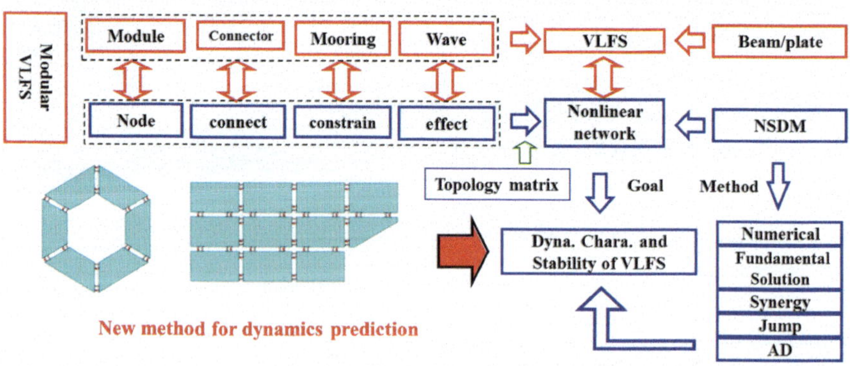

Fig. 2 Idea for modeling and methodology of network structure dynamic method

of dynamic model for modular floating structures with different topological configurations [7], (2) the characteristics of the material and geometric configurations of the flexible connectors are considered in modeling process which can reveal the realistic physical property of the connectors [10] and (3) there is a set of systemic analysis methods, such as amplitude death stability, network syngeneic effect and response jumping [8].

2.3 Experimental Validation of Network Structure Dynamic Method

In order to validate the network structure dynamic method, the numerical results of dynamic responses for a three-modular floating platform model were compared to the experimental data in regular waves [15]. The experimental investigation was conducted in the wave basin at the State Key Laboratory of Coastal and Offshore Engineering, Dalian University of Technology, PR China. The sketch of the wave basin is shown in Fig. 3 and the setup of the experimental facility is shown in Fig. 4.

Figure 5 shows the comparison of four types of results for the motions of heave, roll, pitch and connector loads. As shown in the legend, the results are respectively obtained from the experiment (named as Test), the hydro-elastic method (named as H-method), the linearized network modeling method (named as L-network), and the nonlinear network modeling method (named as N-network). The wave frequency ranges from 0.2 to 1.2 rad/s.

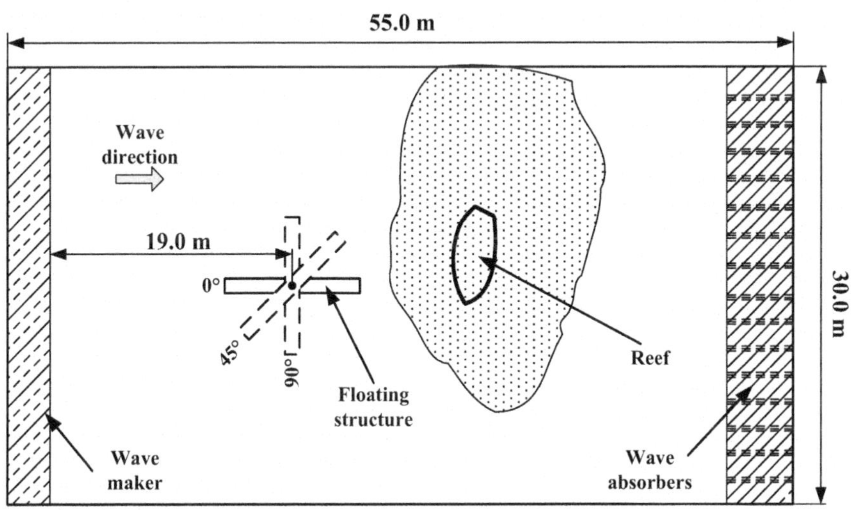

Fig. 3 Sketch of the whole wave basin

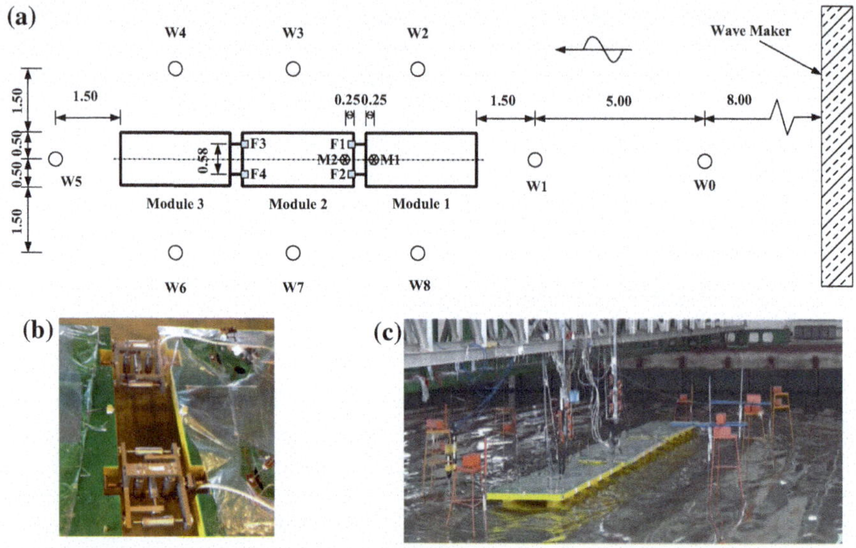

Fig. 4 Setup of experimental facility, **a** layout of sensor locations, **b** photo of connectors, **c** photo of experimental setup

According to the results shown in Fig. 5 and our paper [15], some conclusions can be drawn as follows. First of all, the network modeling methods can simulate the actual system more realistic than the H-method, especially when large motions occur. It is because the characteristics of the material and geometric configurations of the flexible connectors are considered in the modeling process which can reveal the real physical property of the connectors. Next, owing to the limitation of the linear wave theory [16] about the assumption of the constant wetted surface, all three numerical methods may overestimate the responses when large motions occur. Finally, because all the displacements and the geometric parameters are coupled in the connector loads, the outstanding performance of the network modeling method based on the connector loads may more persuasive than simple responses to reflect the usefulness of the method.

3 Amplitude Death Stability of Modular Floating Airport

Unlike the traditional concept of stability, we have proposed a new concept for the stability design of the floating airport based on amplitude death (AD) mechanism [17]. The phenomenon of amplitude death is a special network self-lock stability state which is corresponding to oscillation cease of coupled oscillators in autonomous networks [18] or oscillation suppression in non-autonomous networks [19]. In the following, the phenomenon of AD is illustrated by numerical simulations to better

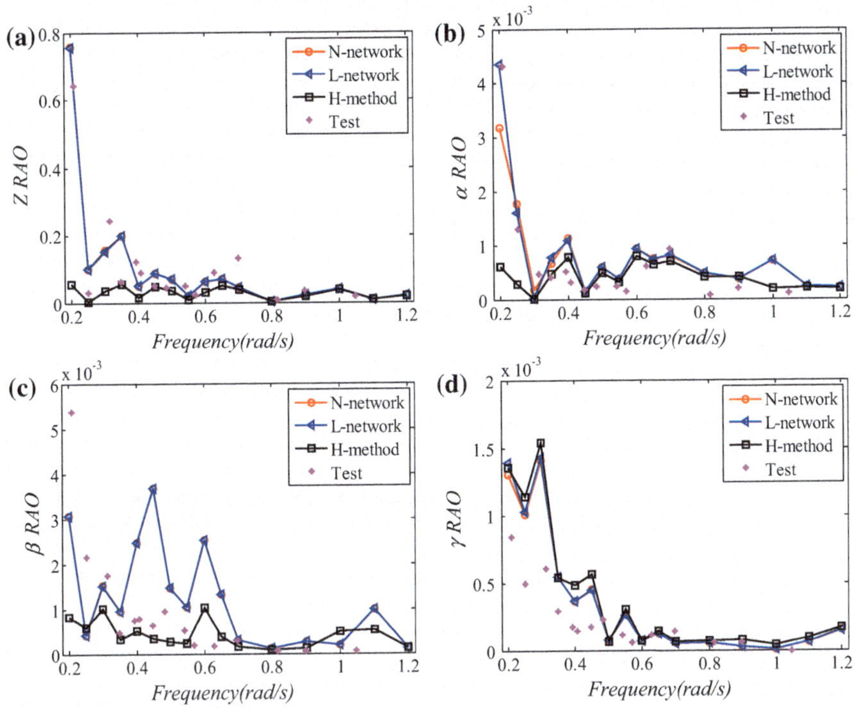

Fig. 5 Comparison between results of numerical methods and experiment for wave incident angle of 45° in **a** heave, **b** roll, **c** pitch, and **d** yaw motions

understand its physical meaning of dynamic stability for the floating airport. The mechanism of AD is investigated using the fundamental solution derived by using an averaging method. At the end of this section, some applications of the amplitude death stability are illustrated. In this section, the model of 5-module floating airport is used and its parameters are given in Ref. [20].

3.1 Amplitude Death Phenomenon of Modular Floating Airport

A chain-type floating airport coupled with promising flexible connectors named as rubber-cable connector as shown in Fig. 6, is employed to illustrate the amplitude death phenomenon. For the rubber-cable connector, the rubber provides the compressive force and the cable restrains departure motions of floating modules. Owing to the different stiffnesses of the rubber and cable, this connector has a piecewise nonlinear property and the cable also exists geometric nonlinearity due to the large size difference between the module and connector.

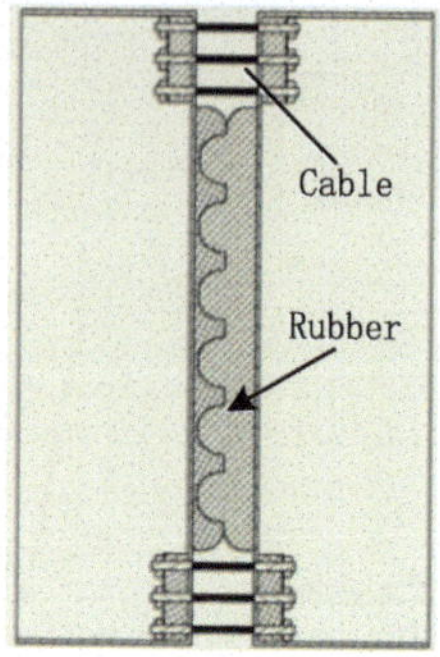

Fig. 6 Sketch of rubber-cable connector

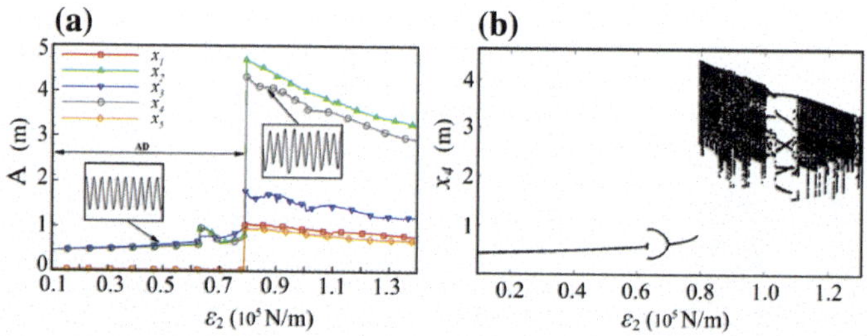

Fig. 7 Variation of dynamic response of floating airport, **a** surge response amplitudes, **b** bifurcation diagram of state variable x_4 of Module 4 [6]

Figure 7a shows the response amplitudes in surge direction as a function of rubber stiffness ε_2. The amplitudes of all modules remain at a very lower level in the interval of $\varepsilon_2 < 7.93 \times 10^4\,\text{N/m}$. The oscillation amplitudes are simultaneously amplified at a critical value of $\varepsilon_2 = 7.93 \times 10^4\,\text{N/m}$. It indicates that there is a jump-up phenomenon that terminates the weak oscillation state. This low-level oscillation state for the interval $\varepsilon_2 < 7.93 \times 10^4\,\text{N/m}$ is regarded as the amplitude death for non-autonomous network systems. In order to examine the oscillation patterns, the bifurcation diagram is shown in Fig. 6b. We find that the weak oscillation state is always corresponding to a simple harmonic motion and its period coincides with the incident wave period exactly. It is worth noting that the large amplitude oscillations after the jump always correspond with high-order harmonics or chaotic motions. Also, the amplitude death state that all modules in every DOFs oscillating with small amplitudes is significant for the safety design of the floating airport.

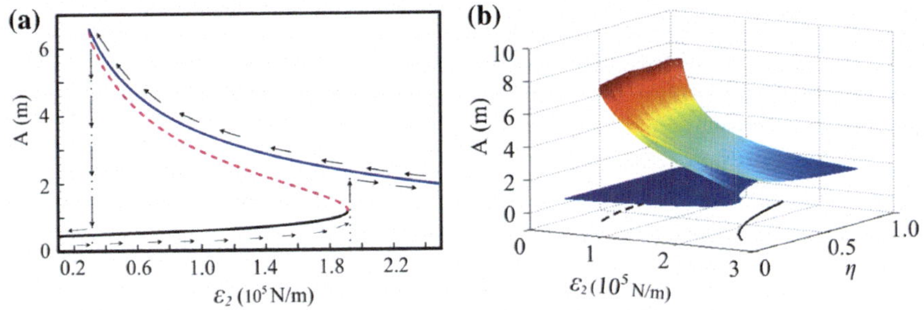

Fig. 8 **a** Fundamental solution versus rubber stiffness, **b** boundaries of bottom surface (deep blue) on parametric space [6]

3.2 *Condition and Mechanism of AD Stability*

Armed with the clue of the amplitude death state and the wave period revealed above, we know that there is a fundamental solution of which the response frequency corresponds with the wave frequency, i.e.

$$\mathbf{X}(t) = \mathbf{u}(t)\cos(\omega t) + \mathbf{v}(t)\sin(\omega t) \tag{2}$$

The above solution must satisfy the network dynamic model (1). By using the simple harmonic average method [21], a set of nonlinear algebraic equations may be derived to obtain the critical condition of AD

$$\Gamma(\mathbf{u}, \mathbf{v}, \varepsilon_1, \varepsilon_2) = \begin{bmatrix} (\mathbf{K} - \mathbf{M}\omega^2)\mathbf{v} - \mathbf{C}\mathbf{u}\omega + \mathbf{q}^1(\mathbf{u}, \mathbf{v}, \varepsilon_1, \varepsilon_2) = 0 \\ (\mathbf{K} - \mathbf{M}\omega^2)\mathbf{u} + \mathbf{C}\mathbf{v}\omega - \mathbf{q}^2(\mathbf{u}, \mathbf{v}, \varepsilon_1, \varepsilon_2) = 0 \end{bmatrix} = \mathbf{0} \tag{3}$$

After solving the nonlinear equation numerically, one obtains the fundamental solution.

Figure 8a shows the fundamental solution under the change of rubber stiffness. From Fig. 8a, we can see that the system has three solution branches in the middle interval of rubber stiffness and sometimes the solution jumps from lower energy branch to a large amplitude branch or drops down from high energy branch to a low-level oscillation branch. It is worth noting that the jump up and drop down phenomena can induce a sudden shift of the response amplitude, which is the mechanism why the amplitude death is terminated suddenly in our non-autonomous network.

This mechanism is very different from that of Hopf bifurcation [22] or Saddle-node bifurcation [23] in an autonomous system. For the safety design of floating airport, the critical condition of amplitude death is important for the design region of the systematic or environmental parameters. Based on the bifurcation theory, the condition can be written as

$$\overline{F}(\mathbf{u}, \mathbf{v}, \varepsilon_1, \varepsilon_2) = \det J(\mathbf{u}, \mathbf{v}, \varepsilon_1, \varepsilon_2) = 0 \tag{4}$$

where J is the Jacobian matrix of function Γ.

Figure 8b plots a 3D surface of the fundamental solution in the parameter plane (ε_2, η), where the symbol $\eta = \varepsilon_1/\varepsilon_2$ and ε_1 is the cable stiffness. The solid line projected on the parameters domain corresponds to the boundary where amplitude death to be vanishing when the coupling strength parameters are set beyond this boundary. To classify the region surrounded by the solid line and the dash line is our central interest for the occurrence domain of AD.

3.3 *Stability Analysis Based on AD Mechanism*

The new concept of amplitude death stability has important significance for the safety design of the floating airport. There are some potential applications based on the amplitude death stability and three examples are illustrated below.

3.3.1 Stiffness Configuration Analysis

Stiffness of flexible connectors has great impacts on responses of multi-modular floating airport. For the rubber-cable connector, to obtain the reasonable stiffness configuration of the rubber and cable is the basis of the engineering design of the connector.

Now we use the stability diagram based on AD in the space of rubber stiffness and stiffness ratio to obtain the reasonable stiffness configuration. In Fig. 9, the white region labeled with C and Pn represents large oscillation states with chaotic or sub-harmonic motions. The red domain labeled with AD indicates the weak oscillation of AD. By using the parameter design of red domain, the resonant frequency of floating airport can avoid wave excitation frequency and improve the stability of floating airport. The stability diagram based on AD provides a visual method for the stiffness configuration analysis.

3.3.2 Connector Configuration Design

The second application of the amplitude death stability is the design of connector configuration which stands for the constraint level in different DOFs. Four types of connections are proposed for assessing the performance of dynamic stability of floating airport.

Figure 10 illustrates the AD regions in the parametric domain spanned by the connector stiffness and the wave period. The red region labeled with 'AD' stands for amplitude death state and the white region indicates the large oscillation state. Viewing the diagram of AD domains for the four connectors, the AD region for the composite connector is the largest in the parametric domain. The red regions for former three connectors shown in Fig. 10a–c are disconnected in comparison with the results in Fig. 10d. This means that there is always a suitable stiffness for the composite connector to ensure the floating airport in an AD state. In comparison with the results analyzed above, we can conclude that the compound connector outperforms in the amplitude death stability of the floating airport. It implies that the ideal connector should have proper stiffness constraint in every degree of freedom, which could provide a theoretical guideline for engineering design.

3.3.3 Stability Control Strategy

From above analysis we know that the AD region is related to the stiffness of the connectors and the wave conditions. Also, a fix stiffness of the certain connector cannot retain a multi-module floating airport in an AD state for all wave frequencies. Therefore, it will be an effective method to adjust the connector stiffness to keep the floating airport in an AD state. Here, we use air-springs as the connections between modules and via changing the air pressure inside air-springs to adjust the stiffness of connectors, as shown in Fig. 11.

We develop a semi-analytical approach to maintain the floating structure always in the AD state base on the fundamental solution (3) derived above [11]. When the wave condition changes, the golden section method permits one to find the optimal stiffness associated with the smallest amplitude from the semi-analytical solution. Then we can adjust the stiffness of air-springs by increasing or decreasing the air pressure to adapt to the wave condition.

Figure 12a shows the case that the wave period changes from 9 to 20 s and the stiffness is initially set to 2.5×10^5 N/m. As time goes on, the optimal stiffness of connectors is determined by an optimization process. We can see that the stiffness is altered in each sampling interval to ensure that the response of the floating structure

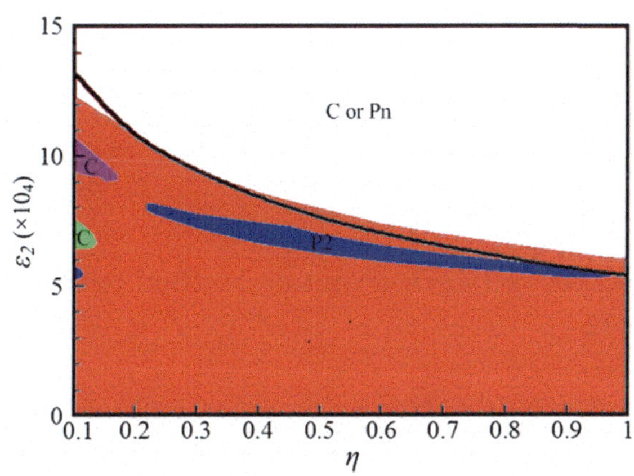

Fig. 9 Stability diagram based on AD in parameter space of stiffness. AD: amplitude death. P2: Period-2 motion. C or Pn: chaotic or high-order periodic motion

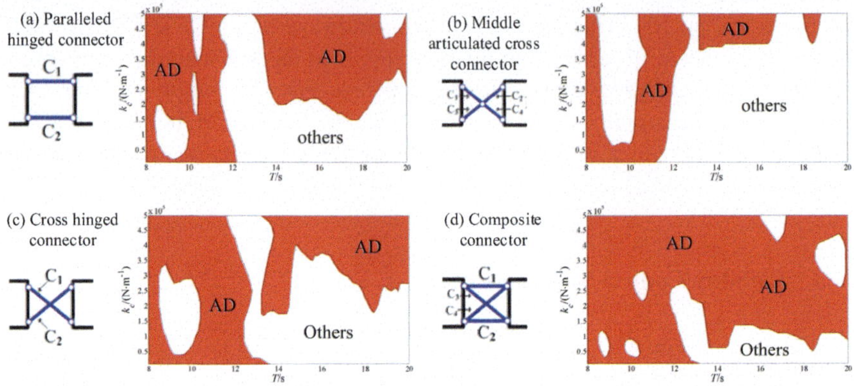

Fig. 10 Stability diagram based on AD in parameter space of stiffness and wave period for different connector configurations

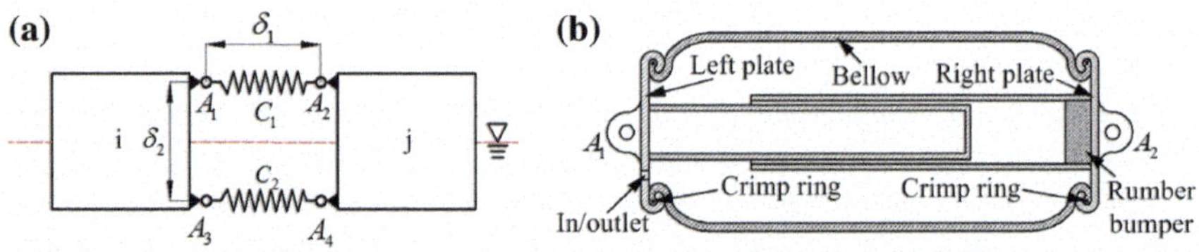

Fig. 11 Sketch for **a** connection between adjacent modules, **b** air-spring connector

is retained in an AD state. Comparing with the peaks of the responses with/without stiffness control, the amplitude drops by more than 50% which indicates that the method is effective.

In Fig. 12b, we consider a more complex case that the wave period and height change simultaneously, as illustrated in the second row diagram. The corresponding response of the floating structure is constantly retained at relatively small oscillatory level via adjusting the stiffness of the connector. It is worth noting that this method may be feeble in a full spectrum of sea conditions, and adaptive strategy should be considered.

4 Applications of Network Dynamic Method in Ocean Engineering

Apart from the remarkable AD phenomenon revealed by the network method, there are many other novel analytical tools which can be applied in ocean engineering. In this section, we will illustrate three typical examples about catastrophe predication, active control and modular configuration.

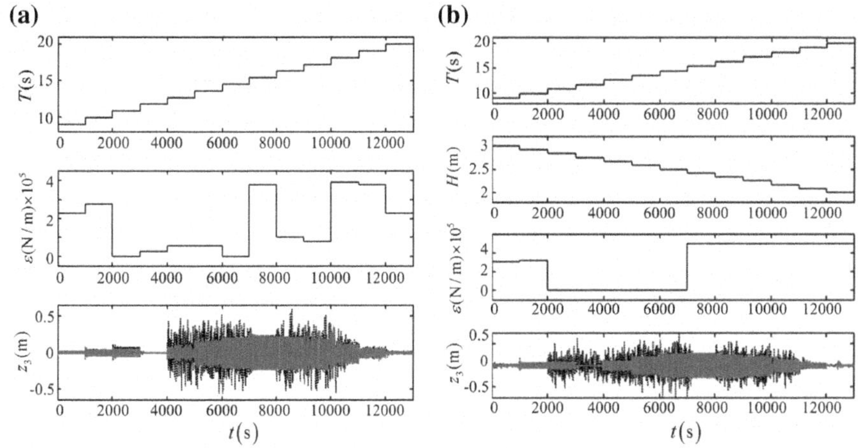

Fig. 12 Retaining system response in a weak oscillatory state when wave condition changes, **a** for change of wave period, **b** for change of wave period and height

4.1 *Catastrophe Predication Method of VLFS*

From above analysis, we know that the responses of the floating airport may jump from a weak oscillation state to a large motion state with multiple amplitude amplification, which can destroy the floating system. We call the jump phenomenon as catastrophe. We proposed a catastrophe predication method based on the reconstruction technique of complex network which can quickly and accurately predict the potential catastrophes [12]. The basic idea is illustrated as follows.

- First, using the sensors to measure the responses of the floating system and transforming the sampling time series of the responses to a certain complex network;
- Second, analyzing the evolution characteristics of the topology indexes (degree, clustering coefficient and so on) and building the catastrophe index based on the topology index;
- Third, predicting the catastrophe of VLFS by the corresponding catastrophe index.

A simple example based on the symbol network method will be illustrated in the following [24]. For a time series $x(t_i)$ sampling from the response of VLFS, the variations r_i of two neighboring data points are calculated. Each variation r_i will be assigned a symbolic element according to a mapping rule f

$$\theta_i = f(\bar{r}_i) = \begin{cases} L \;\; 0 \leq (r_i - \min(r_i))/(\max(r_i) - \min(r_i)) \leq 0.2 \\ l \;\; 0.2 < (r_i - \min(r_i))/(\max(r_i) - \min(r_i)) \leq 0.4 \\ e \;\; 0.4 < (r_i - \min(r_i))/(\max(r_i) - \min(r_i)) \leq 0.6 \\ h \;\; 0.6 < (r_i - \min(r_i))/(\max(r_i) - \min(r_i)) \leq 0.8 \\ H \;\; 0.8 < (r_i - \min(r_i))/(\max(r_i) - \min(r_i)) \leq 1 \end{cases} \quad (5)$$

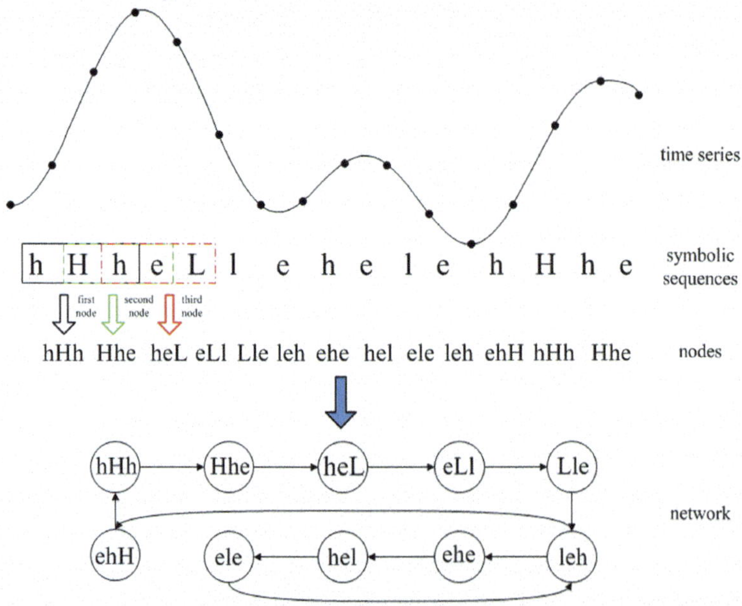

Fig. 13 Process of mapping a time series to a symbolic network

The time series $x(t_i)$ is then transformed to symbolic elements $\theta = (\theta_1, \ldots \theta_i, \theta_n)$ where $\theta_i \in (H, h, e, l, L)$. A symbolic sequence with m consecutive symbols is regarded as a node in the network while the connection is determined by the temporal transformation of nodes. Here, we take $m = 3$ as an example to introduce the procedures required for the construction of symbolic networks as shown in Fig. 13.

With different wave heights H, the floating system will stay in different motion states, as shown in Fig. 14 and the differences will be exhibited in the corresponding virtual networks. In Fig. 14a, a time series of a single period motion is converted into a network with two loops connected at a common node. In Fig. 14b, a time series of a weak sub-harmonic motion results in a network structure of a big loop with two 'shortcuts'. In Fig. 14c, as the harmonic components grow, the corresponding network is configured by a large circle with several inner connections. In Fig. 14d, a chaotic motion leads to a complex network with many inner nodes mutually connected. Figure 14 clearly indicates that there is an obvious relationship between the dynamics of the time series and characteristics of the reconstructed network. The network structure tends to be more complicated when a motion pattern develops from a periodic motion to a chaotic motion.

A Catastrophe Sensitivity Index (CSI) is proposed to measure a qualitative change of topological structures of networks which is defined as follows

$$D = \frac{\upsilon - \mu}{\mu} \times 100\% \tag{6}$$

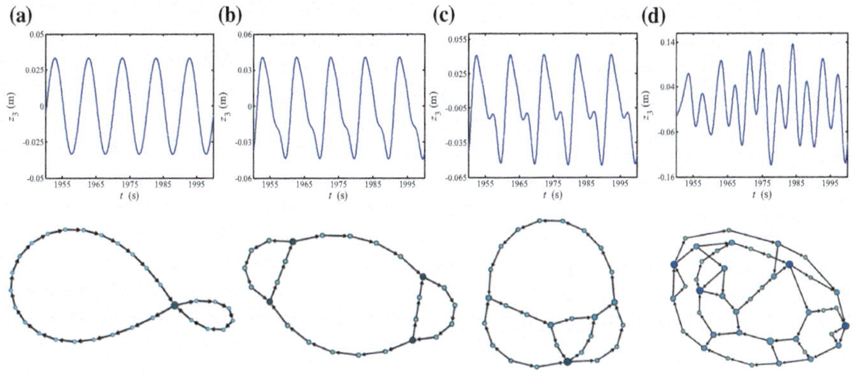

Fig. 14 Response of heave motion and corresponding network graphs generated by time series derived from different wave heights **a** $H = 2.90$ m, **b** $H = 3.20$ m, **c** $H = 3.35$ m and **d** $H = 3.50$ m with mapping parameter $m = 4$

where μ is the average path length of the network from a stable state of an original system and υ is the instant average path length of the network from present evolutionary state of the system.

A case of catastrophic occurrence induced by the variation of wave height is considered. Figure 15a illustrates a linear increase of the wave height with initial value $H = 1$ m. The response of heave motion z_3 is plotted in Fig. 15b, where the response is initially a weak oscillation state with single periodic motion before time of 5930 s, and it clearly turns into a large chaotic motion after time of 6740 s, which indicates the occurrence of floating platform catastrophe. The CSI is depicted in Fig. 15c. It shows that the index remains constant at zero until the time $t = 5931$ s, and then it suddenly drops and fluctuates. The time interval from the drop of the index to commence of catastrophe is regarded as the pre-warning region ($t = 5931$–6740 s) bounded by two red lines in Fig. 15. Note that there is about 800 s of precious pre-warning time for oncoming catastrophic events.

4.2 Active Control of VLFS

A floating airport, flexibly connected by a series of super-scale floating bodies, faces the challenge of tight requirements on the stationarity for operations in random seas. So the active control of the floating airport excited by uncertain waves is necessary. Next, a nonlinear control strategy for the oscillation control of the floating airport excited by uncertain waves is presented [25].

To make the floating system motionless in all degrees of freedom, eight thrusters (propellers) are suggested for installation on each module and a schematic deployment of the eight thrusters is shown in Fig. 16. The first four thrusters are mainly

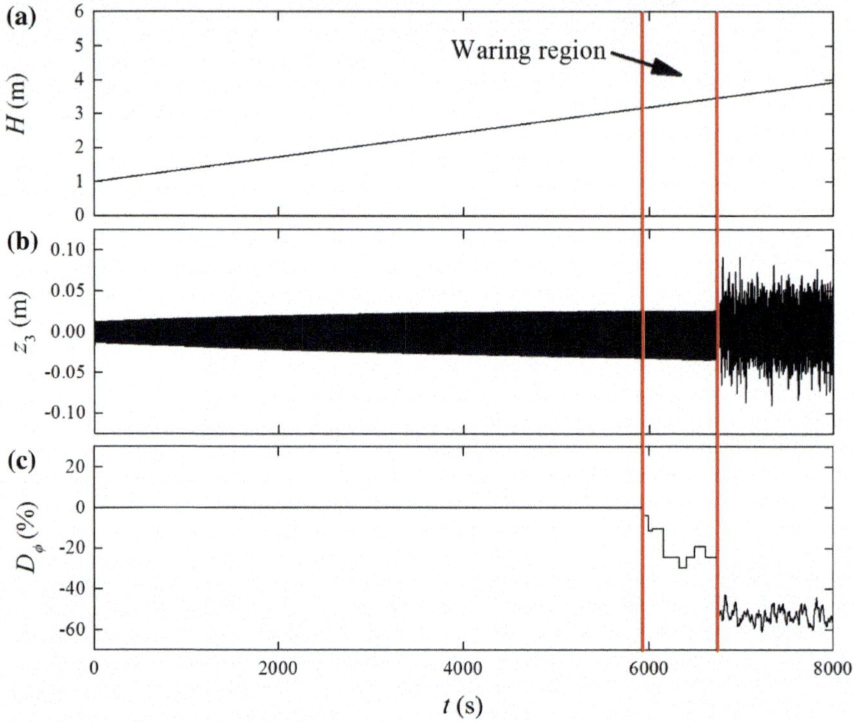

Fig. 15 Variation of **a** wave height, **b** the response of state variable z_3 and **c** CPI of average path length with mapping parameter $\mathrm{m} = 4$

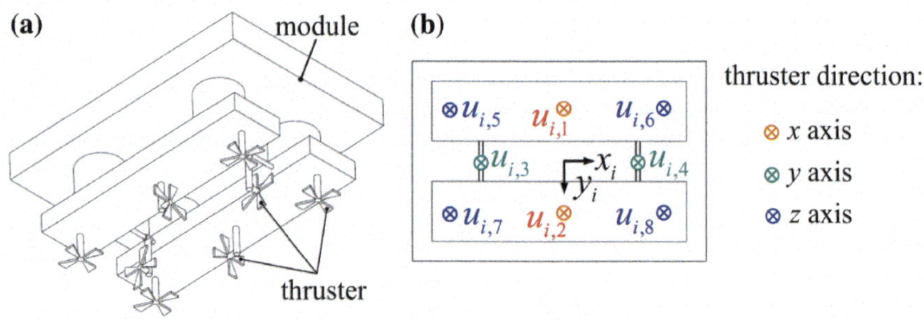

Fig. 16 **a** A schematic deployment of thrusters on a floating module and **b** a bottom view of location of thrusters and their acting directions (red in x, green in y and blue in z)

used to suppress the surge, sway and yaw motions; and the last four thrusters are employed to control the heave, roll and pitch motions. In a controlling process, all the thrusters have to cooperate together to jointly control the flatness of the modularized platform.

By applying the network modeling method, the governing equation of the controlled floating airport can be formed as

$$\mathbf{M}_i\ddot{\mathbf{x}}_i + (\mathbf{S}_i + \mathbf{D}_i)\mathbf{x}_i = \mathbf{f}_{i,w} + \mathbf{f}_{i,g} + \mathbf{f}_{i,u}\; i = 1, \ldots, N \tag{7}$$

On the right hand side of Eq. (7), the wave force $\mathbf{f}_{i,w}$ is treated as unknown and uncertain, considering that the floating system is exposed to a random sea. For the i-th module, the control term $\mathbf{f}_{i,u}$ represents the forces and torques acting on the mass center of the i-th module, so that the vector $\mathbf{f}_{i,u}$ is a function of the control variables $\left[u_{i,1}, u_{i,2}, \ldots, u_{i,8}\right]$.

The backstepping method is employed to design the control law of $\left[u_{i,1}, u_{i,2}, \ldots, u_{i,8}\right]$ in conjunction with a sufficiently smooth projection operator which leads to an adaptive scheme for the disturbance estimator in tracking the actual excitations of uncertain waves [25]. The combined action of the eight thrusters of a module and the estimated wave disturbance are expressed as

$$\begin{aligned}
\bar{\mathbf{f}}_{i,u} &= -\bar{\mathbf{f}}_{i,k} - \bar{\mathbf{f}}_{i,g} - \hat{\bar{\mathbf{f}}}_{i,un} - \dot{\mathbf{x}}_{i,c} - \mathbf{e}_i - \mathbf{e}_{i,2} \\
\dot{\hat{\bar{\mathbf{f}}}}_{i,un} &= \zeta \operatorname{Proj}_d\left(\mathbf{e}_{i,2}, \hat{\bar{\mathbf{f}}}_{i,un}\right) \\
&= \mathbf{e}_{i,2} - \boldsymbol{\eta}_1 \circ \boldsymbol{\eta}_2 \circ \operatorname{Inverse}\left(4\left(\boldsymbol{\varepsilon} \circ \boldsymbol{\varepsilon} + 2\boldsymbol{\varepsilon} \circ \hat{\bar{\mathbf{f}}}_0\right) \circ \hat{\bar{\mathbf{f}}}_0 \circ \hat{\bar{\mathbf{f}}}_0\right) \circ \nabla \mathrm{p}_d\left(\hat{\bar{\mathbf{f}}}_{i,un}\right)
\end{aligned} \tag{8}$$

For controlling the multi-modular floating airport, the magnitude of output forces generated from the thrusters should be as small as possible for energy saving but combined output force has to meet with the condition in Eq. (8). Secondly, the output forces from the thrusters should be distributed as equal as possible to avoid significant difference among outputs. Therefore, an optimization scheme is introduced to determine the eight control thrusts subject to constraints, given by

$$\begin{aligned}
&\min J = \sum_{i=1}^{N}\sum_{k=1}^{8} u_{i,k}^2 \\
&\text{s.t. } \bar{\mathbf{f}}_{i,u}\left(u_{i,1}, u_{i,2}, \ldots, u_{i,8}\right) = -\bar{\mathbf{f}}_{i,k} - \bar{\mathbf{f}}_{i,g} - \hat{\bar{\mathbf{f}}}_{i,un} - \dot{\mathbf{x}}_{i,c} - \mathbf{e}_i - \mathbf{e}_{i,2}
\end{aligned} \tag{9}$$

For N-number of modules, there are 8 N optimization variables. To solve the multi-variable optimization problem with the constraints, the sequential quadratic program (SQP) is employed. This method first determines the search direction based on a sub-problem of quadratic programming (QP) and then determines the search step size by using a proper merit function with the premise of convergence. The method permits the benefit of fast convergence in dealing with multi-variable constrained optimization problems. The SQP problem is solved by a standard software package in MATLAB.

Numerical analysis was conducted to examine the feasibility of the proposed control strategy. The model of a 5-module floating airport is considered and the JON-

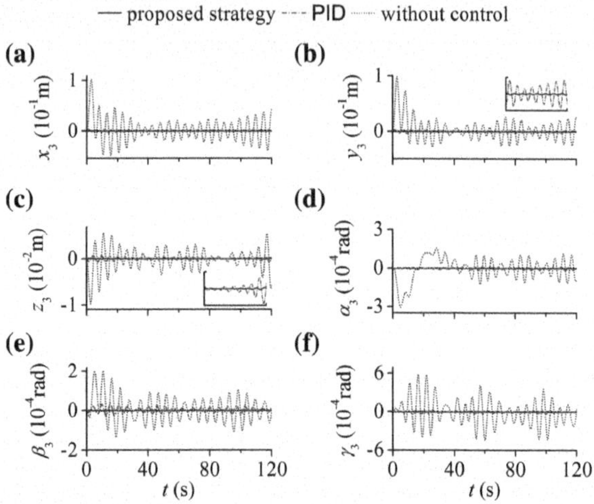

Fig. 17 Motions of third module with/without controls: **a** surge, **b** sway, **c** heave, **d** roll, **e** pitch and **f** yaw

SWAP spectrum with a significant wave $2\pi/5$ is adopted. Figure 17 only shows the six responses of the third module with/without control since the motions of the other modules are similar. The uncontrolled responses (blue-dot-line) fluctuate randomly due to the excitation of irregular waves. On contrary, the responses (black-solid-line) under the proposed control strategy are greatly suppressed. As for a comparison, the responses (red-dash-line) under the PID control scheme are also illustrated. It is clear that both control approaches perform well. Nevertheless our approach outperforms the PID scheme in terms of stationarity quality when examining the detail differences depicted in the amplified sub-plots in Fig. 3b, c.

A benefit of applying the control is that connector loads can be greatly reduced, as shown in Fig. 18. A connector comprises two springs that couple two adjacent modules together. Connector is a key component whose load is a matter of serious concern by design engineers. Here we only present the loads of two springs since the results are quite similar among springs. Figure 18a shows the connection load installed between the second and the third modules. Without control, the spring force reaches to a level near to 500 kN while the spring force with control is reduced to a very small level. Figure 18b shows the force of the connector installed between the third module and the fourth module. Similarly, the spring load is very much reduced during a control process. It is because the control suppresses the motions of modules while the connector load is associated with the relative movements between two adjacent modules. In general, the both control schemes perform well but with slight differences. The connector loads by the PID scheme fluctuate about 20 KN more than that of the proposed method.

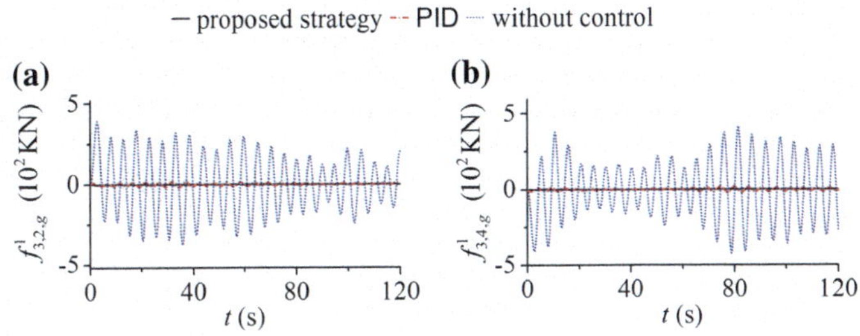

Fig. 18 Spring loads of connectors with/without control

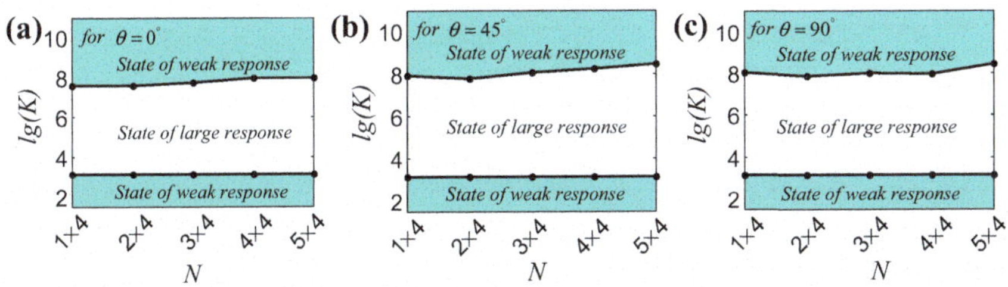

Fig. 19 Critical boundaries of connector stiffness versus quantity of modules for array-type floating system in Sea State 2 with incident wave angles

4.3 *Characteristic Evolution from a Chain-Type to Array-Type Floating Structure*

In this section, we conduct the study from a chain-type toward an array-type floating system, to gain new knowledge and better understanding of the characteristic evolution of scale-extendable islands that expand by attaching additional modules [26]. The array layouts for diverse topologies 1×4, 2×4, 3×4, 4×4 and 5×4 are investigated.

Figure 19 demonstrates the boundaries of connector stiffness versus various array layouts in Sea State 2 with three typical wave angles. We can find that the upper boundary of connector stiffness is rather steady as the platform topology evolves from a chain-type towards an array-type floating system. It suggests that connectors set above the upper boundary may be suitable for a countable scale of array-type floating systems.

Figure 20 depicts the maximum extreme responses versus the wave angle as the layout changes from a chain-type towards array-type floating systems. By the discussions about numerical study in Fig. 20, we can draw a conclusion that the response transitions are not gradual and smooth as the platform layout evolves from a chain-type to array-type floating systems except for the surge motion. In general, the responses of array-type floating systems are greater than that of a chain-type floating system, which is quite opposite to our intuition.

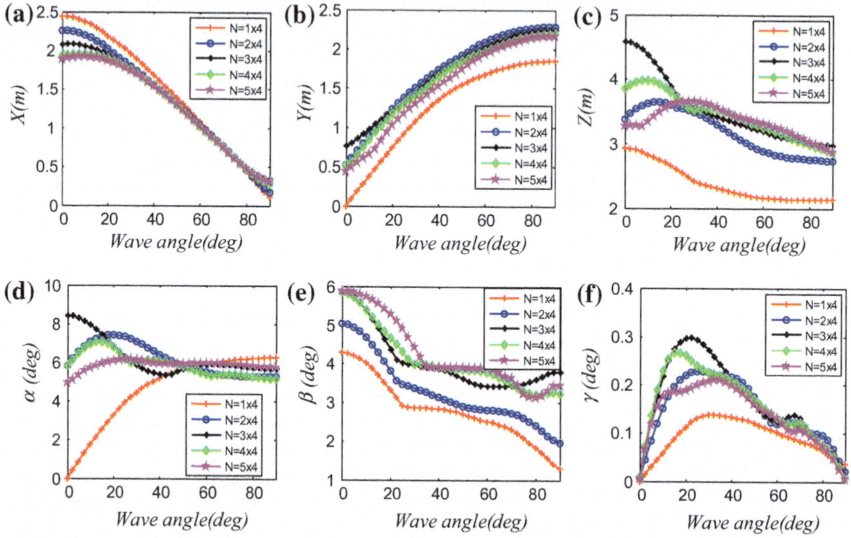

Fig. 20 Maximum extreme responses of array-type floating systems versus wave angle in Sea State 2 (connector stiffness $K = 1 \times 10^{8.7}$ N/m)

Next, we observe the connector loads as the platform layout evolves from a chain-type to array-type floating systems. The array-type floating system is flexibly connected in x and y directions. Figure 21a–c illustrates the maximum extreme loads of connectors deployed in the longitudinal direction. There are three force components, denoted by F_x^L, F_y^L and F_z^L, in x, y and z directions respectively. Similarly, Fig. 21d–f illustrates the maximum extreme loads of connectors deployed in the transversal direction, denoted by the symbols F_x^T, F_y^T and F_z^T respectively. As the scale of floating system is expanded with increasing number of modules, the load level is monotonously increased as well. This is reasonable because a larger scale floating system has to undertake larger hydrodynamic loads which cause the larger interior loads on the connectors. The variation patterns of the maximum loads along the wave angles are similar but the peak values appear in different wave angles.

5 Conclusions

Very large multi-modular floating structure with flexible connectors is a cutting-edge asset in ocean engineering for natural resource development. Based on a view point of network structure, we first introduced the network science to ocean engineering. The modular modeling method is validated via experimental test and new analysis tools are provided. Recent research progress is reviewed herein. It is worth noting that the network structure dynamics method is a new interdiscipline; therefore some basic

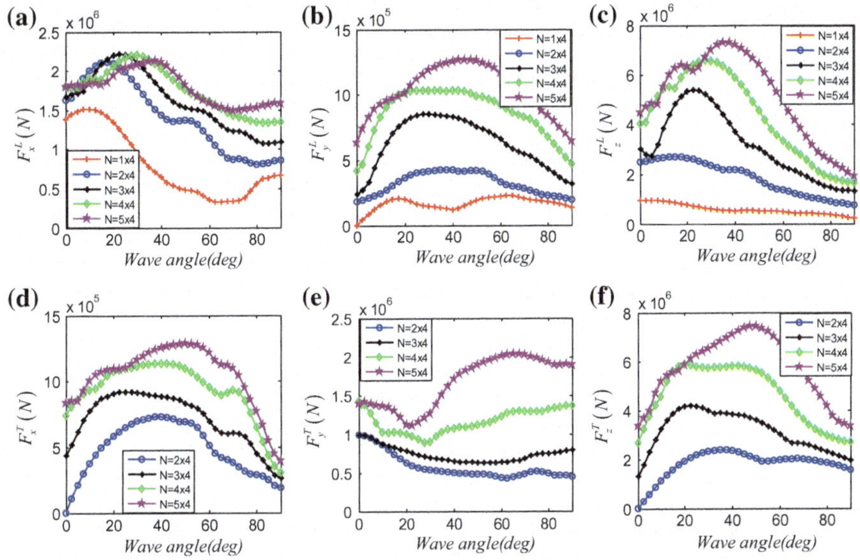

Fig. 21 Maximum extreme connector loads of array-type floating systems versus wave angle for connector stiffness $K = 1 \times 10^{8.7}$ N/m

scientific problems are still to be solved and its potential applications in engineering still needs to be expanded. The ongoing and future works will cover but not limit to the various aspects given below:

(1) Layout and stiffness configuration of flexible connector;
(2) Design of new type connector for VLFSs;
(3) Study on modular configuration of artificial floating island;
(4) Network catastrophe theory and marine engineering application;
(5) Development of application of network dynamics in other engineering fields.

Acknowledgements This research work was supported by the National Natural Science Foundation of China (11702088, 11472100) and the High-tech Ship Research Projects Sponsored by MIIT.

References

1. Watanabe, E., Utsunomiya, T., & Wang, C. M. (2004). Hydroelastic analysis of pontoon-type VLFS: A literature survey. *Engineering Structures, 26,* 245–256.
2. Armstrong, E. R. (1924). *Sea Station*. 1511153.
3. Suzuki, H. (2005). Overview of megafloat: Concept, design criteria, analysis, and design. *Marine Structures, 18,* 111–132.
4. Taylor, R. (2003). MOB project summary and technology spin-offs. In *Proceedings of International Symposium on Ocean Space Utility Technologies* (pp. 29–36).

5. Suzuki, H. (2001). Safety target of very large floating structure used as a floating airport. *Marine Structures, 14,* 103–113.
6. Xu, D., Zhang, H., Lu, C., Qi, E., Tian, C., & Wu, Y. (2014). Analytical criterion for amplitude death in nonautonomous systems with piecewise nonlinear coupling. *Physical Review E, 89,* 42906.
7. Zhang, H., Xu, D., Xia, S., Lu, C., Qi, E., Tian, C., et al. (2015). Nonlinear network modeling of multi-module floating structures with arbitrary flexible connections. *Journal of Fluids and Structures, 59,* 270–284.
8. Zhang, H., Xu, D., Xia, S., & Wu, Y. (2018). Nonlinear dynamics of a non-autonomous network with coupled discrete—continuum oscillators. *Nonlinear Dynamics, 94,* 889–904.
9. Zhang, H., Xu, D., Lu, C., Xia, S., Qi, E., Hu, J., et al. (2015). Network dynamic stability of floating airport based on amplitude death. *Ocean Engineering, 104,* 129–139.
10. Zhang, H., Xu, D., Lu, C., Qi, E., Tian, C., & Wu, Y. (2017). Connection effect on amplitude death stability of multi-module floating airport. *Ocean Engineering, 129,* 46–56.
11. Xia, S., Xu, D., Zhang, H., Qi, E., Hu, J., & Wu, Y. (2016). On retaining a multi-module floating structure in an amplitude death state. *Ocean ngineering, 121,* 134–142.
12. Zhang, H., Xu, D., & Wu, Y. (2018). Predicting catastrophes of non-autonomous networks with visibility graphs and horizontal visibility. *Mechanical Systems and Signal Processing, 104,* 494–502.
13. Watts, D. J., & Strogatz, S. H. (1998). Collective dynamics of "small-world" networks. *Nature, 393,* 440–442.
14. Barabási, A.-L. (2009). Scale-free networks: A decade and beyond. *Science, 325,* 412.
15. Shi, Q., Zhang, H., Xu, D., Qi, E., Tian, C., Ding, J., et al. (2018). Experimental validation of network modeling method on a three-modular floating platform model. *Coastal Engineering, 137,* 92–102.
16. Stoker, J. J. (2011). *Water waves: The mathematical theory with applications*. New York: Wiley.
17. Zhang, H., Xu, D., Xia, S., & Wu, Y. (2017). A new concept for the stability design of floating airport with multiple modules. *Procedia IUTAM, 22,* 221–228.
18. Saxena, G., Punetha, N., Prasad, A., & Ramaswamy, R. (2013). Amplitude death: The cessation of oscillations in coupled nonlinear dynamical systems. *Physics Reports, 1582,* 158–171.
19. Resmi, V., Ambika, G., & Amritkar, R. E. (2011). General mechanism for amplitude death in coupled systems. *Physical Review E, 84,* 46212.
20. Zhang, H., Xu, D., Lu, C., Qi, E., Hu, J., & Wu, Y. (2015). Amplitude death of a multi-module floating airport. *Nonlinear Dynamics, 79,* 2385–2394.
21. Yingli, L., Daolin, X., Yiming, F., & Jiaxi, Z. (2012). Nonlinear dynamic analysis of 2-DOF nonlinear vibration isolation floating raft systems with feedback control. *Chaos, Solitons & Fractals, 45,* 1092–1099.
22. Zhai, Y., Kiss, I., & Hudson, J. (2004). Amplitude death through a Hopf bifurcation in coupled electrochemical oscillators: Experiments and simulations. *Physical Review E, 69,* 26208.
23. Karnatak, R., Punetha, N., Prasad, A., & Ramaswamy, R. (2010). Nature of the phase-flip transition in the synchronized approach to amplitude death. *Physical Review E, 82,* 46219.
24. Yang, G., Xu, D., & Zhang, H. (2019). Catastrophe prediction of multi-modular floating platforms with symbolic networks. Physica A: *Statistical Mechanics* and *Its Applications* (Under review).
25. Xia, S., Xu, D., Zhang, H., & Wu, Y. (2019). Multivariable oscillation control of a modularized floating airport with disturbance of uncertain waves. *Journal of Sound and Vibration, 439,* 310–28.
26. Ding, R., Xu, D., Zhang, H., Zhao, H., Shi, Q., Xia, S., & Wu, Y. (2019). Characteristic transition of scale-extendable floating islands from chain to array. *Marine Structures* (Under review).

Design and Potential Applications of Floating Structures in Singapore

Kok Keng Ang, Jian Dai, Oyvind Hellan, Arnstein Watn and Michael Boon Ing Si

Abstract As an island city-state with about 710 km^2 of land, Singapore treats land as a precious and limited resource. In order to sustain the development growth, Singapore continues to reclaim land from the sea and excavate underground space. In addition, Singapore has also started creating space on the sea by using large floating structure technology. This paper focuses on a multi-purpose floating structure (MPFS) research and development project funded by Land and Liveability National Innovation Challenge (L2 NIC) Directorate and JTC Corporation. The objective of the project is to develop innovative design concepts, optimal structural, and foundation solutions, as well as construction and installation methods for multi-purpose floating structures in Singapore coastal waters. This paper covers three specific applications, namely a floating hydrocarbon storage facility, a floating bridge and a modular multi-purpose floating structure. The technical challenges, conceptual designs, research innovation and key findings will be discussed. The outcomes of this research project may be used as a reference for other potential applications including floating offshore bunker supply bases, LNG regasification facilities, solar plants, desalination plants, piers, shipyards, container port terminals, golf courses, parks and towns/cities.

Keywords Large floating structure · Modular floating structure · Hydrocarbon storage facility · Floating bridge

1 Introduction

As an island city-state with about 710 km^2 of land, Singapore treats land as a precious and limited resource. In order to sustain the development growth, Singapore continues to reclaim land from the sea.

K. K. Ang (✉) · J. Dai · M. B. I. Si
Department of Civil & Environmental Engineering, National University of Singapore, Singapore 117576, Singapore
e-mail: ceeangkk@nus.edu.sg

O. Hellan · A. Watn
SINTEF Ocean, Trondheim, Norway

C. M. Wang et al. (eds.), *WCFS2019*, Lecture Notes in Civil Engineering 41,
https://doi.org/10.1007/978-981-13-8743-2_7

Besides land reclamation, the large floating structure (VLFS) technology is one of the approaches that Singapore is using in creating space on the sea. Building large floating structures is feasible because of Singapore's benign sea state condition and strong offshore construction capability. Towards exploring and implementing large floating structure applications, the Land and Liveability National Innovation Challenge (L2 NIC) Directorate and JTC Corporation have provided S$7.13 million in research grant to the National University of Singapore (NUS) and SINTEF to conduct research on multi-purpose floating structures (MPFS). The objective of the MPFS research and development project is to develop design concepts, innovative and optimal structural, foundation and construction solutions for large floating structures. The project focuses on three specific applications, namely the floating hydrocarbon storage facility (FHSF), the floating bridge (FB) and the modular multi-purpose floating structure (MMFS).

The study on the FHSF focuses on the development of the world's first floating prestressed concrete facility that can accommodate 300,000 m^3 stockpile of hydrocarbon products. The research tasks include evaluation of existing design concepts, development of innovative design concepts, detailed structural and hydrodynamic analyses, providing innovative foundation, material and construction solutions as well as design and conduct of physical model tests.

Similarly, the study on the FB focuses on the evaluation of existing floating bridges and development of innovative design concept and solution of a floating bridge spanning over 500 m of waterbody. Structural and hydrodynamic performances of the proposed floating bridge concept have been investigated. The objective of this research aims to develop a novel and cost-effective design concept for ASEAN's first floating bridge.

The objective of the track on MMFS is to develop innovative solutions for the creation of 'land on sea' of any desired size and shape for generic application by connecting a number of standard modular units. The size and shape of the basic modular units are being carefully investigated and determined that meet a mix of requirements, including constructability and ease of marine installation as well as connectivity and flexibility in meeting the global shape and size of the intended land on sea. A key work scope in MMFS is the development of a few types of module connections that feature ease in connection (and disconnection) between modules in marine condition resulting in the desired type of connection rigidity.

This paper presents the development of innovative design concepts for the FHSF and FB. Numerical simulations on the stability, structural and hydrodynamic performance are conducted and results are discussed. Physical model tests have also been performed to validate the proposed design concept for the FHSF. The research work on MMFS is currently ongoing. Owing to the confidentiality of the studies on MMFS, this paper will only cover the study on FHSF and FB.

2 Floating Hydrocarbon Storage Facility

Hydrocarbons are a primary energy source for Singapore. This research track focuses on evaluating existing design concepts, and developing innovative and optimal structural, foundation and construction solutions for a floating hydrocarbon storage facility. The concept developed shall be suitable for hydrocarbon storage at shallow sea depths. Solutions for large sea depths will require substantial different solutions although some generic technology (material technology, linking systems, etc.) may be adaptable also to other locations and conditions. The generic technology developed through the project should also be applicable to other types of structures and locations and have the potential for world-wide use.

The floating fuel storage facility shall meet the following requirements:

i. The facility shall have a storage capacity of at least 300,000 m^3 of fuel oil and clean petrochemical products (CPP);
ii. The concept will be aimed at oil storage; gas storage requires substantial different solutions and will not be included;
iii. Occupation of the sea space shall be optimized;
iv. The classification rules and safety measures against fire, explosion, oil leakage and collision with small ships shall be satisfied;
v. The design working life shall be 60 years with minimum maintenance;
vi. The facility should facilitate loading/offloading operations and blending of fuel mixtures;
vii. Provision of spaces for Piping/Mechanical/Electrical/Instrumentation (PMEI) system should be allocated in the facility;
viii. The facility must be easily assembled/disassembled so that the facility may be moved elsewhere if required;
ix. The construction and installation of the facility must be cost-effective;
x. Future climate change should not affect the safety/serviceability of the facility;
xi. Existence of the facility shall not disturb/harm the current marine environment;
xii. The concept shall be applicable or adaptable to other environmental conditions including deep sea;
xiii. Environmental impact—sustainable utilization of raw material.

2.1 *Design Concept*

The concept for a floating hydrocarbon storage facility may be categorized into six design components, namely the global structural arrangement (see Fig. 1), storage principle (see Fig. 2), positioning and station keeping system (see Fig. 3), storage tank shape (see Fig. 4), tank wall design (see Fig. 5) and tank connections (see Fig. 6). Based on various design choices for each design component, five distinct design concepts have been proposed that suit Singapore coastal water condition.

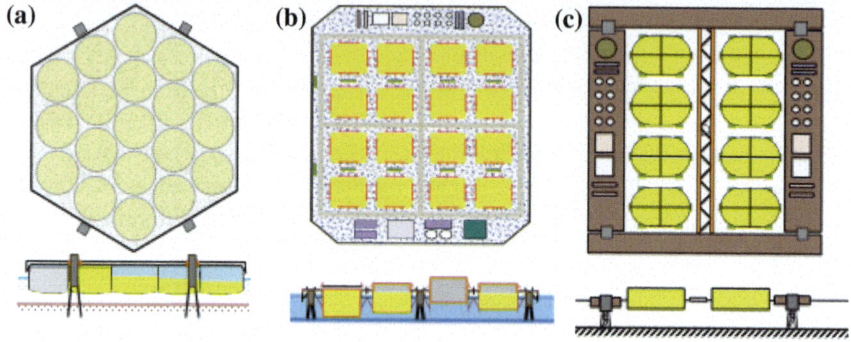

Fig. 1 Global arrangement: **a** monolithic structure, **b** tanks enclosed by platform and **c** tanks enclosed by barges

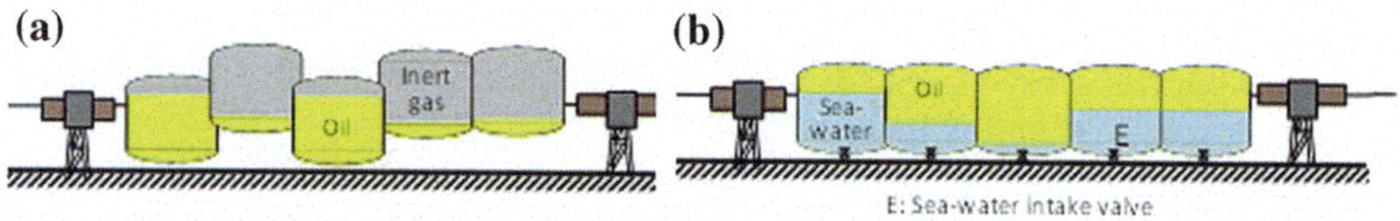

Fig. 2 Storage principle: **a** conventional principle and **b** water displacement principle

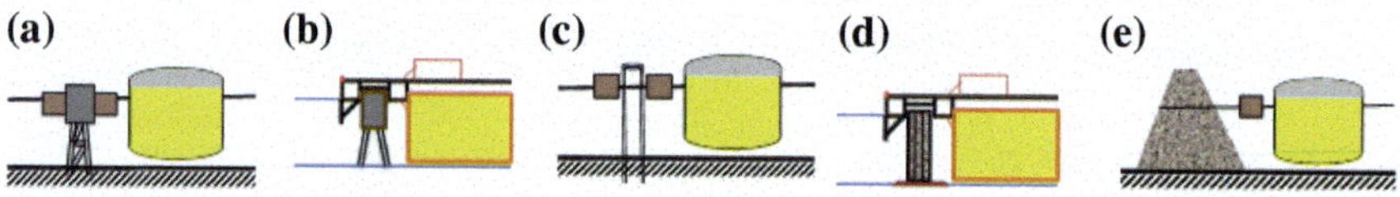

Fig. 3 Positioning and station-keeping: **a** jacket dolphin, **b** jack-up dolphin, **c** mono-pile, **d** sand-filled caisson and **e** sand berm

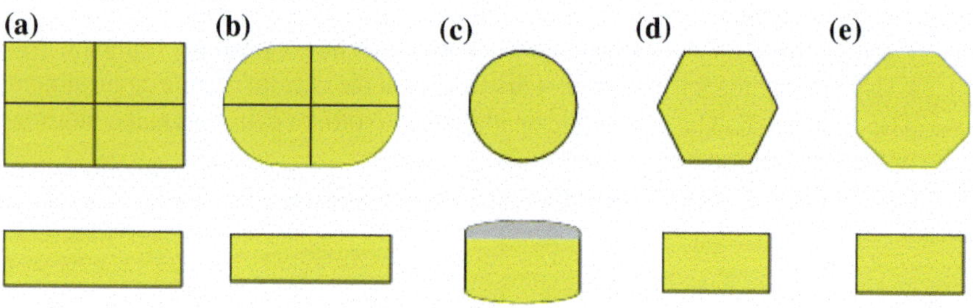

Fig. 4 Storage tank shape: **a** rectangular tank, **b** rectangular tank with rounded corners, **c** cylindrical tank, **d** hexagonal tank and **e** octagonal tank

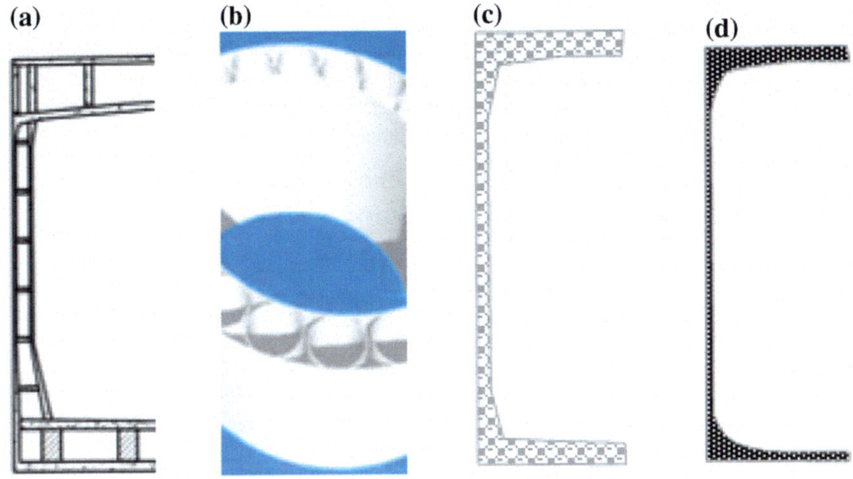

Fig. 5 Tank wall design: **a** double hull wall with horizontal and vertical stiffeners, **b** double hull with vertical stiffeners, **c** lightweight tank wall (lightweight material or wall with voids) and **d** single hull wall

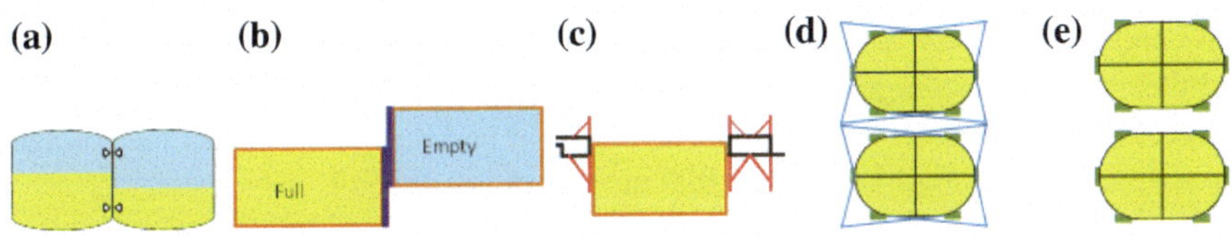

Fig. 6 Tank connection: **a** fixed connection, **b** vertical sliding connection, **c** guide frame connection **d** moored connection and **e** fendered connection

A workshop was organized for the research team to present the proposed design concepts to participants and interact with them to assess these design concepts. The workshop participants included public authorities, universities, research institutions and the offshore and oil industry. From the vibrant discussions between the research team, invited guests and technical advisors, a winning design concept was selected and the team gathered a wide range of valuable feedback from the workshop participants.

Based on the feedbacks from the workshop participants, the research team further improved the selected design concept. Figure 7 shows the conceptual drawings of the proposed FHSF [1]. This concept is self-contained with all essential facilities such as power generation plant, desalination plant, slop and wastewater treatment plant, control room, warehouse, pump rooms, offices and accommodation quarters for workers. It has floating berths on the sides for loading/offloading operations and bunker supply. The floating berths and barges also serve as a protection of the fuel storage modules against waves and ship collision.

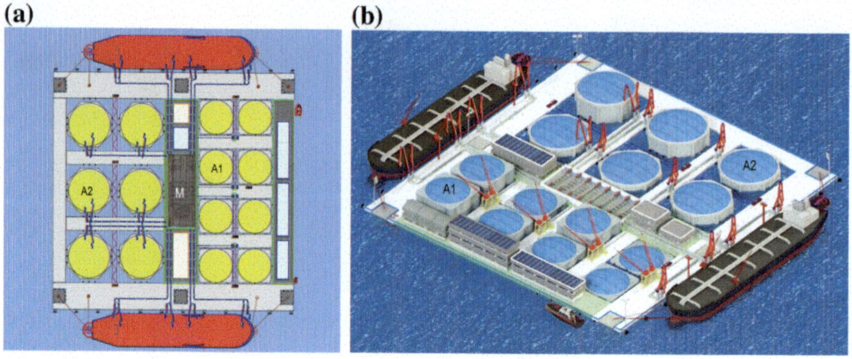

Fig. 7 Floating hydrocarbon storage and bunker facility: **a** plan view and **b** architectural rendering view

The FHSF stores hydrocarbon and provides bunker fuel to ships in harbours. The individual tanks are designed such that they are self-stabilising and the maximum tilt angle under environmental loads does not exceed allowable limits. Depending on the types of product stored, the tanks are categorized into two groups, namely the small tanks and the big tanks. The small tanks are for Clean Petrochemical Products (CPP) and their practical storage capacity ranges from 5000 to about 15,000 m^3 (see A1 in Fig. 1). The big tanks are for crude and fuel oil and they have a practical storage capacity ranging from 20,000 to 35,000 m^3 (see A2 in Fig. 1).

2.2 *Stability and Motion Criteria of Floating Tanks*

Central to the feasibility of the proposed FHSF is the stability and motion of the storage tanks under environmental loads. The floating hydrocarbon storage tanks must be self-stabilising and the displacements of the tank must not exceed limiting values. The latter is required so that loading and offloading processes are not disrupted during operations and the tank does not capsize or damage the surrounding structures and facilities under extreme weather conditions. Currently, there are no available design codes or standards for the stability design of floating tanks and there are no specifications for the limiting values of tank motion. Thus, the recommended limits for design checks on tanks' stability and motion criteria were proposed based on a critical review of existing design guidelines and philosophy for onshore oil tanks and offshore floating vessels [2]:

i. The initial metacentric height GM of moored tanks shall not be smaller than 0.15 m;
ii. In addition to (i), floating tanks that are fully exposed to the open sea need to fulfill all stability requirements as specified by DNV rules for classification of ships [3];

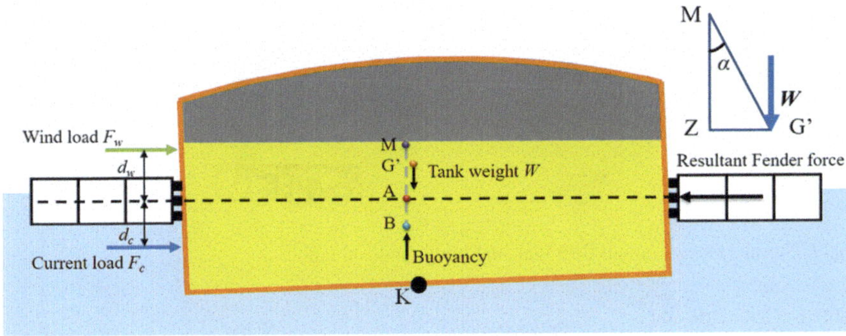

Fig. 8 Floating tank enclosed by barges and subjected to wind and current loads

iii. Rotational (roll/pitch/yaw) motions of moored tanks shall not exceed 2° under working environmental condition;
iv. Sway and surge movements of moored tanks shall not exceed 3 m, and rotational (roll/pitch/yaw) movements of moored floating tanks shall not exceed 5° under extreme environmental condition.

The acceptable limits on the movement of tanks under extreme condition need to be further assessed by detailed hydrostatic and structural analyses results as well as operational considerations. The principle is that these critical movements should not cause damage or loss of integrity and stability to the tank itself and surrounding structures and facilities. Additional motion criteria shall be considered for the construction and installation stage. Furthermore, the freeboard of the tank is assumed to be about 2.5–3 m, which is important to protect the roof structures from direct wave slamming.

Consider a floating hydrocarbon storage tank enclosed by floating barges on the sides. In view of the barges protecting the tanks from incoming waves, we shall assume for the analysis that the tank is mainly subjected to wind and water current loads as shown in Fig. 8. The yellow field in the tank represents the stored hydrocarbon product. The wind and water current speeds are assumed to be constant in the model so that a simple static analysis is possible [4]. In addition, the fenders are designed to restrain the horizontal motion within prescribed limits. Thus, only angular movements, i.e. tilting motion, will be studied.

With the given information on structural configuration and geometry of tank components as well as fuel loading condition, we can estimate the locations of the tank's centre of gravity (CG) and centre of buoyancy (CB) as shown in Fig. 8. The initial metacentric height GM may be calculated from

$$\mathrm{GM} = \mathrm{KB} + \mathrm{BM} - \mathrm{KG} \tag{1}$$

where M, G, B and K are the metacentre, the centre of gravity, the centre of buoyancy and the keel, respectively. In a partially filled condition, there will be a shift in liquid

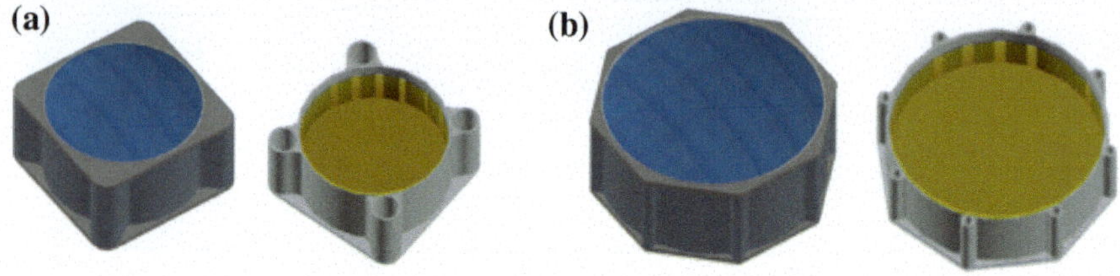

Fig. 9 Storage tanks: **a** small CPP tank and **b** big crude oil tank

Table 1 Geometric and material properties

Parameter	Value
Concrete density	2000 kg/m^3
Seawater density	1025 kg/m^3
CPP density	870 kg/m^3
Crude oil density	1010 kg/m^3
Tank wall thickness	450 mm
Floater wall thickness	350 mm
Roof slab thickness	200 mm
Bottom slab thickness	750 mm

surface given a small tilt on the structure due to the existence of free liquid surface; this is known as Free Surface Effect (FSE) that further reduces the GM, which should be accounted for in the calculation. In addition, the wind and water current loads acting on the structure can be assumed as a uniformly distributed load on the tank wall [5]. The wind and current pressure can be acting in two opposing directions or in the same direction. The maximum case between these two cases that yields the most severe destabilising moment is used for calculations.

The tilting angle of the tank can be calculated based on moment equilibrium about point A, the intersection between resultant fender reaction and the metacentric radius (BM) as shown in Fig. 1. The formula for calculating the tilting angle α is

$$W \times \mathrm{GM} \tan \alpha = |F_w d_w| + |F_c d_c| \tag{2}$$

where W is the weight of the tank, F_w and F_c are the resultant wind load and water current load exerted on the tank walls, d_w and d_c are the lever arms from the centre of the wind load and current load to point A, respectively.

In the FHSF, the tank design features a single hull cylindrical tank with a dome roof on top, a flat slab at the bottom, and several hollow cylindrical floaters attached to the tank wall for both buoyancy and stability, as shown in Fig. 9. This single hull design enjoys nearly balanced hydrostatic pressure on the tank walls. The tank has a diameter D and height H. The dome roof has a common rise to span ratio Hr/D = 1/8. Table 1 lists the key properties of the storage tanks. The wind and water current speed at the selected site in Singapore coastal water are listed in Table 2.

Table 2 Wind and current speeds

Parameter	Value (m/s)
1-year hourly mean maxima wind speed	15.9
100-year hourly mean maxima wind speed	24.0
1-year hourly mean maxima current speed	1.46
100-year hourly mean maxima current speed	1.90

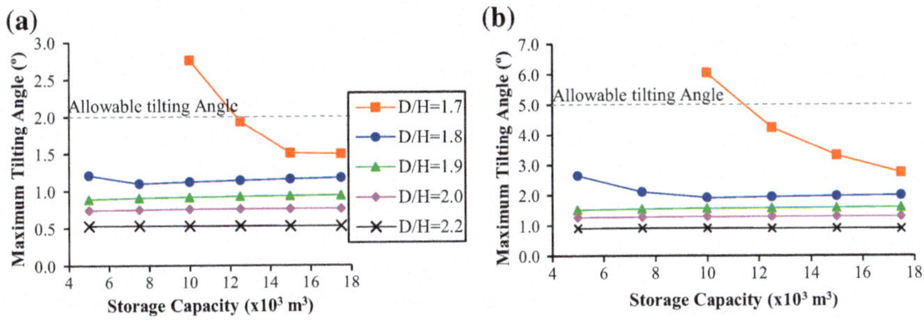

Fig. 10 Maximum tilting angle of small tanks under **a** 1-year design load and **b** 100-year design load

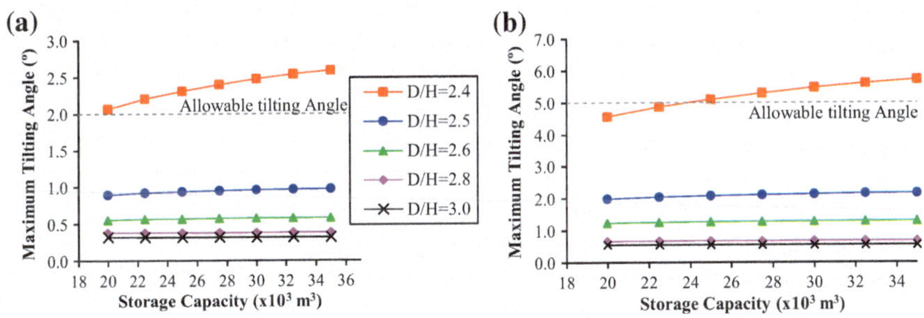

Fig. 11 Maximum tilting angle of big tanks under **a** 1-year design load and **b** 100-year design load

Figure 10 presents the maximum tilting angles of small CPP storage tanks of various storage capacities under wind and current loads at their 1-year and 100-year return periods. It is clear from the figure that larger tank D/H ratios lead to smaller tilting angles (and thus the tanks are more stable). Results show that tanks with $D/H = 1.7$ fulfil the stability and operational requirements on tilting motion when the storage capacity is above 12,500 m^3. When tank D/H reaches 1.8, all the considered tanks satisfy the aforementioned design checks. Similarly, the maximum tilting angles of big oil storage tanks under wind and current loads are presented in Fig. 11. As it can be seen, a D/H ratio exceeding 2.4 is necessary for oil storage tanks to satisfy stability and tilt motion criteria.

2.3 Structural and Hydrodynamic Analyses

Besides the study on hydrostatic stability and quasi-static motion of floating tanks, detailed finite element analysis of the storage tank structure [6] was conducted. Analysis results show that it is possible to avoid the need for vertical prestressing tendons with the introduction of a tapered tank wall at the tank bottom.

Comprehensive hydrodynamic analyses of floating tanks considering fender connection to the surrounding barges [7] and compliant tether mooring lines [8] were carried out. Hydroelastic response analysis of the surrounding barges was also investigated [9]. Detailed model tests on single floating tanks [10], multiple floating tanks and the entire floating facility were conducted. Results show that the hydrodynamic performance of the tanks and barges is satisfactory even with environmental conditions corresponding to a 100-year return period, thereby validating the proposed conceptual design for deployment in Singapore coastal waters.

3 Floating Bridge

Bridges are essential in connecting islands and land parcels separated by a water body to boost economic and leisure activities. When water is very deep and/or the seabed is extremely soft at a location where a bridge is going to be built, conventional piers supporting the bridge become expensive or even impractical. Under these conditions, floating bridges may offer distinct advantages through the use of pontoons to support the bridge deck. The pontoons are supported by natural buoyancy forces and are not dependent on the sea bed condition. More importantly, if the bridge is to be relocated elsewhere when it is no longer needed, a floating bridge allows easy removal as it may be towed away by tug boats.

The second track of MPFS project focuses on evaluating existing design concepts and developing innovative structural, foundation and construction solutions for a floating bridge spanning across waterbodies. The design of the FB should be suitable for shallow water depths and accommodate daily tidal variations at the site which is just next to the Marina Barrage. The design concept shall meet the following requirements:

i. The bridge must have sufficient clearance; at least 15 m over the water surface and clear span for maintenance vessels, commercial river taxis and ad-hoc boats during special events to pass underneath;
ii. The water flow discharged from Marina Barrage (current speed up to 1.5 m/s) shall not endanger the serviceability of the bridge;
iii. The total length of the bridge will be approximately 500 m;
iv. The width of the bridge must be at least 38.6 m in order to accommodate 3 lanes of traffic in both directions;
v. The design service life shall be 100 years with minimum maintenance;
vi. The construction and installation of the bridge must be cost-effective;

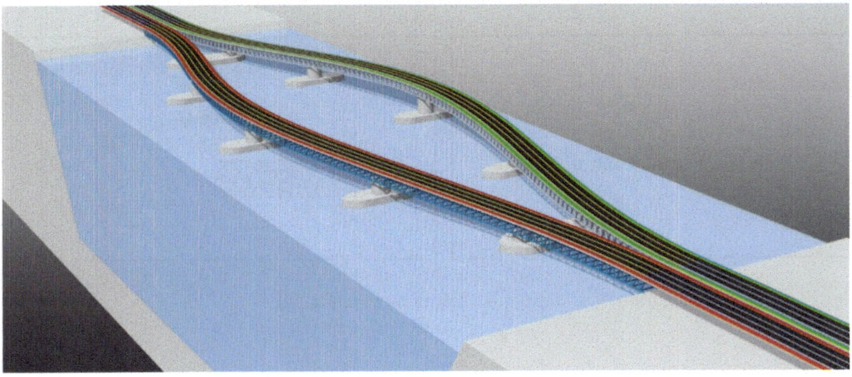

Fig. 12 Aerial view of floating double lateral curved bridge

vii. The bridge design should cater for efficient construction, assembly and decommissioning;
viii. Future climate change should not affect the safety/serviceability of the bridge;
ix. The bridge shall have minimal environmental impact and shall not harm current marine environment;
x. Sustainable utilization of raw material shall be considered;
xi. The bridge should be aesthetically attractive and design should be adapted to the surrounding environment.

3.1 Design Concept

Based on the design requirements specified above, a concept entitled "floating double lateral curved bridge", as shown in Fig. 12, was proposed by the research team. This design concept comprises two oppositely curved pontoon bridges, each carrying a 3-lane roadway and a pedestrian walkway. The bridges are supported by several pontoons spaced at 100–120 m apart. The double arch bridge structures hug an elliptical sea space which may be used to house a suitable structure.

An arch is a much stronger member than a straight beam when subjected to in-plane loadings as it transfers the loadings in axial rather than in bending and shear as in the case of a straight beam. Additionally, an arch will provide more torsional rigidity which improves the resistance to the rolling motion. The double lateral curved bridge achieves not only an aesthetic design but also forms an effective structural system. Hence, the laterally curved configuration is adopted for the bridge deck in order to achieve a mooring free bridge with a clear span across the water body.

However, a drawback of using a laterally curved configuration is the increased reaction at the supports. Figure 13 shows the free body diagrams of a (a) straight beam and (b) simply supported arch that are subjected to an in-plane point load at the

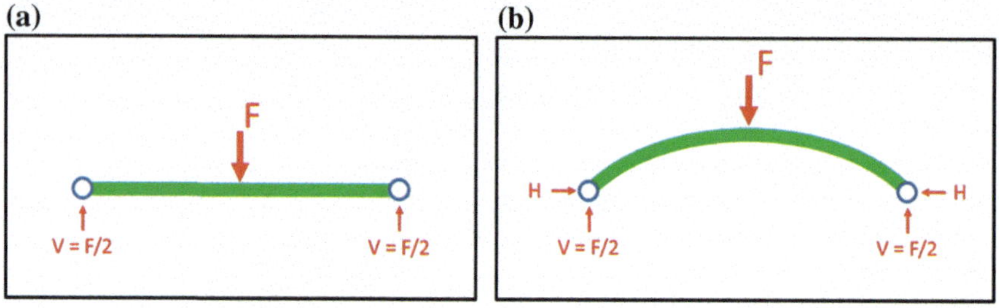

Fig. 13 Free body diagram of **a** straight beam and **b** curved beam

midspan. Clearly, the end reactions of an arch will always be greater than a straight beam due to the existence of horizontal force components. Note that the magnitude of the horizontal reaction component depends on the curvature, stiffness, and opening angle of the arch. According to the practical experience, such a horizontal reaction can be larger than the vertical reaction component in many real-life cases. Thus, a curved bridge will usually require a stronger bridge end support. To overcome this complication, the double curved bridge is proposed to cancel out the undesirable horizontal force components at the points where two bridges join. Figure 14 shows the schematic force diagram of a double lateral curved bridge subject to current and wind loads. As illustrated in Fig. 14, the environmental loads will put one bridge in compression through arch effect and the other in tension via catenary effect. It is expected that the forces will generally be in the same order of magnitude. Consequently, the sum of the horizontal reaction component experienced at the intersections of the two curved bridges is expected to be small. It is thus innovative to adopt a double curved configuration due to the force cancellation effect; thereby resulting in the curved bridge to be similar to a straight bridge as far as support reaction is concerned.

The coastal areas are subject to daily tidal variations that must be accounted for in the design of bridges. The tidal variations impose challenges to the design of bridge abutment and the connection for the floating bridges as the bridge structures are floating up and down with the tides. In the current design proposal, the bridge structure stiffness is specially designed so that the bridge is able to deform with the tidal changes (see Fig. 15).

In summary, the proposed floating bridge design has the following key features:

i. The pontoons are designed to support through buoyancy the self-weight of the bridge as well as the vehicle load.
ii. The double lateral arch bridge floats stably on water.
iii. The overall bridge design ensures that there are minimal lateral motions owing to the arch and catenary effects and hence no moorings are required to hold the pontoons in position.
iv. The curvature of the arch bridge is designed optimally to meet design and operational requirements.
v. The bridge can be designed to have the flexural rigidity to adapt to the tidal variations.

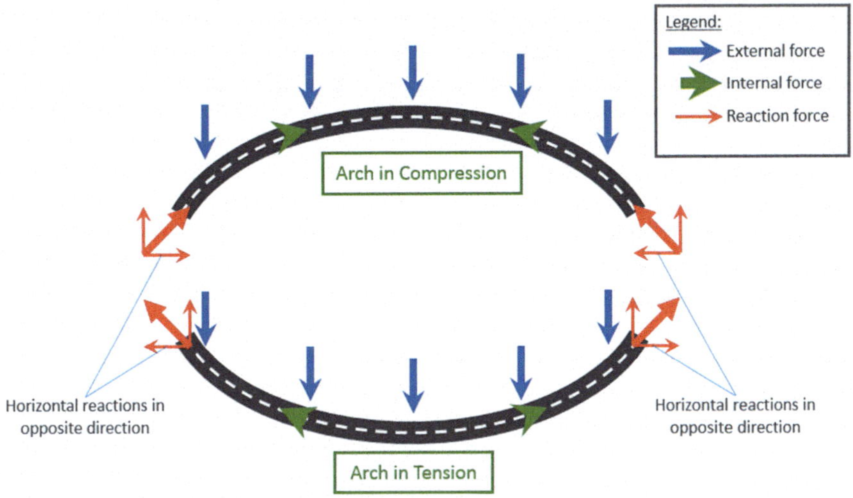

Fig. 14 Force diagram of double lateral curved bridge

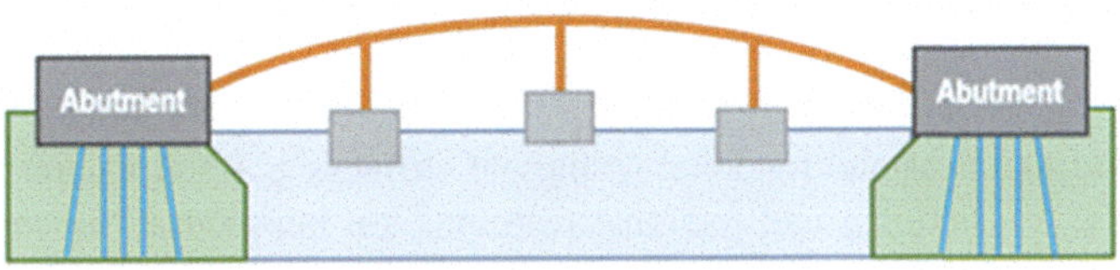

Fig. 15 Principle of bridge to adapt tidal variations

vi. Double-curved geometry is adopted with the intention to cancel the axial reaction force component at the point of intersection of the two curved bridges.
vii. The bridge is modular in design and can be prefabricated in parts offsite to minimize the onsite construction efforts and the impact on the local environment.

3.2 *Analytical Study for Single Curved Floating Bridge*

It is essential to understand the mechanisms corresponding to a floating curved bridge adapted to the daily tidal variations. In addition, the generally soft soil condition in Singapore coastal waters also have implications for the load carrying mechanism and structural performance of the curved bridge. In order to get insights into the physical behavior of the bridge structure, an analytical solution to the responses of a curved bridge in and out of the curvature plane was developed.

Figure 16 shows a beam model of the curved bridge. The bridge has a radius of curvature R and an overall length L. The angle subtended at the centre of curvature is θ. In the global Y-direction, the beam is discretely supported by pontoons, which

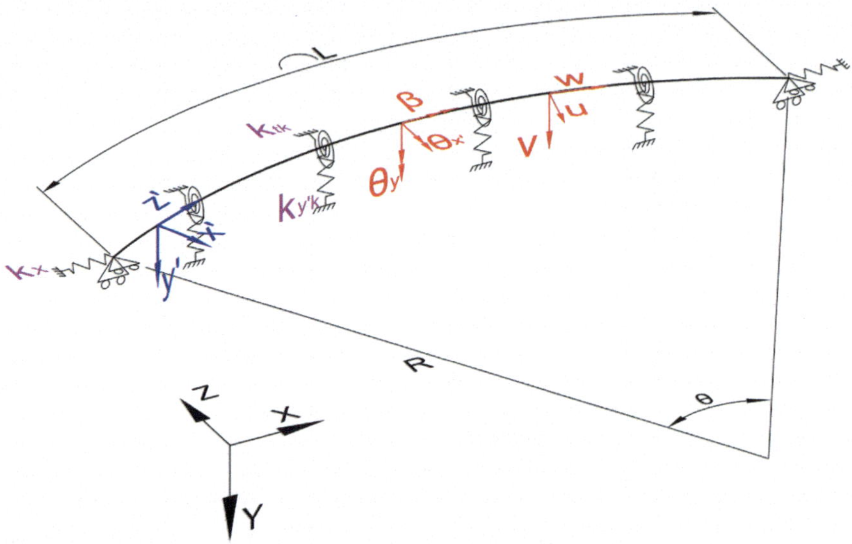

Fig. 16 Coordinates system and boundary conditions of curved beam

are modelled as linear and torsional springs of stiffness $k_{y'k}$ and k_{tk}, respectively. Additionally, translational and torsional restraints are imposed at the ends of the bridge to prevent the beam from displacing in global *Y*-direction as well as rotating about its local *z*' axis. In the global *X*-direction, linear springs, $k_{s,X}$ are added at both ends to represent the effect of soft soil foundation. Similar boundary conditions should also be applied to the beam in the global *Z*-direction. However, the beam is assumed to be simply supported at two ends instead. This is because the reaction forces in *Z*-direction are always statically determinate and have no effect on the internal forces. Therefore, it is reasonable to adopt this simplification. The movement of the abutment in the global *Z*-direction is a rigid body motion which can be easily computed once the foundation stiffness is known. According to the classical theory [11], the in-plane responses of a curved beam are decoupled from its out-of-plane responses. Hence, the solutions to the in-plane and out-of-plane responses can be derived independently.

In the curvature plane, the bridge is subject to wave and current forces acting on the pontoons. The spring deformations representing the soft soil foundation restraint at the bridge ends are denoted as δ_H. According to Young and Budynas [12], the horizontal displacement of the end support may be assumed to be caused by (a) a vertical concentrated force along the beam or (b) a horizontal concentrated load acting at roller end as shown in Fig. 17a, b, respectively. The horizontal deformation at beam end for case (a) is given by [12]

$$\delta_{Ha,i} = -\frac{R^3}{EI_{y'}}(LP_{Ha}) \tag{3}$$

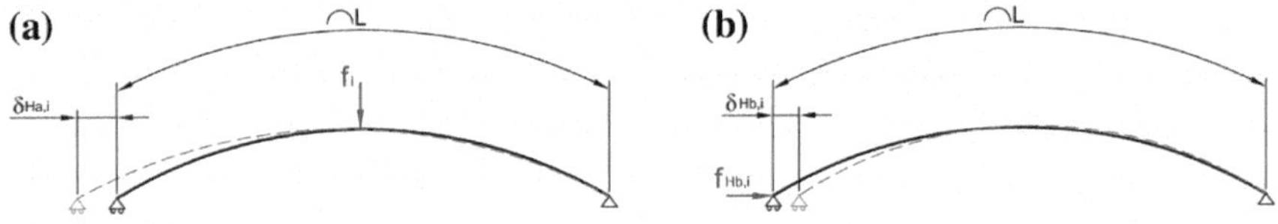

Fig. 17 Pin-roller supported arch subject to **a** vertical and **b** horizontal point loads

where LP_{Ha} is the loading terms which takes into account the geometry and properties of the beam, as well as the location and magnitude of the applied load, which may be expressed as

$$LP_{Ha} = f_i\left[\frac{\theta}{2}\sin\frac{\theta}{2}\cos\frac{\theta}{2} - \emptyset\sin\emptyset\cos\frac{\theta}{2} + \frac{k_1}{2}\left(\cos^2\frac{\theta}{2} - \cos^2\emptyset\right) + k_2\cos\frac{\theta}{2}\left(\cos\frac{\theta}{2} - \cos\emptyset\right)\right] \tag{4}$$

where f_i is the ith external transverse load applied to the bridge; θ and ϕ are the subtended angle of the curved beam and the angle measured counter-clockwise from the mid-span of the beam to the position of load, respectively; and k_1 and k_2 are the correction factors for shear and hoop stress, respectively. Similarly, for case (b), the displacement of the roller support can be computed from

$$\delta_{Hb,i} = -\frac{R^3}{EI_{y'}}(LP_{Ha}) \tag{5a}$$

$$LP_{Hb} = f_{Hb,i}\left[\theta\cos^2\frac{\theta}{2} + \frac{k_1}{2}(\theta - \sin\theta) - k_2\sin\theta\right] \tag{5b}$$

It is clear from Eq. (5b) that $f_{Hb,i}$ can be used to represent the reaction force of a spring shown in Fig. 16. A strain compatibility relationship can be employed to obtain the actual spring deformation due to the ith external transverse load. By setting up the equilibrium state of the curved beam, the internal member forces can be directly evaluated.

Unlike the support condition of the curved beam for the in-plane case, the beam is discretely supported at the locations of the pontoons owing to the buoyancy out of the curvature plane, as shown in Fig. 16. By neglecting the inertia and viscous damping terms, the out-of-plane governing equations may be written as [11]

$$\frac{\partial^2}{\partial z^2}\left(EI_x(z)\frac{\partial^2 v}{\partial z^2} - \frac{1}{R}\beta\right) - \frac{GJ(z)}{R}\left(\frac{1}{R}\frac{\partial^2 v}{\partial z^2} + \frac{\partial^2 \beta}{\partial z^2}\right) + \sum_{k=1}^{Np} k_{yk}(v - H_T - H_e)\delta(z - z_k) = \rho A(z)g \tag{6a}$$

$$\frac{EI(z)}{R}\left(\frac{\beta}{R}-\frac{\partial^2 v}{\partial z^2}\right)-GJ(z)\left(\frac{\partial^2 \beta}{\partial z^2}+\frac{1}{R}\frac{\partial^2 v}{\partial z^2}\right)+\sum_{k=1}^{Np} k_{tk}\beta\delta(z-z_k)=0 \quad (6b)$$

where v and β are the vertical displacement and torsional deformation, respectively; k_{yk} and k_{tk} the vertical and torsional hydrostatic stiffness, respectively; Np refers to the number of pontoons; H_T and H_e are the tide-induced surface elevation and the water surface elevation in the equilibrium state.

In view of the boundary conditions and the relationship between v and β, both vertical and torsional deformations of the curved beam can be expressed as the summation of a series of sinusoidal functions as

$$v=\sum_{i=1}^{n} q_{vi} sin\frac{i\pi z}{L},\ \beta=\sum_{i=1}^{n} q_{\beta i} sin\frac{i\pi z}{L} \quad (7)$$

where q_{vi} and $q_{\beta i}$ denote the generalized coordinates of the ith mode; and n the number of modes. To solve the coupled differential equations, Galerkin's approach is adopted to formulate the weighted residual forms of the governing equations, which leads to ith mode governing equations

$$\sum_{j=1}^{n} a_{ij}q_{vj}+\sum_{j=1}^{n} b_{ij}q_{\beta j}=\rho g\int_0^L A(z)\sin\left(\frac{i\pi z}{L}\right)dz \quad (8a)$$

$$\sum_{j=1}^{n} c_{ij}q_{\beta j}+\sum_{j=1}^{n} d_{ij}q_{vj}=0 \quad (8b)$$

where the coefficients are given by

$$a_{ij}=\int_0^L\left[EI_x(z)\left(\frac{i\pi}{L}\right)^4+\frac{GJ(z)}{R^2}\left(\frac{i\pi}{L}\right)^2\right]\sin\left(\frac{i\pi z}{L}\right)\sin\left(\frac{j\pi z}{L}\right)\mathrm{d}z$$
$$+\sum_{k=1}^{Np} k_{yk}\sin\left(\frac{i\pi z_k}{L}\right)\sin\left(\frac{j\pi z_k}{L}\right) \quad (9a)$$

$$b_{ij}=\frac{1}{R}\left(\frac{i\pi}{L}\right)^2\int_0^L(EI_x(z)+GJ(z))\sin\left(\frac{i\pi z}{L}\right)\sin\left(\frac{j\pi z}{L}\right)\mathrm{d}z \quad (9b)$$

$$c_{ij}=\int_0^L\left[\frac{EI_x(z)}{R^2}+GJ(z)\left(\frac{i\pi}{L}\right)^2\right]\sin\left(\frac{i\pi z}{L}\right)\sin\left(\frac{j\pi z}{L}\right)\mathrm{d}z$$
$$+\sum_{k=1}^{Np} k_{tk}\sin\left(\frac{i\pi z_k}{L}\right)\sin\left(\frac{j\pi z_k}{L}\right) \quad (9c)$$

$$d_{ij}=\frac{1}{R}\left(\frac{i\pi}{L}\right)^2\int_0^L(EI_x(z)+GJ(z))\sin\left(\frac{i\pi z}{L}\right)\sin\left(\frac{j\pi z}{L}\right)\mathrm{d}z \quad (9d)$$

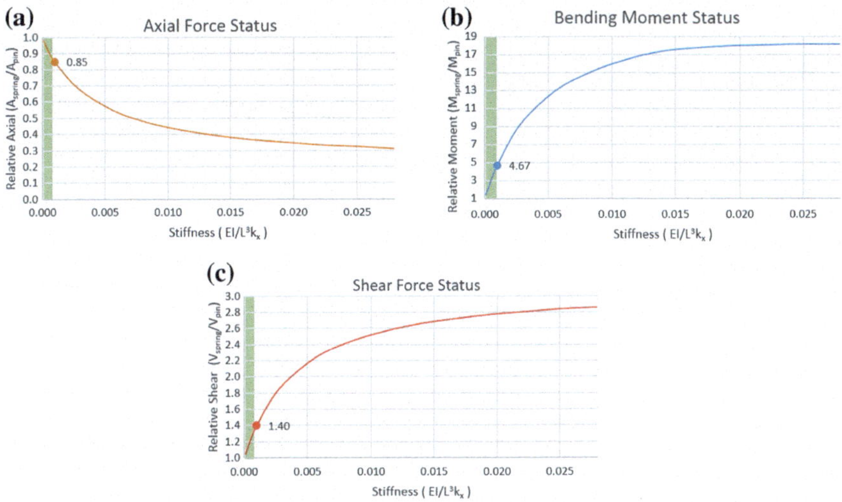

Fig. 18 Bridge in-plane response

The accuracy and computational efficiency of the analytical solutions have been documented in the authors' earlier publications [11, 13]. To study the effect of soft soil foundation on the in-plane response of a curved floating bridge, a numerical investigation was carried out by conducting a series of parametric studies with different stiffness ratios of the bridge's flexural rigidity, $EI_{y'}/L^3$ to the foundation stiffness at bridge ends, k_x. Figure 18 presents the ratio of the forces in a spring supported curved bridge to those of a pin supported curved bridge. Different bridge structural stiffness to soft soil foundation stiffness ratios are considered. It can be seen that for this particular load arrangement, the reduction in the axial force can be up to 70% which in turn amplifies the bending moment by 18 times as the foundation stiffness reduces. Hence, it is crucial that a sufficient rigidity of the end abutment is retained so as to ensure that the membrane action of the curved bridge is effectively activated. For example, at least 85% of the axial force accompanied by 367% increment in the bending moment can be achieved by a stiffness ratio EI/L^3k_x smaller than 1×10^{-4}.

The effect of tide-induced water surface elevation is next investigated. Figure 19 shows the vertical displacement and torsional deformation of the bridge when it is subjected to a 2 m low tide. Note that the range of the out-of-plane stiffness of the curved bridge is practically selected based on the sectional properties of the Norwegian Bergsøysund bridge. It is observed from Fig. 19 that the vertical displacement and rotation angle of the bridge start to converge when the out-of-plane second moment of inertia exceeds 10 m^4. When the bridge stiffness is too low, i.e. $I_x = 1$ m^4, the bridge spans between two adjacent pontoons exhibit noticeable deflections due to the self-weight of the bridge. Such magnitude is definitely not acceptable as it will not only hinder the serviceability of the bridge but it may also lead to rupture of the superstructure and the bridge deck. On the other hand, if the bridge stiffness

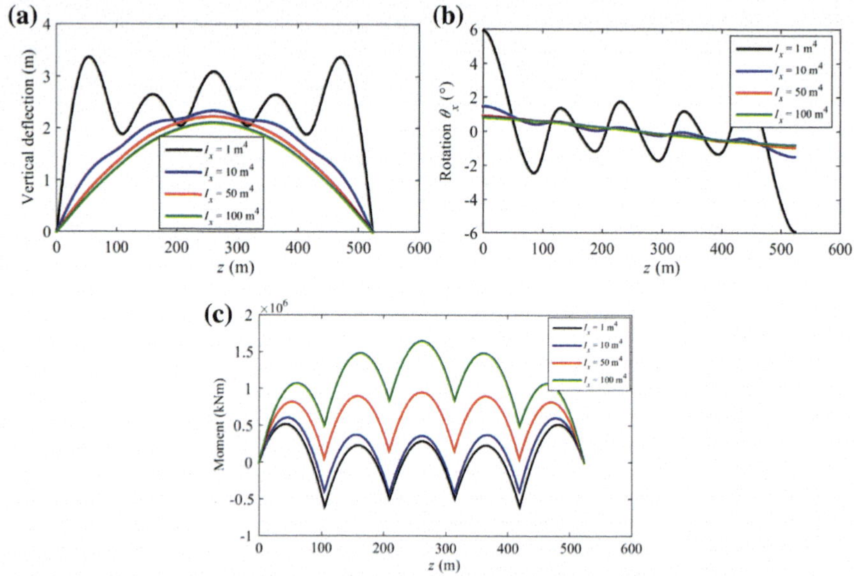

Fig. 19 Curved bridges with different flexural rigidities subject to tidal variations

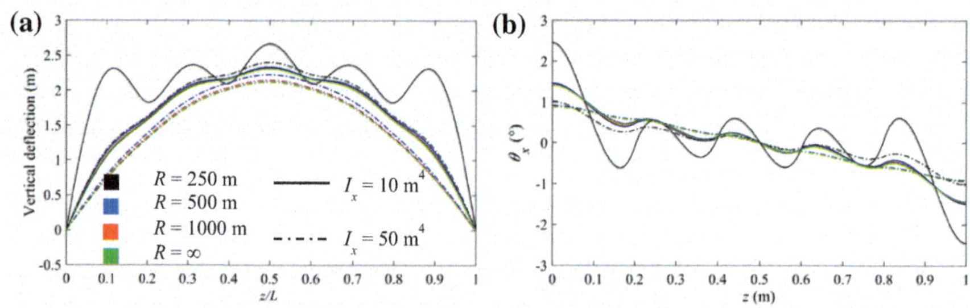

Fig. 20 Curved bridges with different radii and flexural rigidities subject to tidal variations

is too large, i.e. $I_x = 100\ \text{m}^4$, the bridge will be subject to huge bending moments although the deflection is restrained. In view of Fig. 19, it may be reasonable to design a bridge with an out-of-plane second moment of inertia in the range of 10–50 m^4 which results in a well-restrained deflection under high/low tides and does not attract large bending moment acting on the bridge.

Figure 20 shows the effect of both structure stiffness and radius on the out-of-plane responses of the curved bridge under a 2 m tide. As it can be seen, when the radius of the bridge is small, the vertical displacement is large. This is due to the fact that the bridge span is also large. However, when the radius exceeds 500 m, the effect of the bridge radius is found to be small. When this is read in conjunction with the in-plane responses of a curved bridge, one may conclude that the optimum bridge radius ranges from 500 to 1000 m.

3.3 Hydrodynamic Study of Double Curved Floating Bridge

Section 3.2 is concerned with a static analytical study of a single curved bridge. In order to investigate the dynamic response of a double lateral arch bridge and the effect of force cancellation at the bridge ends, a more detailed numerical model was constructed. In the numerical model, the bridge structure is modeled as flexible beams supported by elliptical cylindrical pontoons that float on water surface. Parametric study of the bridge structural rigidity was performed to investigate its effect on the structural and hydrodynamic responses. White noise, regular and irregular wave simulations were carried out in the numerical investigation. The analysis results demonstrate that the floating bridge concept is valid when the structural rigidity is well chosen. For more details, one may refer to the published results in [14].

4 Conclusions

This paper presents the development of innovative design concepts for a floating hydrocarbon storage facility (FHSF) and a floating bridge (FB). Numerical simulations on the stability, structural and hydrodynamic performances were conducted. The results show that constructing large floating structures in Singapore coastal waters is very doable owing to the benign sea state. Besides FHSF and FB, other potential applications for VLFS include floating LNG regasification facilities, desalination plants, piers and even towns or cities.

Acknowledgements This research is supported in part by the Singapore Ministry of National Development and the National Research Foundation, Prime Minister's Office under the Land and Liveability National Innovation Challenge (L2 NIC) Research Programme (L2 NIC Award No. L2 NICTDF1-2015-2.). Any opinions, findings, and conclusions or recommendations expressed in this material are those of the author(s) and do not reflect the views of the Singapore Ministry of National Development and National Research Foundation, Prime Minister's Office, Singapore. We thank the JTC Corporation for their support.

References

1. Wang, C. M., Ang. K. K., Dai, J., Lim, B. K., Magee, A. R., & Watn, A, et al. (2017). *Floating Hydrocarbon Storage and Bunker Facility*. Singapore Provisional Patent 10201702591X.
2. Ang. K. K., Dai, J., Lim, B. K., & Wang, C. M. (2017). Floating hydrocarbon storage and bunker facility. In *Proceedings of the 10th International Research Conference of KDU*. Ratmalana, Sri Lanka, 3–4 Aug 2017.
3. DNV. (1995). *Rules for classification of ships. Part 3 Chapter 4: Stability and watertight integrity*. Norway: Det Norske Veritas.
4. DNV. (2011). *DNV-OS-C301: Stability and watertight integrity*. Norway: Det Norske Veritas.
5. DNV. (2010). *DNV-RP-C205: Environmental conditions and environmental loads*. Norway: Det Norske Veritas.

6. Jiang, D., Tan, K. H., Dai, J., Ong, K. C. G., & Heng, S. (2018). Structural performance evaluation of innovative prestressed concrete floating fuel storage tanks. *Structural Concrete, 20,* 15–31. https://doi.org/10.1002/suco.201800032.
7. Wan, L., Han, M., Jin, J., Zhang, C., Magee, A. R., Hellan, O., et al. (2018). Global dynamic response analysis of oil storage tank in finite water depth: Focusing on fender mooring system parameter design. *Ocean Engineering, 148,* 247–262.
8. Wan, L., Zhang, C., Magee, A. R., Jin, J., Han, M., Ang, K. K., et al. (2018). An innovative mooring system for floating storage tanks and stochastic dynamic response analysis. *Ocean Engineering, 170,* 361–373.
9. Dai, J., Ang, K. K., & Zhang, C. (2019). Hydroelastic analysis of modular floating barges for hydrocarbon storage facility. In *Proceedings of the 1st Vietnam Symposium on Advances in Offshore Engineering*, Hanoi, Vietnam, 1–3 Nov 2018.
10. Zhang, C., Dai, J., Ang., K. K., & Magee, A. R. (2019). Comparison study on bottom plate effect on single hydrocarbon storage tank through decay test. In *Proceedings of the 1st Vietnam Symposium on Advances in Offshore Engineering*, Hanoi, Vietnam, 1–3 Nov 2018.
11. Dai, J., & Ang, K. K. (2015). Steady-state response of a curved beam on a viscously damped foundation subjected to a sequence of moving loads. *Proceedings of the Institution of Mechanical Engineers, Part F: Journal of Rail and Rapid Transit, 229,* 391–442.
12. Young, W. C., & Budynas, R. G. (2002). *Roark's formulas for stress and strain* (7th ed.). Darby, USA: McGraw-Hill.
13. Dai, J., Ang, K. K., & Lim, B. K. (2017). Response of a floating curved pontoon bridge subjected to tide induced water surface variation: an analytical approach. In *Proceedings of the 10th International Research Conference of KDU*. Ratmalana, Sri Lanka, 3–4 Aug 2017.
14. Wan, L., Magee, A. R., Hellan, O., Watn, A., Ang, K. K., & Wang, C.M. (2017). Initial design of a double curved floating bridge and global hydrodynamic responses under environmental conditions. In *Proceedings of the 36th International Conference on Ocean, Offshore & Arctic Engineering*. Trondheim, Norway, 35–30 June 2017.

Hydrodynamic Responses and Loads of a Model Floating Hydrocarbon Storage Tank System for Concept Validation and Numerical Verification

Chi Zhang, Allan R. Magee, Nuno Fonseca, Øyvind Hellan and Kok Keng Ang

Abstract An innovative floating hydrocarbon storage facility (FHSF) has been proposed to utilize the shielded near-shore area for countries with large demand on the land space such as Singapore and Japan. The concept comprises 14 floating hydrocarbon storage tanks (FHST) and several surrounding floating barges. All the modular designed FHSTs are loosely connected to the barges through a soft mooring system so as to reduce the loads, and the entire system is free to float to reduce the tidal influence. The single FSHT has been proven to have moderate hydrodynamic responses in previous studies, but there still exist concerns on the influence of potential resonances in the narrow gaps and the strong hydrodynamic interactions. The loads on the specially designed soft mooring system have to be checked. The complete system is complex and difficult to analyze. So, experimental studies were performed on both a simplified system and the complete system to ensure the quality and reduce the uncertainty in the experiments. The simplified system consists of two FHSTs and a surrounding floating barge frame. The experiments were performed in the ocean basin in SINTEF Ocean. A series of random, wide-band and realistic random wave tests were carried out to generate benchmark data to verify numerical analysis tools. This paper will focus on this simplified system that represents the complete system's behavior. A frequency domain numerical model of the simplified system was established based on potential theory. Empirical coefficients were used to account for viscous damping. The numerical results are comparable to the experimental results in general. The statistical responses of the FHST in the design sea states are also within the acceptable range even with the hydrodynamic interactions. However, further improvement on the system such as a better design of the floating barge is necessary.

Keywords Hydrodynamic interactions · Gap resonance · VLFS · Floating structures · Hydrocarbon storage

C. Zhang (✉) · A. R. Magee · K. K. Ang
Department of Civil and Environmental Engineering, National University of Singapore, Kent Ridge, Singapore
e-mail: ceezhchi@nus.edu.sg

N. Fonseca · Ø. Hellan
SINTEF Ocean, Trondheim, Norway

C. M. Wang et al. (eds.), *WCFS2019*, Lecture Notes in Civil Engineering 41,
https://doi.org/10.1007/978-981-13-8743-2_8

1 Introduction

Very large floating structure (VLFS) provides a solution for ocean space utilization. It might also be an alternative solution to the land reclamation. There exists disputation on the negative environmental impacts from land reclamation [1]. Several innovative concepts on VLFS have been proposed, tested or applied, for example, the Mega float project [2], the floating performance platform in Singapore [3] and the floating bridge in Norway [4]. Lamas-Pardo [5] reviewed recent applications of VLFS.

The floating hydrocarbon storage facility (FHSF) is a recently developed special modular VLFS concept. The FHSF aims at providing hydrocarbon storage for logistics around the Singapore Strait. It is partially inspired by the successful application of the floating oil storage bases in Japan (see Fig. 1a that shows the floating oil storage base in Kamigoto Island). The conceptual design of the FHSF is shown in Fig. 1b. Its total design storage capacity is 300,000 m^3. The physical test model of the FHSF is shown in Fig. 1c. Two types of FHST have been designed as shown in Fig. 1d. The storage capacity of the smaller FHST and larger FHST are 12,500 and 33,500 m^3 respectively. The entire system is made of concrete and has a 60-year service life. The 14 FHSTs are moored to the interconnected surrounding floating barges through an innovative soft mooring system [6]. The floating barges are restrained by mooring dolphins, but they are allowed to move freely in the vertical direction in order to reduce the tidal loads. With a modular design, the system is scalable for future expansion needs. A detailed system description can be found in [6]. The hydrodynamic responses of the FHSF under the sea states around Singapore is the focus that determines the feasibility of the concept. The comprehensive numerical and experimental studies have proven a very good hydrodynamic performance of the single FHST in the design sea states [7, 8]. However, it is still not easy to fully understand the hydrodynamic properties of the entire system due to its complexity. A literature review was performed to figure out the method to solve the hydrodynamic responses of VLFS. The method to solve the dynamic responses of VLFS can be different according to its type. VLFS can be categorized into pontoon/mat type and semisubmersible type [9]. The dynamic responses of the pontoon/mat type VLFS can be evaluated through hydroelastic theory [10, 11]. The semisubmersible type VLFS and other modular type VLFS have the advantage of reduced internal loads with a properly designed connection. The hydrodynamic responses of the modular VLFS can be investigated through generalized mode method [12, 13] or rigid module flexible connection model (RMFC) [14, 15]. Based on RMFC model, Zhang et al. [16] has investigated the dynamics of VLFS through the nonlinear network theory for the modular VLFS, and the numerical model is validated by a model test [17].

Compared to the studies on VLFS in the literature, there exist several complexities on the hydrodynamic analysis on the FHSF. Firstly, the modules of FHSF are placed in proximity with narrow gaps. The width of these gaps is around 5–8 m to allow the horizontal motions of FHST. The fluid oscillation may exist in the narrow gaps between adjacent barges [18–20]. The gap resonances can affect the motions of floating structure significantly. Secondly, the modules are connected by relatively

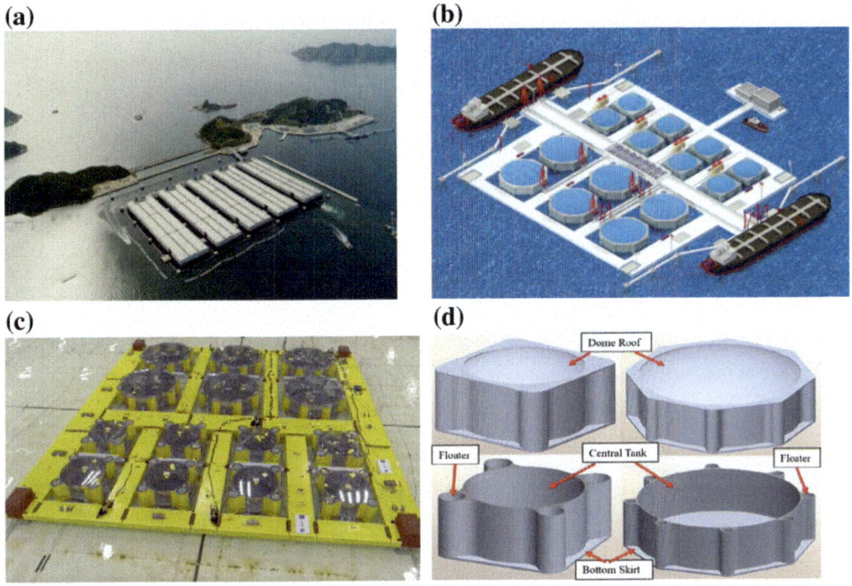

Fig. 1 **a** Floating oil storage base in Kamigoto Island; **b** Conceptual design of the FHSF; **c** Physical models of the FHSF at SINTEF Ocean; **d** Conceptual design of the smaller and larger FHST. Scaling ratios of the two FHSTs are different in (**d**) for ease of viewing

soft mooring system. In the literature study, the modules are connected either through hinges or a relatively strong connection. In the FHSF system, the hydrodynamic and mechanical coupling among modules cannot be ignored in the system, which makes numerical analysis time-consuming. Thirdly, there exist a large number of FHSTs with internal tanks. They are also designed in complex geometrical shapes. One example is the thin bottom skirts attached to the bottom plate [21]. It is important to describe the FSHT in the numerical model accurately and efficiently. These features challenge the existing numerical tools. The difficulties in the experimental study are also increased significantly.

In order to reduce the complexity and uncertainty, a simplified system comprising two tanks and a surrounding barge frame is first studied. The simplified system contains the most critical features of the entire system. Both numerical and experimental studies start with the simplified system and progress to the more complicated numerical model on the entire system. The experimental work on the simplified system and the entire system were completed in June 2018 at the SINTEF ocean. Numerous experimental cases involving different loading conditions, environmental conditions and even the accidental conditions were studied. The test results are used for both conceptual valuation and numerical validation. The numerical model is done for the simplified system in WAMIT that uses the higher order boundary element method (HOBEM). The HOBEM can represent an abrupt change in body geometry by using higher-order interpolation functions and it is more robust and efficient [22]. A time domain model will be constructed for the simplified system based on the validated model in the future to account for the nonlinearities. In all these models, both FHST and the barges are assumed as rigid bodies and their connections are also properly modeled.

The aim of this paper is to do a preliminary validation of the numerical model as well as the conceptual validations based on the simplified system. The latter will be done based on the experimental results at this stage. The paper is organized as follows. The description of the simplified system is introduced in Sect. 2. The experimental work is summarized in Sect. 3 followed by the numerical model in Sect. 4. The numerical and experimental results will be compared in Sect. 5. Moreover, the statistical responses in the extreme sea states will be discussed. The main findings and future work will be discussed in Sect. 6.

2 Simplified Two Tank System Description

The simplified FHSF system is taken from one corner of the entire system. The key features of the simplified system are the same as the entire system. The simplified system comprises two smaller FHSTs and a barge frame. The details on the design of the FHSTs is described in detail in [7]. The larger FHSTs are expected to have better hydrodynamic performance and thus it is not selected in the simplified system. The sketch of the simplified system configuration is shown in Fig. 2, and the dimensions are summarized in Table 1. The FHST is connected to the barge frame through 8 mooring ropes. The ropes are pre-tensioned so that the ropes will not be slack before the FHST can touch the barge frame. The anchor points located at the corners of the barge frame. The barge frame is also restrained by 4 mooring dolphins as in the entire system. The barge frame can move freely in the vertical direction. The mooring dolphins will be connected to pile groups in the seabed. In this system, the key features of the entire system are kept such as the narrow gaps and the multi-body interactions. Thus, the study on the simplified system can furnish a good understanding of the performance of the entire system.

Table 1 Geometrical dimensions of FHST (Unit: m)

Dimension	Prototype	Model
Total height	22.60	0.4144
Length/width (edge to edge)	37.63	0.8362
Overall diameter	33.90	0.7533
Internal clear diameter	33.00	0.7333
The diameter of the floater	7.980	–

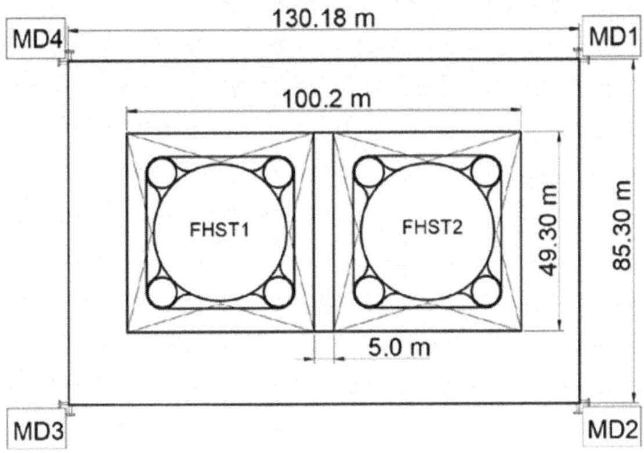

Fig. 2 Sketch of simplified FHST system proposed for numerical and experimental studies

3 Description of Experiments

The experiment performed in SINTEF Ocean is divided into two phases. Phase-1 focuses on the experimental work on the simplified system while Phase-2 focuses on the dynamic responses of the entire system in random sea states. The test facility is the ocean basin in SINTEF Ocean. It has a length of 80 m, a width of 50 m, and a depth of 10 m. The depth of the basin can be adjusted from 0 to 0.88 m by moving a steel bottom. In this model test, the water depth is fixed to 0.40 m which corresponds to 18 m in the prototype. The double-hinged wave paddle locates at the short end of the basin. It can generate long crest waves with a maximum wave height of 0.5 m. This is enough to cover the current design sea states. Passive wave absorbers on the two ends of the basin ensure less than 5% reflection of the incoming waves. The model is placed in the middle of the basin 35.26 m. This ensures no serious reflection can be generated by the model.

The models were designed under a Froude scaling factor of 1:45 for both phases. The dimensions of the FSHT model and barge frame model are listed in Tables 1, 2, 3 and 4. The mooring ropes are modeled by linear springs but with the same configuration in the prototype. 4 mooring dolphins are fitted to the corners of the barge frame. A single mooring dolphin is embedded by two load cells so that the global loads can be recorded. Different wave headings can be tested by rotating the models. Three wave headings (0°, 45°, and 90°) were tested. The definition of these wave headings is shown in Fig. 3a. They are denoted as the head sea, oblique sea and beam sea condition, respectively. The loading conditions include empty, 20% filling ratio and 100% filling ratio of the designed storage capacity. These loading conditions will be denoted as FL00, FL20 and FL100 cases. Some simplifications have been made in the model test. The internal liquid is simulated by fresh water which has the scaled mass from the hydrocarbon products from the prototype. The roof of the FHST is assumed flat in the model test. These simplifications are not significant. In the model test, the origin of the body fixed coordinates locates at the center of each body's water plane. The x-axis points towards the wave maker, z-axis

Table 2 Geometrical dimensions of barge frame (Unit: m)

Dimension	Prototype	Model
Length (exterior)	130.20	2.8933
Length (interior)	100.20	2.2267
Width (exterior)	85.30	1.8956
Width (interior)	49.30	1.0956
Depth	6.00	0.1333

Table 3 Mass properties of FHST

Parameter	Prototype	Model	Unit
Mass (empty)	7,494,737	79.0–79.5	kg
Rxx/Ryy (empty)	15.06	0.318	m
Rzz(empty)	16.33	0.406	m
Vertical COG	8.17	0.1675	m
Mass (20%)	9,669,737	103.0–103.5	kg
Mass (100%)	18,119,737	193.5–194.0	kg

Table 4 Mass properties of barge frame

Dimension	Prototype	Model	Unit
Total mass	24,938,634	267	kg
Vertical COG	3.42	0.076	m
Rxx	44.35	0.9856	m
Ryy	30.85	0.6856	m
Rzz	42.47	0.9439	m

Table 5 1-year and 100-year sea states at a specific location around Singapore

Return period	H_s (m)	T_p (s)
1 year	1.0	5.0
100 year	1.8	7.0

points upwards and *y*-axis can be determined by the right-hand screw rule. The origin of the global coordinates is fixed at the center of the water plane of the barge frame.

Key parameters such as motions of the FHST, global loads on the mooring dolphins, tension loads on the mooring lines and the free surface in the narrow gaps have been monitored in the model test. The 6 D.O.F motions of the FHSTs are monitored through OQUS motion tracking system. The forces on the mooring dolphins were measured by two load cells inside the dolphin. The tension loads on 4 out of 8 mooring lines are obtained. 11 wave probes are utilized to monitor the free surface elevation in the basin, in the narrow gaps and inside the tanks. 2 more wave probes are installed in the basin to monitor the wave field in the basin. There are three cameras mounted in the front and on the side of the model and under water. One more camera was mounted on the barge frame to monitor the local free surface motions. All the systems are synchronized to give an output with a sampling frequency at 20 Hz.

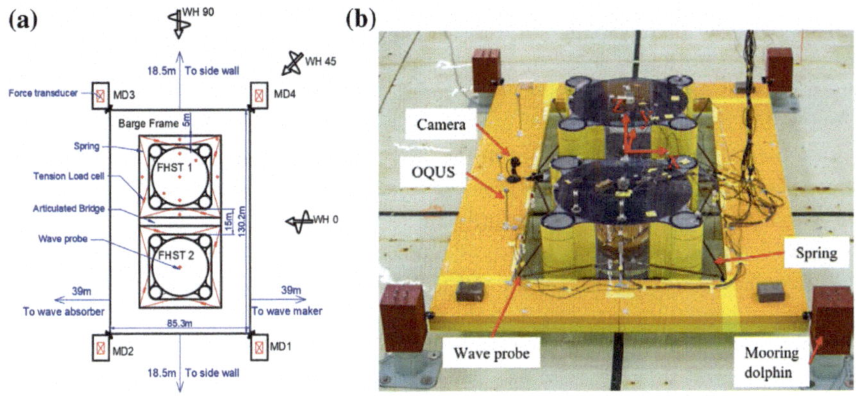

Fig. 3 **a** Layout of simplified system in ocean basin (Phase-1); **b** Physical model of simplified system in the ocean basin

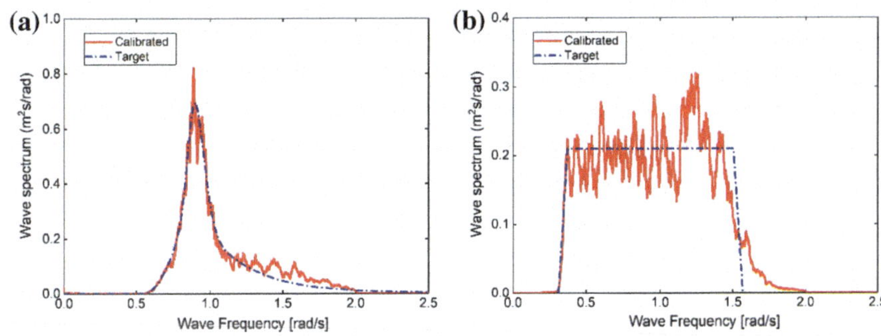

Fig. 4 **a** Incident wave spectrum of 100-year wave; **b** Incident wave spectrum of white noise

White noise test is performed to obtain the RAOs of the system for the numerical validation. The significant wave height H_s of the white noise is 2 m in full scale. The wave periods of the white noise are from 4 to 20 s; thereby covering the wave frequency range. The sea states with return periods of 1 year and 100 years have been defined as the operational and extreme conditions for the FHSF. The H_s and peak period T_p of the corresponding 1-year and 100-year waves are given in Table 5. In this paper, we will focus on the extreme response in the 100-year waves. Combined wave and current tests but will not be discussed herein. JONSWAP spectrum is assumed for the 100-year waves. Both white noise and the 100-year wave tests last for more than 1800 s in model scale corresponding to 3 h in the prototype. All the incident waves were calibrated by the iteration procedure. The target and calibrated incident wave spectrum of the white noise and 100-year waves are plotted in Fig. 4a, b.

Based on the results from the white noise tests, the RAOs are calculated by

$$\boldsymbol{RAO} = \sqrt{S_r/S_I} \tag{1}$$

in which S_r is the response spectrum and S_I the incident wave spectrum. Both spectra are derived from the recorded time history when the motion of the model reaches the steady state. Threshold has been set for data processing. When the wave energy is below the threshold, the RAOs are not used to control the quality of the results. The statistical results are taken from the 100-year wave tests. The noise has been filtered before taking the statistics. One should note the "extreme responses" herein are based on one realization. It is sufficient to verify the concept at this stage. The extreme response analysis requires more realizations of different random seeds. This will be done once the numerical model is validated. Both RAOs and the statistical results have been transferred back to the full scale. The results presented herein are all in full scale unless otherwise specified.

4 Description of Numerical Model

The numerical model was established based on the linear potential theory. The velocity potential φ governed by Laplace's equation in the fluid domain can be solved with the properly defined boundary conditions. The detailed definition of the boundary value problem (BVP) for a floating body can be found in [23]. The velocity potential is normally decomposed into incident wave potential $\emptyset_I$, scattered potential $\emptyset_s$, radiation potential in N D.O.Fs., i.e.

$$\emptyset = \emptyset_{\mathrm{I}} + \emptyset_{\mathrm{s}} + \sum_{i=1}^{N} \emptyset_i \tag{2}$$

The incident wave potential and scattered potential are usually combined as diffraction potential $\emptyset_{\mathrm{D}}$.

For a single body, $N = 6$ in Eq. (2). For a multibody system, it is straightforward to solve the radiation problem by extending N from 6 to $6 \times N$ [12]. The generalized modes such as flexible modes of structure can also be included. For solving the boundary value problems, Green's theorem is applied to derive integral equations. By using HOBEM, B-splines are utilized to describe each patch of the body. This is also in an equivalent manner to the representation of the velocity potential. As commented by Lee and Newman [22], the higher-order method is more efficient and accurate in most cases. The bottom skirts which a thin plate on the bottom as shown in Fig. 5a, have been proven to improve the hydrodynamic responses. The bottom skirts are modeled by thin submerged elements with the 'dipole' method. At this stage, the effects of fluid in the internal tanks are also considered, but the sloshing can only be considered in the linear range. The details on these advanced functions can be found in [22]. Once the radiation and diffraction potentials are solved, the added mass and damping coefficient and the first-order wave excitation forces can be expressed as

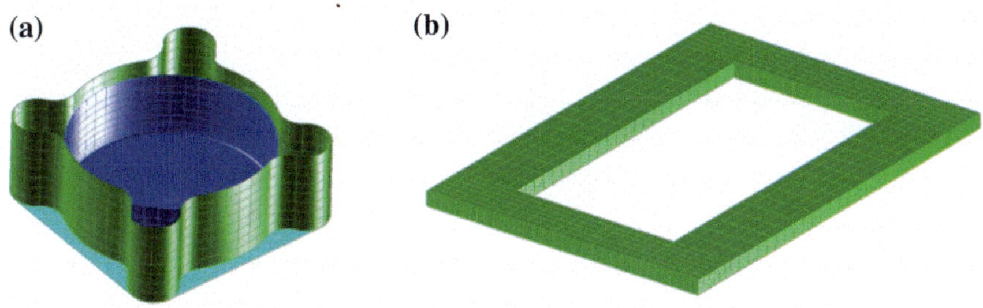

Fig. 5 **a** Hydrodynamic model of FHST with an internal tank for WAMIT; **b** hydrodynamic model of barge frame. Scaling ratios are different in (**a**) and (**b**) for ease of viewing

$$\boldsymbol{F}_i = -\mathbf{i}\rho\omega \iint_S \varnothing_D \boldsymbol{n}_i d\boldsymbol{S} \tag{3}$$

$$\boldsymbol{A}_{ij} - \mathbf{i}\rho\omega\boldsymbol{B}_{ij} = \rho \iint_S \varnothing_j \boldsymbol{n}_i d\boldsymbol{S} \tag{4}$$

where $\boldsymbol{F}_i$ is the wave excitation force in the i-th mode of motion, $\boldsymbol{A}_{ij}$ and $\boldsymbol{B}_{ij}$ are the added mass and radiation damping in i-th mode induced by the motion in j-th mode, S is the wet surface of the floating body, $\boldsymbol{n}_i$ is the normal vector of the wet surface, $\boldsymbol{\rho}$ the water density and ω the wave frequency.

For a rigid floating system with zero forward speed in waves moving in the steady state, the equation of motion due to first-order wave excitations can be described in the frequency domain as

$$-\omega^2(\boldsymbol{M} + \boldsymbol{A}(\omega))\boldsymbol{X}(\omega) + \mathbf{i}\omega(\boldsymbol{B}(\omega) + \boldsymbol{B}_{ext})\boldsymbol{X}(\omega) + (\boldsymbol{C} + \boldsymbol{C}_{ext})\boldsymbol{X}(\omega) = \boldsymbol{F}(\omega) \tag{5}$$

where ω is the wave frequency;$\mathbf{M}$ the structural mass matrix;$\mathbf{A}(\omega)$ and$\mathbf{B}(\omega)$ are the frequency dependent added mass and damping matrix respectively; $\boldsymbol{C}$ is the restoring matrix; $\boldsymbol{B}_{ext}$ and $\boldsymbol{C}_{ext}$ are the additional damping matrix and external restoring matrix;$\mathbf{X}(\omega)$ is the motion vector; $\mathrm{F}(\omega)$ the external force vector. All the vectors and matrices are expressed in the frequency domain. So, the motion RAOs in linear range can be solved efficiently.

The numerical model was built in the potential flow code WAMIT v7.2 [22] with the same setup as in the model test. Higher order boundary element method (HOBEM) was adopted for the simplified system due to its complexity. The bottom skirts were simulated by thin elements provided in WAMIT. The patches of the FHST and barge frame are shown in Fig. 5a, b. In total, 32 patches were used for the FHST and 28 patches were used for the barge frame. 3rd order B-spline was used to describe all the patches. Additional stiffness was added in the surge and sway in the numerical model to simulate the mooring system. The equivalent stiffness is 1200 kN/m in the prototype. In order to damp the unrealistic motions in near resonance frequencies, 10% of critical damping was introduced in pitch and roll. 10% of critical damping in heave was also introduced. The damping ratios were estimated from the decay tests. As there exists narrow gaps and sloshing, the frequency interval was set to be 0.01 rad/s from 0.2 to 2 rad/s so as to capture the resonant phenomenon accurately. No free surface damping was introduced in the numerical model at this stage.

5 Results and Comparison STUDIES

Numerous cases have been tested in the experiments as there exist various combinations of environmental conditions, loading conditions and directions of wave and winds. To validate the numerical model, a simple case in the head sea when the two FHSTs are empty was selected. Larger responses of the FHSTs in the random waves were expected as the inertia of the FHST was smaller than in other cases. After the preliminary conceptual and numerical verification, more complicated cases will be performed in the future.

5.1 Comparison of Numerical and Experimental Results in Head Sea

5.1.1 Comparison of Motion RAOs in Head Sea

The RAOs of the 6 D.O.F motions of the barge frame from numerical simulation and the experiments are compared in Fig. 6a, b. FL00 in these figures denotes that the FHSTs are empty. Good agreements of the RAOs can be seen from these figures. The numerical model tends to overpredict some peaks on the RAO of heave. This is relevant to the gap resonances which have been overestimated in the potential flow as no free surface dissipation was introduced. To solve this problem, the generalized modes were introduced in the numerical model. Alternatively, the energy dissipation term could be introduced on the free surface boundary conditions. The interesting finding is the knockout period on the RAOs of the barge motion. The knock period is around 6.5 s either in heave or pitch. They are relevant for cancelation of the wave excitations on the front and the backside barge components. From the RAOs, we can find that the knockout period is very close to the peak period of the 100-year waves. In general, the RAOs of the barge motion in heave and pitch is small. This indicates the barge frame is well-designed.

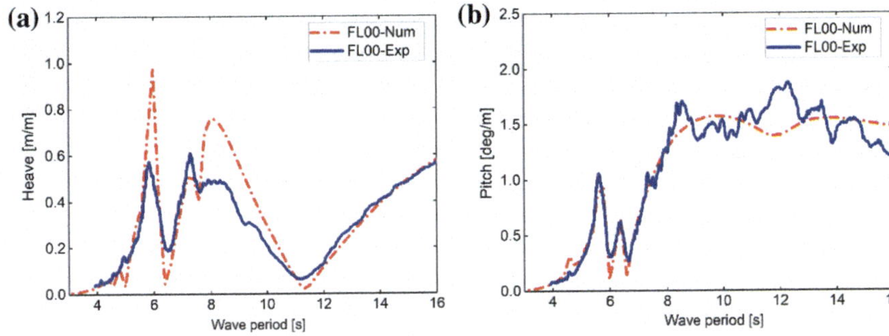

Fig. 6 **a** Numerical and experimental RAOs in heave of barge frame; **b** Numerical and experimental RAOs in pitch of barge frame

5.1.2 Comparison of Motion RAOs of FHSTs in Head Sea

The RAOs of the motion in the surge, heave, pitch and roll of the FHSTs are plotted in Fig. 7a–d. The first three are expected to be the dominating motions in the head seas. As the system is symmetrical, the motion RAOs of the two FHSTs are the same in the numerical results. Very close motion RAOs of the two FHSTs are given by experiments. Fairly good agreement can be found between the numerical and experimental results in surge, heave, and pitch. The good agreement on pitch and heave indicate that the damping ratio used in these two D.O.Fs in the numerical model are proper. The roll motions are excited in the head seas due to the hydrodynamic interactions. The overall trends of the RAOs in roll are well predicted by the numerical model, but it underestimates the roll motions especially in the long waves. It is hard to justify the reason for this underestimation, but one suspects that it is due to the gap resonances. In addition, there exist several peaks on the RAOs from the numerical simulations such as the peak at around 5.5 s. These peaks can be confirmed to be relevant to the gap resonances later. There also exists sway motion in this head sea condition. The RAOs in sway has a similar trend as in roll, so it is not plotted here. The yaw motion is also not plotted as the RAOs are very small. This can be found through the statistics of yaw motion later. The natural periods of surge and pitch are all outside the wave frequency range. The heave motion is also small when the incident wave period is between 4 and 7.5 s. These results verify the proper design of such a complicated system. Through these comparisons, it also verifies that the numerical model based on HOBEM has very good accuracy although the FHST has a very complex shape and the system is also complicated. There remains some additional work such as introducing free surface damping. These works will be performed in the future.

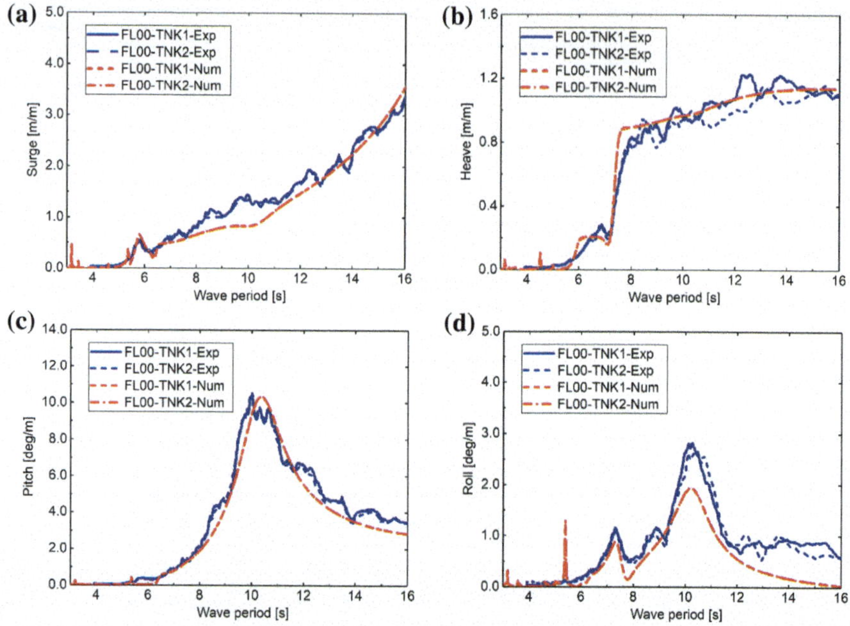

Fig. 7 **a** Numerical and experimental RAOs in surge of barge frame; **b** Numerical and experimental RAOs in heave of barge frame; **c** Numerical and experimental RAOs in pitch of barge frame; **d** Numerical and experimental RAOs in sway of barge frame

5.2 *Statistical Responses in Extreme Wave Conditions*

Statistical results show the performance of the system in the realistic sea states. As mentioned earlier, the dynamic responses of the system in the FL00 under the 100-year storm is of interest. The statistical responses, the body motion, the free surface elevation, and tension load on the mooring line and the global loads in the horizontal plane are discussed. These will enable one to verify the performance of the simplified FHSF system in extreme conditions.

5.2.1 Statistical Results of Body Motions in 100-Year Waves

Figure 8a, b show the results of the barge frame motion in heave and pitch. The maximum heave and pitch are around 0.7 m and 1.4 deg. The mild global motions of the barge frame in the 100-year storm are critical to the on-site facilities. However, we should note that with a combination of heave and pitch, the vertical motion of the barge components can be larger than 1 m. The freeboard of the barge frame is only 2 m. The green water cannot be avoided on the deck. To solve this problem, the freeboard must be increased. Figure 8c, d show the results of the translational

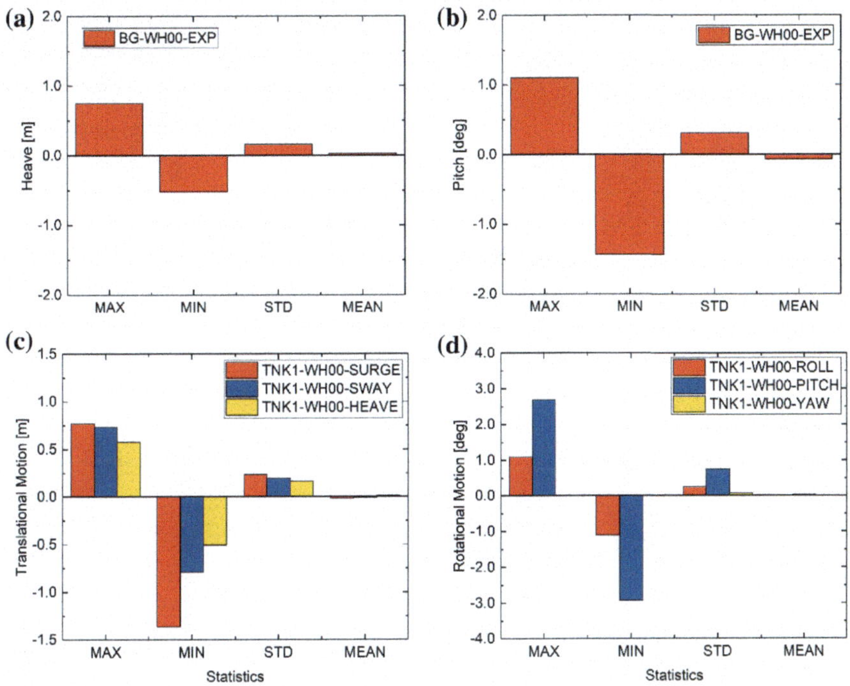

Fig. 8 **a** Statistics in heave of barge frame; **b** statistics in pitch of barge frame; **c** statistics of translational motions of FHST; **d** statistics of rotational motions of FHST

motions, i.e. surge, sway and heave and the rotational motions, i.e. roll, sway, yaw. It can be found that the maximum displacement in the horizontal plane is around 1.3 m in the 100-year storm. Considering the 5 m gap between the FHST and the barge frame, no collisions will happen. The maximum pitch is less than 3° which is smaller than 5° which has been set as the limit state of the FHST by the project team. In this system, the yaw motion is almost zero. This supports that the innovative soft mooring system can control the yaw motion very well. An interesting find is the maximum motion in sway in the head sea condition can be half of the surge motion. The strong hydrodynamic interactions must be considered in the design of such a system. These minor motions in the extreme sea states validate the employment of this concept in Singapore's waters.

5.2.2 Statistical Results of Free Surface Elevation in 100-Year Waves

The free surface elevation in the narrow gaps is one of the concerns for this design concept. The statistics of the free surface elevations are plotted in Fig. 9. The 5 wave probes were installed in the narrow gaps as shown in Fig. 3. W08 is on the left-hand side of the FHST, W09 at the front side corner of the free surface, W10 and W12 are

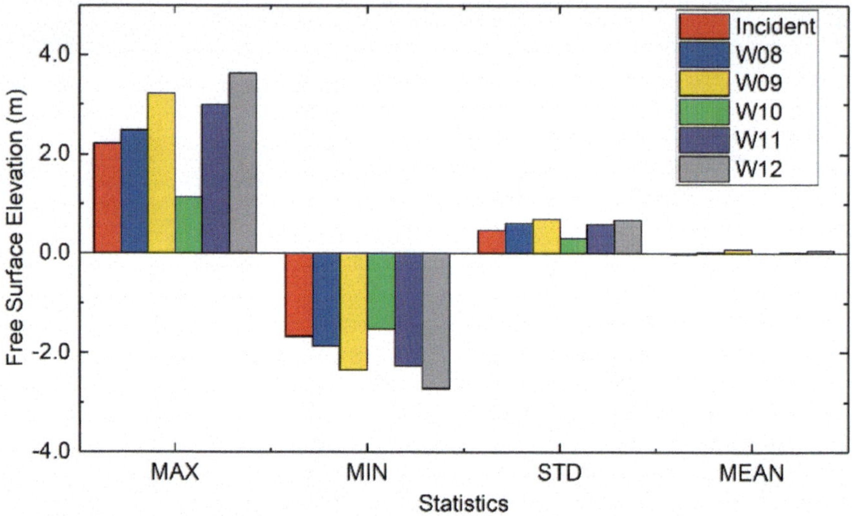

Fig. 9 Statistics of free surface elevation in narrow gaps

placed behind and in front of the FHST, W11 is close to the center of the free surface. The statistics of the calibrated incident wave (when the model is not installed) are also plotted for ease of comparison. It can be found that the maximum free surface elevation occurs at W12 at around 3.8 m. It can be around 2 times that of maxima of the incident waves. This position is on the weather side which is strongly influenced by the incident waves. On its opposite position W10, the free surface oscillation is much smaller because of the shielding effects from the barge frame and the FHST. The maxima at this position are only half of the maxima of the incident waves. The second large maximum free surface elevation appears at W09. This position is also on the weather side. W11 also experience relatively large free surface oscillations. All these large free surface elevations are relevant to the resonance of the fluid. These results also indicate the barge frame design should be improved to avoid serious green water induced by the gap resonances.

5.2.3 Statistical Results of Tension Loads in 100-Year Waves

The tension loads were measured by tension load cells on the selected mooring lines. As the system is symmetrical, the statistics of the tension forces on four of the mooring lines are plotted in Fig. 10a. These mooring lines connect the FHST-1 to the barge frame. T1 to T4 represents the loads on the mooring lines on four sides of the FHST-1. One can find the position of the load cells in Fig. 3. As the motion in sway is not small, the displacement of the FHST is along the diagonal direction of itself. This can explain why the maximum tension loads are larger in T1 and T3. The maximum loads on the mooring lines are smaller than 300 kN. Normal offshore

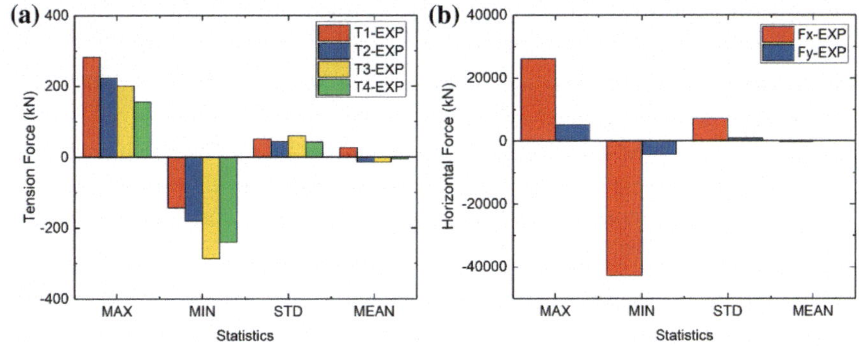

Fig. 10 **a** Statistics of mooring loads on selected members of mooring system; **b** statistics of global loads in horizontal plane

ropes/wires can resist the loads. The statistics of the global horizontal loads on the system are shown in Fig. 10b. The maximum force in the x-direction is around 40 MN, and the maximum force in the y-direction is around 5 MN. The latter is generated by hydrodynamic interactions. The global loads are not small for this simplified system. However, we cannot conclude that the loads on the entire system will be too large as there will be cancellation of forces in the entire system. The maximum horizontal load measured from the model test phase-3 is only 92.6 MN. More details on this will be reported in the future.

6 Conclusion Remarks

This paper summarized the recent research work on the hydrodynamics of the innovative floating hydrocarbon storage facility. A simplified two-tank system for the FHSF facility is proposed for numerical and experimental studies. The experimental work on the simplified system and the entire system is briefly introduced. The numerical model on this simplified system is also presented. Numerical and experimental results are compared. The statistics of the hydrodynamic responses and loads are discussed. Below, we summarize the main findings of this study.

(1) A numerical model based on potential flow through HOBEM can give acceptable results on hydrodynamic responses of the simplified system. There is some overestimation on the heave motion of the barge frame and underestimation on the roll motion of the FHST. These are suspected to be caused by fluid resonances in the narrow gaps as no free surface damping is introduced in the numerical model.
(2) RAOs of the 6 D.O.F motions of the FHST and the heave and pitch motions of the barge frame are small in the wave frequency range. The natural period of surge and pitch of the FHST have been shifted away from the wave frequency

range. It verifies that the system has been properly designed for the given sea states.

(3) The statics of the motions of the barge frame and FHST are all mild in the 100-year storm in the head sea condition. The maximum pitch of the FHST is smaller than 3° and the maximum surge is smaller than 1.5 m. Both are within the acceptable range. The loads on the mooring lines are also at an acceptable level. The minor motions and loads support that the FHSF concept is very doable under the given sea states.

In the future, more cases should be verified based on the validated numerical model. For example, the case with internal sloshing effect will be further investigated. The severity of the gap resonances and occurrence of green water should be analyzed especially when there exist combined wind, wave, and current actions. The performance of the system under such combined environmental conditions should be further verified.

Acknowledgements This research is supported in part by the Singapore Ministry of National Development and the National Research Foundation, Prime Minister's Office under the Land and Liveability National Innovation Challenge (L2 NIC) Research Programme (L2 NIC Award No L2 NICTDF1-2015-2). Any opinions, findings, and conclusions or recommendations expressed in this material are those of the author(s) and do not reflect the views of the Singapore Ministry of National Development and National Research Foundation, Prime Minister's Office, Singapore. The authors acknowledge WAMIT for providing educational license.

References

1. Wang, C. M., & Tay, Z. Y. (2011). Hydroelastic analysis and response of pontoon-type very large floating structures. In *Fluid structure interaction II* (pp. 103–130). Heidelberg: Springer.
2. Miyajima, S., Seto, H., & Ohta, M. (2003). Hydroelastic responses of the Mega-Float Phase-II model in waves. *International Journal of Offshore and Polar Engineering*, *13*(04).
3. Wang, C. M., & Wang, B. T. (2015). *Large floating structures*. Singapore: Springer.
4. Cheng, Z., Gao, Z., & Moan, T. (2018). Hydrodynamic load modeling and analysis of a floating bridge in homogeneous wave conditions. *Marine Structures, 59,* 122–141.
5. Lamas-Pardo, M., Iglesias, G., & Carral, L. (2015). A review of Very Large Floating Structures (VLFS) for coastal and offshore uses. *Ocean Engineering, 109,* 677–690.
6. Wan, L., Zhang, C., Magee, A. R., Jin, J., Han, M., Ang, K. K., et al. (2018). An innovative mooring system for floating storage tanks and stochastic dynamic response analysis. *Ocean Engineering, 170,* 361–373.
7. Zhang, C., Magee, A. R., Wan, L., Wang, C.M., & Hellan, Ø. (2017). Experimental study of hydrodynamic responses of a single floating storage tank with internal fluid. In *ASME 2017 36th International Conference on Ocean, Offshore and Arctic Engineering.* Trondheim, Norway, 25–30 June 2017. OMAE2017-61867.
8. Wan, L., Han, M., Jin, J., Zhang, C., Magee, A. R., Hellan, Ø., et al. (2018). Global dynamic response analysis of oil storage tank in finite water depth: Focusing on fender mooring system parameter design. *Ocean Engineering, 148,* 247–262.
9. Watanabe, E., Utsunomiya, T., & Wang, C. M. (2004). Hydroelastic analysis of pontoon-type VLFS: A literature survey. *Engineering Structures, 26*(2), 245–256.
10. Newman, J. N. (1994). Wave effects on deformable bodies. *Applied Ocean Research, 16*(1), 47–59.

11. Ding, J., Tian, C., Wu, Y. S., Wang, X. F., Liu, X. L., & Zhang, K. (2019). A simplified method to estimate the hydroelastic responses of VLFS in the inhomogeneous waves. *Ocean Engineering, 172,* 434–445.
12. Newman, J. N. (2001). Wave effects on multiple bodies. In M. Kashiwagi (Ed.), *Hydrodynamics in ship and ocean engineering* (Vol. 3, pp. 3–26), RIAM.
13. Taghipour, R., & Moan, T. (2008). Efficient frequency-domain analysis of dynamic response for the multi-body wave energy converter in multi-directional wave. In *The Eighteenth International Offshore and Polar Engineering Conference*. Vancouver, Canada, 6–11 July 2008. ISOPE-I-08-269.
14. Riggs, H. R., Ertekin, R. C., & Mills, T. R. J. (2000). A comparative study of RMFC and FEA models for the wave-induced response of a MOB. *Marine Structures, 13*(4–5), 217–232.
15. Fu, S., Moan, T., Chen, X., & Cui, W. (2007). Hydroelastic analysis of flexible floating interconnected structures. *Ocean Engineering, 34*(11–12), 1516–1531.
16. Zhang, H. C., Xu, D. L., Xia, S. Y., Lu, C., Qi, E. R., Tian, C., et al. (2015). Nonlinear network modeling of multi-module floating structures with arbitrary flexible connections. *Journal of Fluids and Structures, 59,* 270–284.
17. Shi, Q. J., Zhang, H. C., Xu, D. L., Qi, E. R., Tian, C., Ding, J., et al. (2018). Experimental validation of network modeling method on a three-modular floating platform model. *Coastal Engineering, 137,* 92–102.
18. Faltinsen, O. M., & Timokha, A. N. (2015). On damping of two-dimensional piston-mode sloshing in a rectangular moonpool under forced heave motions. *Journal of Fluid Mechanics,* 772.
19. Molin, B., Zhang, X., Huang, H., & Remy, F. (2018). On natural modes in moonpools and gaps in finite depth. *Journal of Fluid Mechanics, 840,* 530–554.
20. Zhao, W., Pan, Z., Lin, F., Li, B., Taylor, P. H., & Efthymiou, M. (2018). Estimation of gap resonance relevant to side-by-side offloading. *Ocean Engineering, 153,* 1–9.
21. Zhang, C., Dai, J., Ang, K. K., & Magee, A. (2018). Comparison study on bottom plate effect on single hydrocarbon storage tank through decay test. In *Proceedings of the 1st Vietnam Symposium on Advances in Offshore Engineering* (pp. 544–550). Hanoi, Vietnam, 1–3 November 2018.
22. Lee, C. H., & Newman, J. N. (2013). *WAMIT User Manual*, version 7.0. Chestnut Hill, MA: WAMIT, Inc.
23. Faltinsen, O. (1990). *Sea loads on ships and offshore structures*. Cambridge, UK: Cambridge University Press.

Wave Induced Motions of a Floating Mega Island

William Otto, Olaf Waals, Tim Bunnik and Coline Ceneray

Abstract Floating mega islands can provide an attractive solution for creating temporal or more permanent space in coastal areas with a high demand for real estate. Also at open sea in the vicinity of wind farms, fish farms or logistical cross points, a floating mega island could be used as a hub, eliminating costly transfers. One of the aspects which needs to be understood is the wave induced motion of such a floating mega island. A piece-wise flexible island has been model tested at MARIN. The motion behavior in mild and severe sea states has been investigated. In this paper, the motion behavior is described and explained by comparing model test results with numerical simulations. An interesting aspect in this is the relative importance of wave diffraction, wave radiation and the dissipation of energy in the construction. The wave drift loads on the island that consists of 87 interconnected triangular pontoons are calculated and analyzed.

Keywords Mega-Floater motion behavior · Multi-body diffraction

1 Background and Applications

With an increasing population living mostly in coastal regions there is an increasing need for building space in coastal areas. Nowadays, this is mostly provided by land reclamation projects in densely populated areas, such as Singapore and the Netherlands. Some coastal regions also suffer from sea level rise and local subsidence of the soil. For example, in Jakarta additional measures have been taken to protect the city from the ocean. This paper discusses the possibility to use floating platforms as part of city development.

W. Otto (✉) · O. Waals · T. Bunnik
MARIN, Wageningen, Netherlands
e-mail: w.otto@marin.nl

C. Ceneray
ENSTA, Brest, France

C. M. Wang et al. (eds.), *WCFS2019*, Lecture Notes in Civil Engineering 41,
https://doi.org/10.1007/978-981-13-8743-2_9

Fig. 1 Village on lake Titicaca in Peru (left) and a fishery village in Ha Long Bay in Vietnam

Fig. 2 Left: A 1000 m floating runway in Tokyo bay, Right: Osaka Airport on reclaimed land

Also at open sea in the vicinity of wind farms, fish farms or logistical cross points, a floating mega island could be used as a hub, eliminating costly transfers. Mega islands can offer temporal or more permanent living and working space for:

- Developing, generating, storing, and maintaining sustainable energy
- Loading and transhipping cargo in coastal areas where there is little infrastructure;
- Cultivating food, such as seaweed and fish;
- Building houses and recreation close to the water.

The use of floating elements for living space is not new. There are small communities that have been living on floating islands for a long time. This is mostly in protected waters or on lakes. Two examples are shown in Fig. 1.

Floating constructions have been tested for large city infrastructures such as airports. One example is the 1000 m floating runway was tested in Tokyo bay [1]. In Japan, there is also a large experience with airport construction on reclaimed land (Fig. 2).

For city construction, one could list the most important design aspects as shown in Table 1. In this table, a comparison is made between the design aspects for a floating and a reclaimed island. Although there is much more experience with land reclamation projects the potential application area for floating islands is much larger.

Table 1 Design aspects

Floating	Land Reclamation
• Relatively new technology • Motions • Larger water depth • Mooring loads • Risk of sinking • Design for tsunami • Cost • Impact on environment • Legal issues • Modular • Redeployable	• Proven technology • No motions • Limited water depth • Seawall loads • Risk of flooding • Design for earthquakes • Cost • Impact on environment • Legal Issues

Fig. 3 Artist impression of a large floating island

In limited water depths, one could also combine a fixed sea wall with a floating part. This has for example been done in the Semarang airport in Indonesia, where they are building a floating terminal as part of the airport extension.

Concepts for a floating city have been proposed by Quirk [2] and Roeffen [3]. These studies show the benefits of living at floating island by stipulating the risk of coastal flooding and the access to new sources of nutrition. Floating (air)ports consisting of an assembly flexibly connected modules have been studied by Kikutake [1] and Zhang [4], with the focus on a number of modules in the order of ~2 to 10. The response of very large flexible floaters has been investigated by Utsunomiya [5]. Murai [6] investigated the interaction with the local bathymetry.

Watanabe [7] has presented a concise literature overview of the work that was done on very large floating structures. A study on the effect of air cushions on the motions of large floaters was done by Van Kessel [8]. In the present study, we are investigating the response of coupled large triangular floaters. An artist impression is shown in Fig. 3.

Fig. 4 Model of the floating island in the basin (87 modules)

2 Piecewise Flexible Island

In this paper the wave induced motions of a piecewise flexible island consisting of triangular rigid bodies are investigated. This island has been tested in MARIN's Offshore Basin as described by Waals et al. [9], see Fig. 4. For most of the applications of large floating islands it is a design objective to minimize the motions and building cost. In general the motions of floaters become smaller with increasing floater size. On the other hand larger floaters are more expensive and impractical to build. Very large floaters may show significant deflection and modal response and cannot be considered as one rigid body. Owing to the oscillating nature, the resulting strain may lead to local fatigue damage of the structure. By using a piecewise flexible floater the aim is to design the island such that the majority of the bending strain is in the connections between the modules. This reduces the internal loads in the island modules compared to a design where the mega structure is built in one piece. For the present island a system of interconnected triangular floaters was selected. The triangular shape was selected to restrict the least degrees of freedom of motion as possible for each individual floater. By connecting each pontoon on three sides an island surface is obtained that can bend in several directions. This allows for oblique wave conditions to pass the island with similar load levels as for the head on cases.

3 Wave Induced Motions

In [9], the motion Response Amplitude Operators (RAO's) of an 87 body linear diffraction calculation have been compared to the measured motions in the model scale tests. The initial comparison was poor. The authors suggest that the most likely explanation of the poor comparison is that the mechanical fenders and lines interconnecting the triangular modules were not taken into account in the numerical simulations. Also, the only source of damping was the wave radiation damping from potential flow. In order to better understand the numerical calculations, first a single body diffraction calculation was performed which is described in Sect. 3.1.

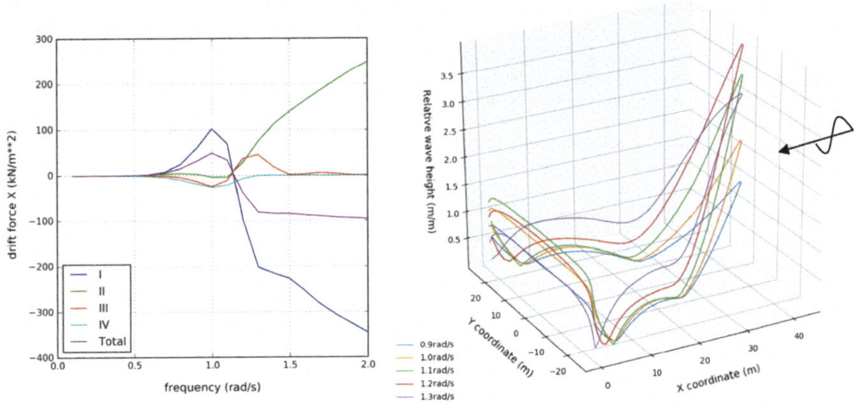

Fig. 5 Left, contributions to mean drift force on a single body triangle. Right, relative wave height along contour of triangle in unit wave amplitude

The improved results of the 87 body diffraction calculation is described in Sect. 3.2. A physical interpretation of the results is discussed in Sect. 3.3, the influence of the floater motions on the mean wave drift force of the assembly is discussed in Sect. 3.4.

3.1 Single Body Diffraction on Triangular Body

Before performing the diffraction calculation on the assembled island, a diffraction calculation was performed on one single triangle. These calculations showed interesting behavior in the drift forces. This is shown in Fig. 5 on the left for head waves, being waves traveling towards the point of the triangle. The total drift force is shown by the purple line, the four individual mathematical terms which add up to the total drift force are shown as well. It can be seen that around 1.0 rad/s, the sum of the drift force has a positive value. Physically, the interpretation of this is that the single triangle in waves is drifting towards the incoming waves. This peculiar result is counterintuitive compared to experience with other floaters and in the opinion of the authors unlikely for a blunt body in a viscous flow.

To better understand this behavior of the drift force, the four individual contributions to the mean drift are plotted as well. The full derivation of the four contributions can be found in [10], and the physical meaning is explained below;

- I = force due to first order relative wave elevation
- II = pressure drop due to the first order velocity
- III = force due to angular motion and inertia force
- IV = pressure due to gradient first order pressure and motion

As it can be seen, the peculiar drift force on a single body triangle in waves around 1 rad/s is mainly caused by the first contribution, which is associated with the relative

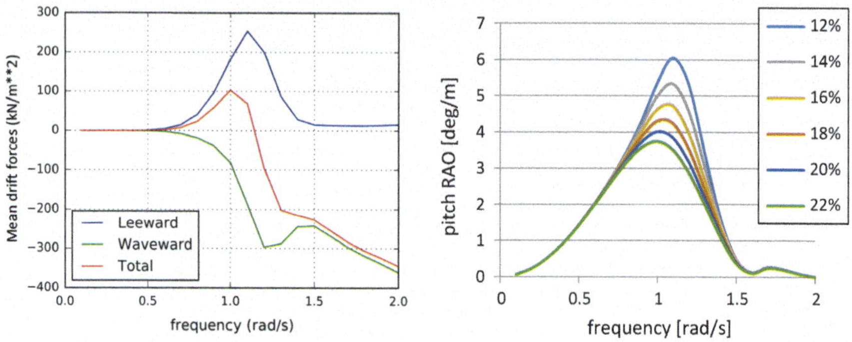

Fig. 6 Left, first contribution of mean drift force, Eq. (1) over waveward sides and leeward side. Right, pitch RAO for different values of critical damping

wave height along the module. Mesh sensitivity studies have shown consistent results for different mesh parameters. The relative wave heights are therefore plotted in Fig. 5 on the right for the relevant frequencies. It can be seen that the solution is smooth, and there are no discontinuities which might imply numerical instabilities. Note that the relative wave height is a result from the undisturbed wave, the diffracted wave, the radiated wave and the body motion, as calculated by the diffraction code. Given these relative heights, the first contribution of the drift force can be calculated by the integral of the relative wave height along the waterline WL (1);

$$F^{(2)} = -\frac{1}{2}\rho g \int_{WL} \eta_r^2 \vec{n} dl \tag{1}$$

In which η_r denotes the relative wave height and n the outward normal vector to the waterline. This equation applies to the waterline of all three sides of the triangle. Note that in X-direction, the normal vector of the two waveward sides have the same normal vector leeward of +0.5, the leeward side of the triangle has a normal vector of −1. The integral over the two waveward sides are shown in green in Fig. 6 on the left, the integral over the leeward side is shown in blue.

It can be observed that the positive drift force is originating from the relative wave height at the leeward side of the triangle, the relative wave height is pushing the triangle at the leeward side against the wave. The relative wave height at this side of the triangle is mainly due to the motion response of the triangle, as the undisturbed wave and the diffracted wave are cancelling out each other to a great extend. The fairly straight lines in Fig. 5 on the leeward side of the triangle support this, the relative wave height at the leeward side is dominated by the pitch motion of the triangle. This leads to the hypotheses that the pitch response of the triangle is overestimated. Note that these calculations are performed with a potential flow method, without viscosity. The pitch response in reality is damped by radiation damping as well as by viscous damping. For this particular floater, the radiation damping around 1.0 rad/s is 12% of the critical damping. Additional linearised damping can be added to the diffraction

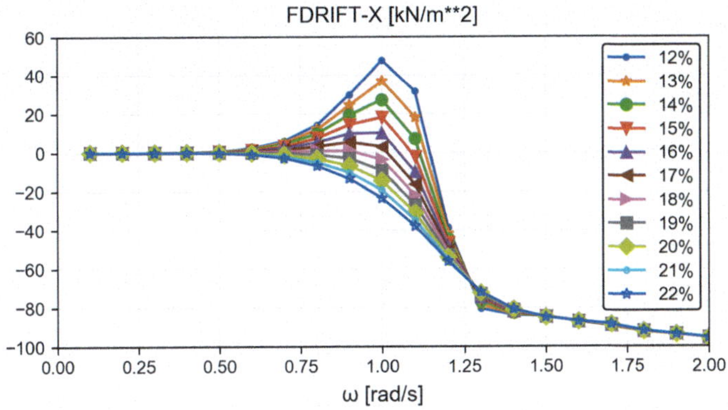

Fig. 7 Total drift force on single body triangle for different pitch damping values

calculation in order to allow for viscous effects. The sensitivity of added linear damping to the pitch motion RAO is shown in Fig. 6 on the right. It can be seen that for 12% critical damping (only radiation), the peak response is 6.0 deg/m, and drops to 4.5 deg/m for 18% of critical damping (6% viscous damping added). The sensitivity of the total mean drift force, the sum of all contributions, to the pitch damping is shown in Fig. 7. It can be seen that the positive drift force does not appear for damping levels higher than 18%, which corresponds to 6% added linearised viscous damping. Future work will show how realistic this is, by comparing the calculated motion response with single body model tests and, if feasible, viscous CFD calculations. When making this future comparison, it should be noted that viscous damping in general has a quadratic nature and care has to be taken on how to linearise it. For the present paper, it is assumed that adding 6% of linearised viscous damping is the most realistic but also most conservative scenario as this is the least amount of damping which needs to be added in order to eliminate the positive drift forces.

3.2 87-Body Motion Response Calculations

The motion response of the 87-triangle assembly is calculated in the frequency domain by using the results from a diffraction calculation as described in [9], a linearised viscous damping contribution as described in Sect. 3.1 and the stiffness and damping from the mechanical fenders and lines. The determination of the global stiffness matrix and damping matrix is a tedious task; their size is 522×522 (6 dof times 87 bodies), and they have contributions of not only the 87 floaters but also from the 256 fenders and 512 lines used to keep the assembly together. This makes their determination by hand not only time consuming but also prone to errors. In order to fill the damping and stiffness matrices in a convenient way a new functionality was added to aNySIM-XMF, which is an in-house MARIN tool dedicated

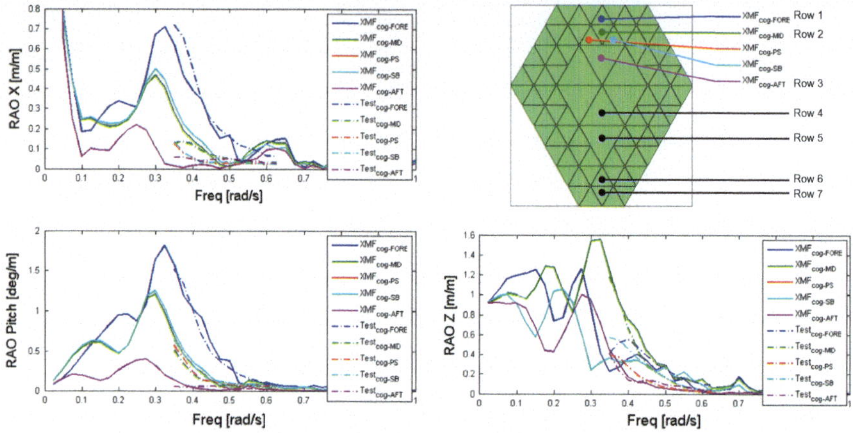

Fig. 8 Motion RAO of 87-body frequency domain calculation compared to model tests

to time domain simulations. By design aNySIM-XMF is a multi-body simulation tool in which mechanical joints and lines can be connected between n-number of bodies. Although designed for time domain simulations, the new functionality uses imposed motions to construct the stiffness matrix and imposed velocities to construct the damping matrix. Each body is consecutively moved by a small offset (1 mm in this case) and the resulting restoring force from all aNySIM components are stored in the linearised global stiffness matrix. This includes the hydrostatic as well as the mechanical restoring. Note that especially for the central triangles, a motion of one triangle results in a force on multiple others. The stiffness matrix therefore has a lot of non-diagonal interaction terms, in total 10,829 elements of the matrix are filled with non-zero entries. The damping matrix is filled in a similar manner.

The comparison between the calculated motion RAOs and those derived from the model tests as described in [9] is shown in Fig. 8. In general there is a good agreement between the tests and XMF calculations in the frequency range measured. During the tests, the motions of only five triangles were measured. Triangle denoted FORE is on the first row on the waveward side, the triangles MID, PS and SB are on the second row and AFT is the larger triangle on the third row. A trend of decreasing motion response when moving further away from the waveward row is clearly visible in the calculations as well as in the tests.

As it can be seen in Fig. 8, the frequency with the most response is around 0.325 rad/s. The motion response at this frequency is shown in Fig. 9. On the left is the real part of the motion RAO, which can be interpreted as a snapshot in time at t0, on the right the imaginary part of the motion RAO, which occurs ¼ wave period later. The phasing is with respect to the undisturbed wave at the CoG AFT, which is where the turret mooring is located in the tests as described in [9]. The colors represent the vertical motion, the horizontal motions are amplified 25 times in order

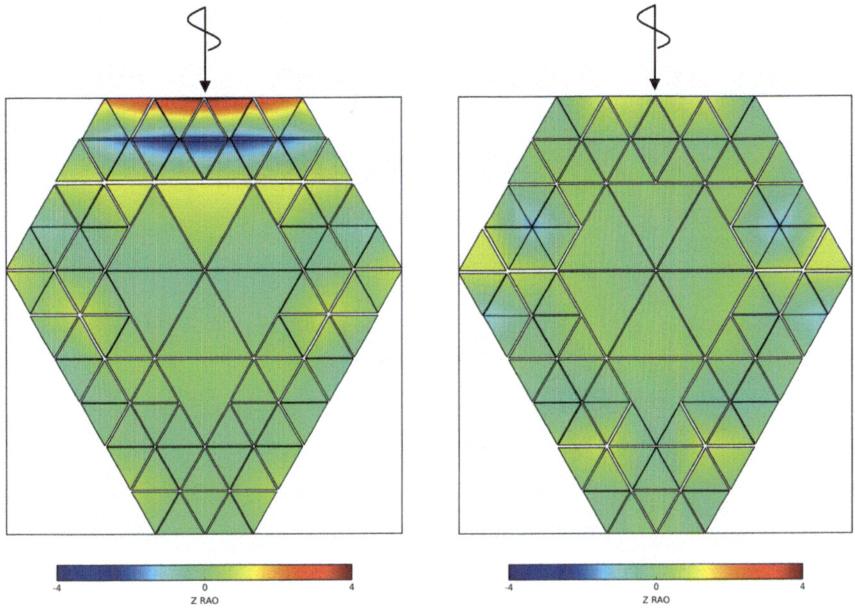

Fig. 9 Motion RAO at 0.325 rad/s of 87-body frequency domain calculation, horizontal RAO motions amplified 25 times. Real (left) and imaginary part (right)

to make them more visible. Note that because of this the first and second row on the waveward side are partially overlapping as there is compression in the fenders.

Although the motions of only five triangles were measured in the model tests, the video cameras confirm the general impression that the first and second rows on the waveward side move while the remainder of the assembly stays relatively still. This can be a convenient property for applications which have strict motion restrictions such as floating cities. In order to be able to design and optimize this behavior, the driving physics behind this are investigated in the next section.

3.3 Driving Physics Behind Motion Response

From both model tests and numerical calculations, the waveward triangles show the largest motion response and the response decreases when going further to the leeward side of the assembly. To better understand what is causing this behaviour, the pressure and response are plotted in Fig. 10. Note that in linear diffraction theory, the total wave field is reconstructed out of three wave fields; the undisturbed wave, the diffracted wave and the radiated wave. The wave excitation is defined as the resulting force of the combined undisturbed and diffracted wave field. On the top row of Fig. 10 the real part of the pressure fields and response are shown. As it can

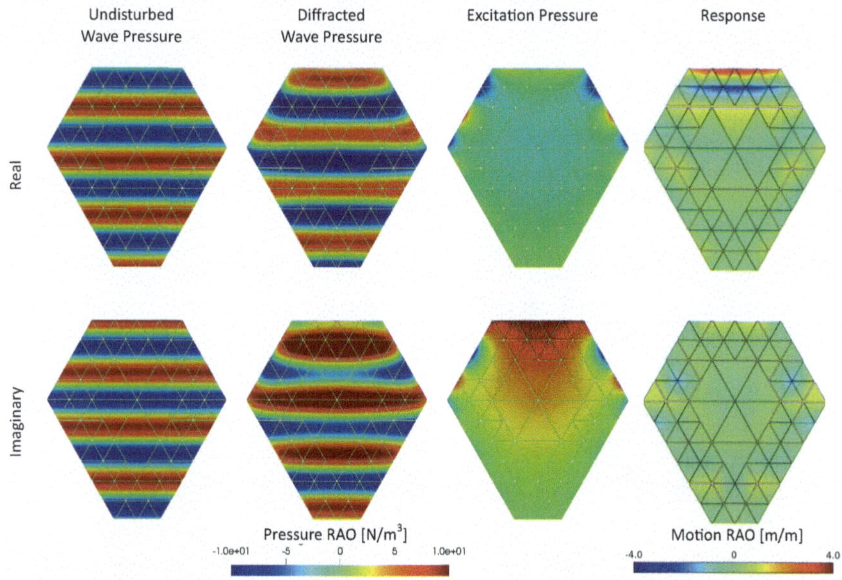

Fig. 10 Wave pressure and motion response per meter wave amplitude at 0.325 rad/s, top row real part, bottom row imaginary part, result of 87-body diffraction calculation

be seen, the undisturbed wave and the diffracted wave field are to a large extend in counter phase, cancelling out each other, which results in a reduced excitation. In the bottom row of Fig. 10 the imaginary part of the pressure fields are shown (¼ wave period later than the real part). Here the diffracted wave is cancelling out the undisturbed wave only at the leeward side of the assembly. At the waveward side of the assembly, there is a net excitation force on the front.

The motion response is largely explained by the wave excitation, the first and second row of triangles on the waveward side have the largest motion response and also the largest wave excitation. Note that the response on the waveward side is mainly real, while the excitation is mainly imaginary. The phase lag between response and excitation is caused by the inertia of the structure. The wave excitation further to the leeward side of the assembly is decreasing to almost zero as the incoming wave is being diffracted by the first rows. It can be interpreted as that the leeward pontoons are sheltered by the waveward pontoons, the first two rows act like a wave deflector.

Another interesting overview is shown in Fig. 11. Here the calculated pitch and heave RAO of the central triangle of each row (see Fig. 8) is plotted. Around 0.275 rad/s, it is clear that the first two rows show the largest response, however the rows more to the leeward side also pitch and heave considerably. The wave excitation pressure at 0.275 rad/s is shown in Fig. 12. The wave excitation at this frequency shows a similar pattern as shown in Fig. 10, there is even less wave excitation at the leeward side at 0.275 rad/s than at 0.325 rad/s. Still the response at the leeward side around this frequency is larger as shown in Fig. 11.

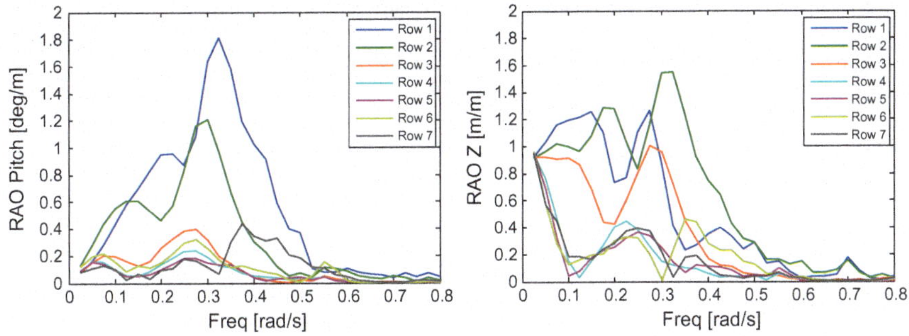

Fig. 11 Left; Pitch RAO Right; Heave RAO; central triangles of consecutive rows from full computation

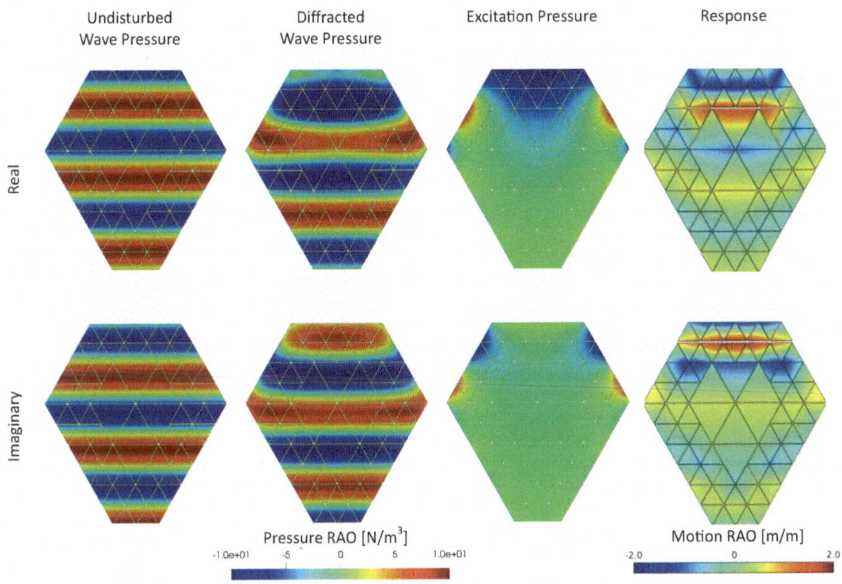

Fig. 12 Wave pressure and motion response per meter wave amplitude at 0.275 rad/s, top row real part, bottom row imaginary part, result of 87-body diffraction calculation

The response at 0.275 rad/s is also shown in Fig. 12. Behind the first two rows, a repetitive wave pattern becomes visible. It appears that the leeward side of the assembly is moving along with the undisturbed wave, despite the fact that the wave excitation indicates that the incoming wave is already diffracted at the waveward rows (the undisturbed wave is cancelled out by the diffracted wave). The contours of the motion response also show a bended, u-shaped profile in this plot.

In order to increase the understanding of this motion response, the same calculation has been performed without body interaction terms in the added mass and damping matrices. The physical meaning of this is that the force of a wave radiated

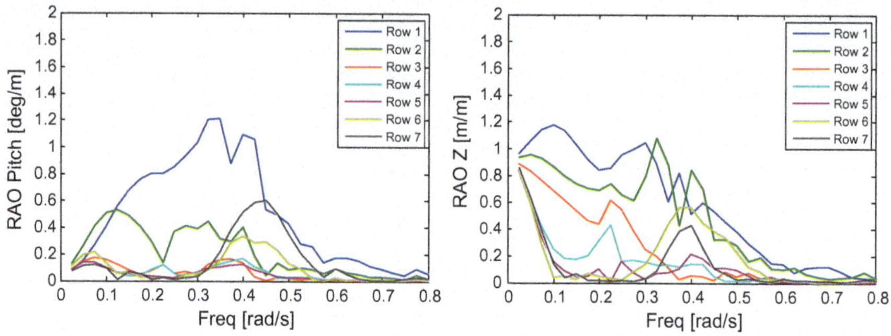

Fig. 13 Left; Pitch RAO Right; Heave RAO; central triangles of consecutive rows from computation without the radiating wave interaction between bodies

by a body is only felt by the body itself and not felt by the surrounding bodies. The resulting pitch RAO is shown in Fig. 13. It can be seen that the hump around 0.275 rad/s of the leeward rows (5, 6, 7) has almost disappeared.

Note that this plot has no further physical meaning. The only purpose of this plot is to investigate what is causing the response of these rows. It appears that at this particular frequency the leeward triangles are responding to the radiated waves from the waveward triangles. As radiated waves propagate in concentric circles, this might also be an explanation of the u-shaped contourlines in the motion plot. Note that these radiated waves are by definition caused by the motions of these bodies and they are proportional to their motion amplitude. In other words, the motions of the leeward triangles are hydrodynamically coupled to the motions of the waveward triangles by the radiated waves. This can be valuable insight when designing and optimizing a segmented island for a benign motion response, as the motions of the whole assembly can be altered by altering the motions of the first rows. This might for instance be done by varying the draft (inertia) of the triangles or by the stiffness of the connections. A sensitivity study of the motion response to these parameters will be part of future work.

Another interesting phenomenon in Fig. 11 appears around 0.45–0.50 rad/s. Although in general the response at this frequency is smaller than the response at lower frequencies, the last row shows a significant pitch response when compared to the first rows. The wave pressures and response at these frequencies are shown in Figs. 14 and 15. It is interesting to see that while at the frequencies 0.275–0.325 rad/s the excitation in the leeward islands was reduced to almost zero, the wave excitation at the leeward triangles around the frequencies 0.45–0.50 rad/s is in the same order of magnitude as on the waveward islands.

Further analysis of Figs. 14 and 15 shows that these excitation pressures are a result of the interference between the undisturbed and the diffracted wave. It seems that the diffracted waves form a frequency dependent pressure pattern underneath the island. The antinodes of this pattern is shown in the third picture from the left. The higher the frequency, the more complex the pattern becomes.

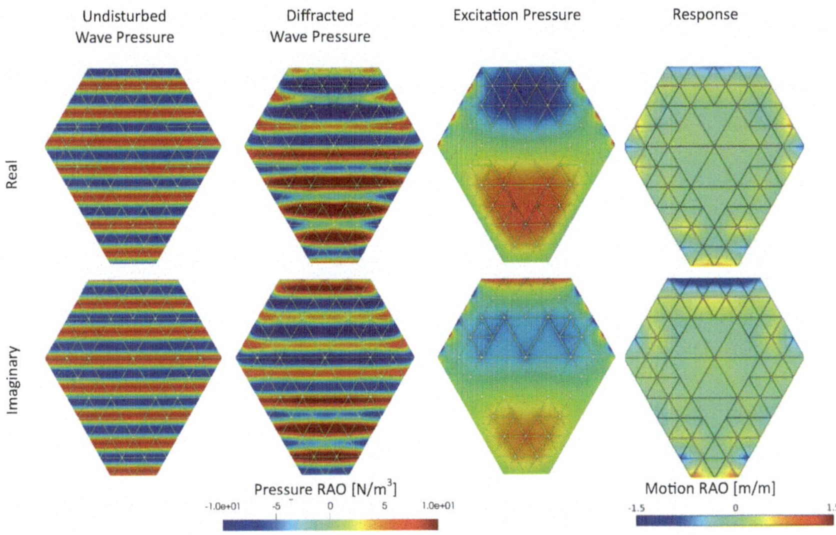

Fig. 14 Wave pressure and motion response per meter wave amplitude at 0.450 rad/s, top row real part, bottom row imaginary part, result of 87-body diffraction calculation

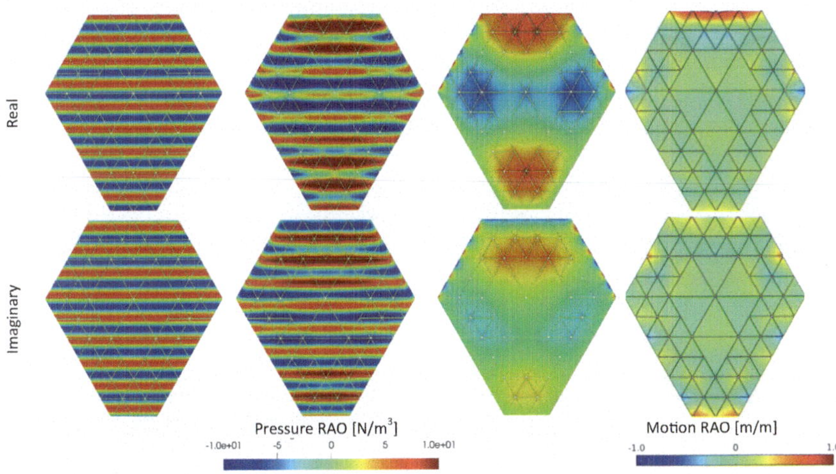

Fig. 15 Wave pressure and motion response per meter wave amplitude at 0.500 rad/s, top row real part, bottom row imaginary part, result of 87-body diffraction calculation

For this specific assembly the excitation at these frequencies does not lead to a large response because of the inertia of the individual modules. For design optimizations however this insight can be important when choosing the optimum ballast weight of the modules.

3.4 87-Body QTF Drift Forces

The first-order pressures, water velocities and motion responses resulting from the linear diffraction analysis can be used to compute Quadratic Transfer Functions (QTFs) for the mean and low-frequency wave drift forces using the direct pressure integration approach proposed by Pinkster [10]. In order to suppress unrealistic water motions in the gap between the triangles damping was added to the linearized free surface condition. In the frequency domain:

$$\frac{\partial}{\partial z}\varphi(\underline{x}) - (1 - i\varepsilon)\frac{\omega^2}{g}\varphi(\underline{x}) = 0 \quad (2)$$

This equation is enforced by panels on the free surface inside the gaps between the triangles. A damping value $\varepsilon = 0.03$ was used, which is based on experience with side-by-side offloading simulations, see Bunnik [12].

The wave drift forces are responsible for the mean offset and low-frequency motions of the island and are the governing the loads in the mooring lines. The drift forces on each individual triangle were computed. In this paper, only the mean drift force is considered. Figure 16 shows the surge wave drift force QTF on the entire island in head seas. The QTF has been computed with and without connection springs and added damping (see discussion in Sect. 3.2 on the effect of the motion response).

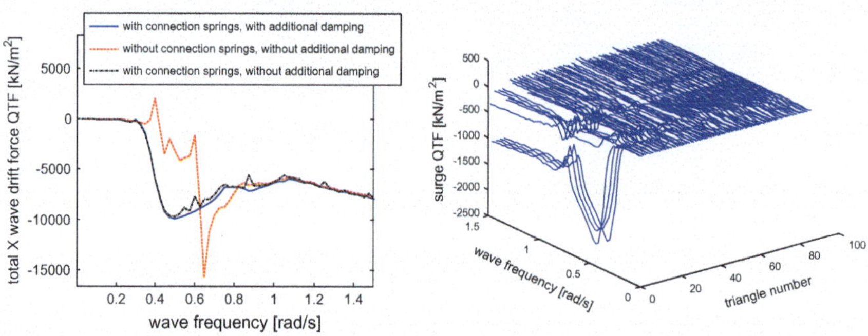

Fig. 16 Left; Mean surge wave drift force on all the 87 triangles combined. Effect of connection springs and added damping is shown. Right; Mean surge wave drift force on all the individual 87 triangles

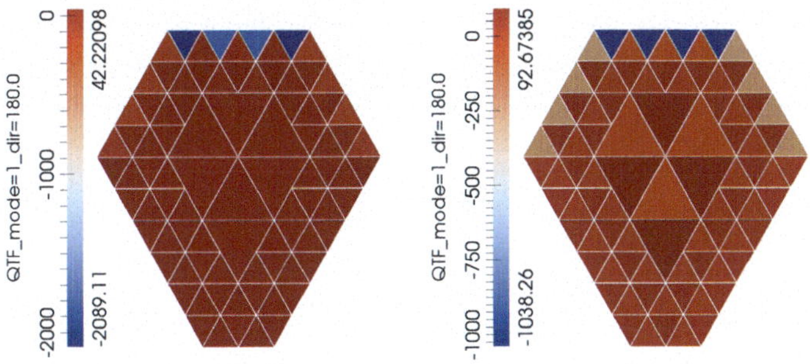

Fig. 17 Mean surge wave drift force per triangle. Left; 0.45 rad/s, Right; 1.25 rad/s

The effect of the connection springs on the QTF (through the motion response) is very large. The positive drift force disappears and the peak shifts to a lower frequency. The effect of adding the damping is to smooth the small oscillations that are present in the QTF. The connection spring and damping have no effect at higher frequencies because the triangles are not moving in short waves. The QTF is shown up to a frequency of 1.5 rad/s, which covers most of the wave energy of interest for the island. For higher frequencies, the present panel distribution was found to be too coarse to provide realistic results.

The mean drift force on each individual triangle is shown in Fig. 17 for two frequencies. For 0.45 rad/s, where the QTF is at a maximum, it can be seen that the drift force is mainly acting on the first waveward row. This insight can help in the design of mooring and connection, as the distribution of the loads is such that the leeward islands experience a benign drift force. The individual triangle with the largest positive drift force in this assembly is 42 kN/m^2, which is a small value compared to the largest negative value.

4 Conclusions and Further Work

In this paper, the wave induced motions of a piecewise flexible island consisting of 87 triangular rigid bodies have been investigated. This investigation is the first step towards a feasibility study of such a construction, with the obtained insights the safe operable limits and connection loads can be examined in future work. A linear diffraction method provided wave excitation and radiation values, the response is calculated in frequency domain with the incorporation of the mechanical connections in the global stiffness and damping matrix. Calculations on a single triangle led to the inclusion of an additional linear pitch damping in order to get realistic drift forces. For the assembly of 87 triangles, the calculated motion response shows a good similarity with the model tests for the triangles and frequencies measured.

Interesting wave interference patterns result in different motion response at different frequencies. At the frequencies with the largest motion response, the wave excitation at the leeward modules is close to zero as the waveward islands diffract the incoming wave. This results is a benign motion response of the leeward islands, making those suitable for motion sensitive applications such as living. At slightly lower frequencies there is however motion response of the leeward islands, despite that there is no wave excitation acting on them. They are responding to the radiated waves of the waveward triangles. For higher frequencies the interference patterns result in wave excitation on both the waveward and leeward triangles. The motion response on this excitation is however small due to the inertia of the triangles.

The wave drift force has been calculated on the 87-triangle assembly as well. The drift force is sensitive to the motion of the triangles and therefore it is relevant to take the global stiffness and damping matrix, including mechanical connections, into account in the drift calculation. In this case, especially the stiffness matrix has a large effect on the drift force. The drift force is mainly acting on the waveward triangles, the leeward triangle experience almost no drift force.

Further work will include a more in-depth study to the viscous damping contribution on the triangles to better understand the assumptions in this paper. Also further work will be done to improve the wave response of the island by varying global design parameters such as draft, assembly shape, module shape and module size. As the present study shows that the waveward triangles are efficient in sheltering the leeward islands, the design optimizations could also include a separate breakwater, either floating or fixed.

Acknowledgements This project has received funding from the European Union's Horizon 2020 research and innovation programme under grant agreement No. 774253. The opinions expressed in this document reflect only the author's view and in no way reflect the European Commission's opinions. The European Commission is not responsible for any use that may be made of the information it contains.

References

1. Kikutake, T. (1998). A Mega-Float Airport, The State of the Art. In *Proceedings of the 17th International Conference on Offshore Mechanics and Artic Engineering*, OMAE 1998. Lisbon, Portugal.
2. Quirk, J. (2017). *Seasteading: How ocean cities will change the world*. New York: Simon & Schuster.
3. Roeffen, B., Dal Bo Zanon, B., Czapiewska, K. M., & de Graaf, R. E. (2013). Reducing global land scarcity with floating urban development. In *Conference International Water Week*, Amsterdam.
4. Zhang, H. (2015). Nonlinear network dynamic characteristics of multi-module floating airport with flexible connectors. In *Proceedings of the 25th International Offshore and Polar Engineering Conference*, ISOPE2015, Kona, Hawaii, USA.

5. Utsunomiya, T., & Okafuji, T. (2006). Wave response analysis of a VLFS by accelerated Green's function method in infinite water depth. In *Proceedings of the 16th International Offshore and Polar Engineering Conference*, ISOPE2006, San Francisco, California, USA.
6. Murai, M., Inoue, Y., & Nakamura, T. (2003). The prediction method of hydroelastic response of VLFS with sea bottom topographical effects. In *Proceedings of the 13th International Offshore and Polar Engineering Conference*, ISOPE2003, Honolulu, Hawaii, USA.
7. Watanabe, E., Utsunomiya, T., & Wang, C. M. (2004). Hydroelastic analysis of pontoon-type VLFS: A literature survey. *Engineering Structures, 26*, 245–256.
8. Van Kessel, J. (2007). The effect of aircushion division on the motins of large floating structures. In *Proceedings of the 26th International Conference on Offshore Mechanics and Arctic Engineering*, OMAE2007, San Diego, California, USA.
9. Waals, O. J. (2018). Model tests and numerical analysis for a floating Mega Island. In *Proceedings of the ASME 2018 37th International Conference on Ocean, Offshore and Arctic Engineering*, OMAE1018, Madrid, Spain.
10. Pinkster, J. (1980). Low frequency second order wave exciting forces on floating structures. PhD thesis, TU Delft.
11. Utsunomiya, T., Watanabe, E., & Nakamura, N. (2001). Analysis of drift force on VLFS by the near-field approach. In *Proceedings of the 11th International Offshore and Polar Engineering Conference*, ISOPE2001, Stavanger, Norway.
12. Bunnik, T. (2009). Hydrodynamic analysis for side-by-side offloading. In *Proceedings of the 19th International Offshore and Polar Engineering Conference*, ISOPE2009, Osaka, Japan.

Fish Farming in Floating Structures

Tor Ole Olsen

Abstract The world will see an increased use of floating structures for various purposes as well as an increased harvesting of food from the sea. Farming of salmon has become a major contributor to food. Norway has excellent conditions for farming salmon; a long coastline (100,000 km including islands) with nice and fresh water. Norway produces 1.3 million tons of salmon per year (the weight of all Norwegian people is 0.3 million tons). The export value of salmon is the second largest after oil and gas. Starting from a small scale some 50 years ago, entrepreneurial fishermen have developed salmon farming into a huge business. Traditional salmon farming is performed in open nets. The open net solution is inexpensive, and suitable when placed in pure and clean water with ample current that provides for changing the water, which is important for salmons. However, there are challenges for the open net approach. The nets are vulnerable, and a broken net allows farmed salmons to escape, and possibly mingling with the wild salmons. The open net allows feces to fall through, and polluting the sea. Sickness may be spread by toxic water, and sea lice may enter and infest the fish. The obvious remedy to these challenges in "crowded" areas is to farm in closed buckets. One prototype bucket was designed, built and installed on the west coast of Norway, with excellent results. No salmon lice were found in the bucket. The salmon "liked" the closed bucket because one can provide a current that gives the salmon exercise. Salmons farmed in this way are larger and better fit, and thereby achieve a higher selling price. There are many activities around farming that will be described in the paper, and they are linked to the experience of marine concrete structures in general.

Keywords Fish farming · Floating structures · Design of floating structures · Design for fish welfare

---

T. O. Olsen (✉)
Dr.techn.Olav Olsen, POB 139, 1325 Lysaker, Norway
e-mail: too@OlavOlsen.no

C. M. Wang et al. (eds.), *WCFS2019*, Lecture Notes in Civil Engineering 41,
https://doi.org/10.1007/978-981-13-8743-2_10

1 Introduction

As a designer of concrete structures you need to know stresses and strains, behavior of the reinforcement, the loads, and many other things. As a conceptual designer of concrete structures, you also need to understand the purpose of the structure. All structures have a purpose. When designing a bridge conceptually, you have a number of handbooks to form your basis for design. Every country has road authorities that produce a number of handbooks. How do you proceed when designing for fish welfare? The fish, obviously, does neither speak nor write handbooks. You need to rely on experience and other disciplines, like biology.

2 Present Technology

Industrialized fish farming started with entrepreneurial fishermen throwing nets out from their seaside properties, and farming fish in them. Fish farming depends on local and national legislations, as well as relations to environmental restraints, local people, unions and costs and economics. I shall refer to fish farming in my country, Norway, that I know the best.

There are 4000 open net cages for fish farming in Norway, the largest 160 m in circumference. This size is tuned to the maximum number of fish allowed in one cage, 200,000. The reason behind this limitation is the consequence of failure of the net and farmed fish mingling with wild fish. Norway produces 1.3 million tons of salmon a year. The weight of all Norwegian people, large and small, is 0.3 million tons! In 2018, Norway exported seafood, not only salmon, for 10 billion Euros, which is more than the export value of wine from France. If every Norwegian should eat all the exported seafood, they would have 7 seafood meals a day, every day of the year. So seafood is important for Norway. But there are challenges!

Norway has a lot of gifts from nature: clean water, proper temperature, long coastline (100,000 km including islands), protected fiords, and long traditions of mastering the sea, which has been so important for the country.

So what are the challenges?

- *Lice* certainly influence the welfare of the salmon. In addition it is costly to handle, and lice is a quick adaptor to new treatments. Many years ago it was believed that the cure was close at hand, but the opposite is the case.
- *Escaping salmon* is, of course a loss, but it is also a danger to the wild salmon, through mixed breeding.
- *Feces*: Fish live in the ocean, and the ocean can take care of feces. The problem is when the feces is concentrated, as it is in many Norwegian fiords.
- *Overfeeding* means that some of the feeding goes straight through the net. Neighboring fish have no limits to intake of food, and becomes deformed.
- *Medicines* where doses do not only go to the needed ones.
- *Illness* speaks for itself, and it becomes a large problem in congested areas.

It is obvious that most of the problems are associated to "crowded" areas. It may sound a bit spoiled to say this, Norway has a land area of 533 times that of Singapore, and approximately the same human population. But many of the fiords are "crowded" by fish farming cages. The result of all this, in Norway, is that production of salmon has not increased since 2012, in spite of the wishes from politicians and others.

3 Cure

The simple cure, in a "crowded" area, is a watertight cage. Obvious it might seem, but there are some hesitations. Mostly because it is a novel way of fish farming, and it is a way of fish farming that reduces the benefits of clean and properly tempered water. And there is a cost issue. However, the benefits against external threats are clear. In floating closed cages, one can collect the water from a depth beyond the lice. The cost of pumping is not so large, as the water level inside the closed cage is almost the same as outside. The water needs to be changed every 60–90 min. One can collect the feces, and dry it and use it for fertilizing, or even fuel. As an example, cement producers are looking for alternatives to fossil fuel, to reduce the environmental impact of cement production.

Another cure is moving the farming offshore, to a less "crowded" area. An open net may be used, but it needs to be protected against ships and propellers. Yet another cure is placing the farming on land. This is, as of now, a more expensive solution, and will not be discussed herein.

4 Marine Concrete Structures

Fortunately, we know how to design and build marine concrete structures. In particular, over the past 50 years, about 50 marine structures have been successfully designed and built for the oil and gas industry. One of the more spectacular ones is the Troll A Platform, operated by Equinor and others. Sitting on the sea bed 303 m below the water surface, in 30 m waves, it has an overturning mudline moment of 100,000 MNm, on soft, yogurt like clay. Parts of the structure is subjected to 303 m water pressure, about 1000 times the load on an ordinary floor. Figure 1 shows the Troll A platform with other structures. The small one is the City Hall of Oslo, Norway. Katie Melua sings in the bottom of the platform.

Figure 2 illustrates a tow out of the Troll A platform, with its 22,000 t deck 150 m above sea level. The Troll A platform displaced 1 million tonnes when towed out. The Gullfaks C platform displaced 1.5 million tonnes as the heaviest object ever moved by man over a distance of some 400 km (see Fig. 3). These are the two beauties of the sea—buoyancy and the ability to transport on it.

Well-designed and well-built concrete structures are robust and sustainable in the marine environment. Core tests of offshore concrete platforms for the oil and gas

Fig. 1 Troll A and other structures, and Katie Melua giving a concert 303 m below sea level

Fig. 2 Tow out of Troll A

Fig. 3 Gullfaks C, its final destination, built in, and transported from Stavanger

Fig. 4 Challenges at sea

industry located in the harsh North Sea indicate a life of more than 200 years. Figure 4 illustrates some of the challenges at sea.

The experience from the marine structures for the oil and gas industry is vital. But this is not enough. The oil and gas industry has rigorous regimes for safety, QA, and codes and standards for design and construction. That is all fine, but the level of cost is accordingly. Other structures, in particular fish farming structures, are at a different level, and most importantly, the fish farming structures are not heavily loaded. Design has to account for this.

Marine concrete structures is not a novel phenomenon. Concrete ships and barges have been built for more than a century, particularly in war times, when steel was scarce. Lambot built his concrete canoe almost two centuries ago (see Fig. 5). Also in Fig. 5, one can see the concrete Mulberry Harbor built and installed by the British during the Second World War to transport military vehicles from the ships to the Normandy beach.

Fig. 5 Lambots canoe from 1848, and Mulberry Harbor from the Second World War

Fig. 6 Sea bath in Oslo

Fig. 7 Floating apartments, concept, Norway

In Norway, with 100,000 km coastline and few people, floating concrete structures have been built for recreation, Fig. 6 shows a very popular sea bath installed in Oslo.

Also in cities, floating structures provide possibilities for pleasant and creative dwellings, Fig. 7.

Other examples of marine concrete structures are floating wind generators (see Fig. 8), and submerged floating tunnels for strait crossings (see Fig. 9).

References [1, 2] give many more examples of large floating structures.

The international concrete federation fib has recognized the importance of marine concrete structures for the future, and works actively with the issue.

Fig. 8 Floating wind generator

Fig. 9 Submerged floating tunnel

5 Design

When doing conceptual design it is important to understand the purpose of the structure, the loading, and the governing load conditions: in other words "first principles". It is also important to understand the structural response. This is important because it is not possible to evaluate all loading situations. In many respects, this calls for a general oversight of the design.

The conceptual design phase is the time to make the good choices that will influence the success of the structure. The experience from the oil and gas industry is

Fig. 10 Heavily loaded gas platform Troll A and an almost not loaded fish farming bucket

Fig. 11 Shell structure

valuable and useful for a future with different kinds of floating structures. However, the designer of different structures should be aware of the changing circumstances.

In Fig. 10, the left structure Troll A is extremely stressed. When designing it, the gas delivery from the platform was already sold, and the delivery date was set. The fish farming "bucket" to the right (explained in detail in the next section) is practically not stressed at all, and it competes with a net. Design is fascinating!

One of the obvious challenges of design is weight; heavy structures do not float. And heavy structures are often the result of poor design, consequently good design is essential. Shell structures are efficient structures to carry distributed loads, illustrated in Fig. 11, and they are therefore often used.

From a design and cost point of view, there are many loads acting on a floating structure. Waves from all directions, ballasting changing over time, and live loads changing over time (Fig. 12).

Another important element of the marine structure is the method of construction, as illustrated in Fig. 13; a sea crane lifting a small structure from land to sea. Larger structures need other means, such as a dry dock. The principles of constructing large structures are illustrated in Fig. 14. It is very important to integrate construction and design closely. Also important: to float, marine structures are made as slender as

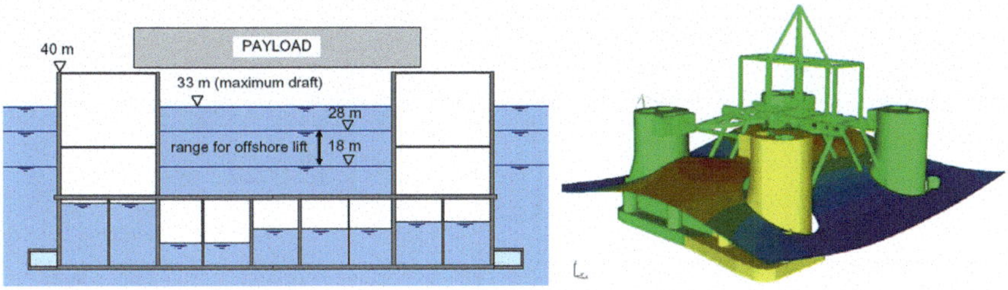

Fig. 12 Loads on a floating structure

Fig. 13 Small sea structure lifted from dock to sea

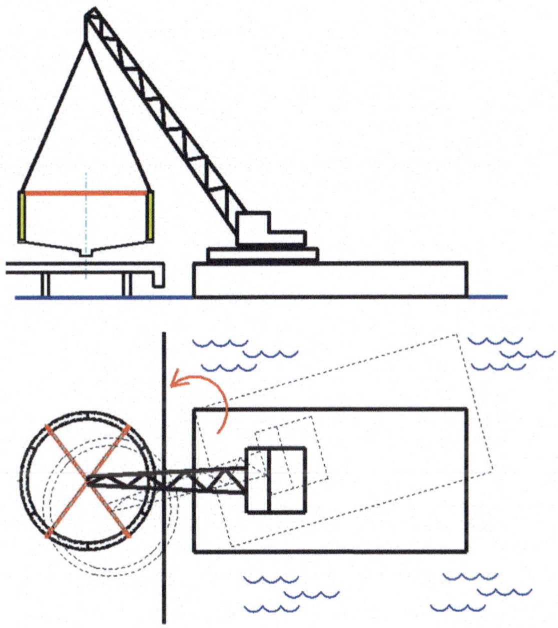

possible. This means that the content of reinforcement is high, calling for careful design and construction planning. Needless to say the weight control is an important part of the marine structure project.

To handle all the loads and all the code checking, programs that automate design are required.

Whilst design automation to calculate loads and perform code checks is part of everyday life in a design office, the eyes and reviews of an experienced engineering team is still a prerequisite for successful design deliverables.

Both linear and non-linear finite element programs are commercially available. Practical experience when modelling and selecting element types is a critical aspect of the linear and non-linear FE analyses employed for concrete design. Reinforced concrete will demonstrate non-linear behavior and methods such as the Modified Compression Field Method and the Modified Compression Field Theory, devel-

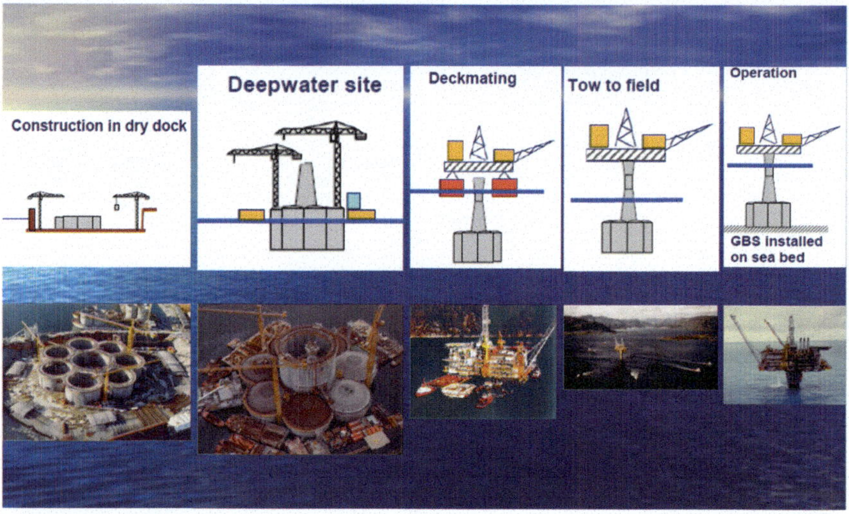

Fig. 14 Principles of constructing large marine concrete structures. The Draugen Platform is the example here. First dry dock, so float out and completion before deck mating. The deck is built in parallel with the concrete structure. Finally tow out, installation and production of, in this case, oil

oped by Professor Collins and his team at University of Toronto, provide a rational approach to estimating shear capacity beyond the standard design codes. This is especially important for shear, as shear is often handled empirically, and marine concrete structures do not resemble laboratory test specimens. In our practice, we have developed software for about 40 years, and are still refining it. We call it ShellDesign.

Traditionally large Finite Element (FE) analyses are based on the principle of linearly material behavior. As we know, reinforced concrete does not behave linearly, Fig. 15 is a simple illustration of this, and shows that the assumption of linearly elastic behavior may be unsafe in some cases and too safe/expensive in other cases.

As a consequence we have developed a nonlinear scheme; first run a linearly elastic FE analysis, then code check sections based on ordinary non-linearly principles. Then we compare stiffness from the point checks with the stiffness's assumed in the FE analysis, and upgrade as required. This is an iterative process, possible even for very large global FE analyses. The scheme is illustrated in Fig. 16. The Modified Compression Field method is incorporated in ShellDesign.

It is thus possible to handle a wide range of analyses of reinforced concrete structures. The linear and effortless transition from simple conventional concrete design, to advanced non-linear, triaxial concrete analyses equips the end user with a highly efficient design tool. The result is lighter and safer structures. This analysis made it possible, for the Sea bath described previously, to reinforce some of the walls with only one layer of reinforcement, saving a lot of weight due to strict requirements to cover of reinforcement for structures in the sea. A detailed description of ShellDesign, with examples, is given in Ref. [3].

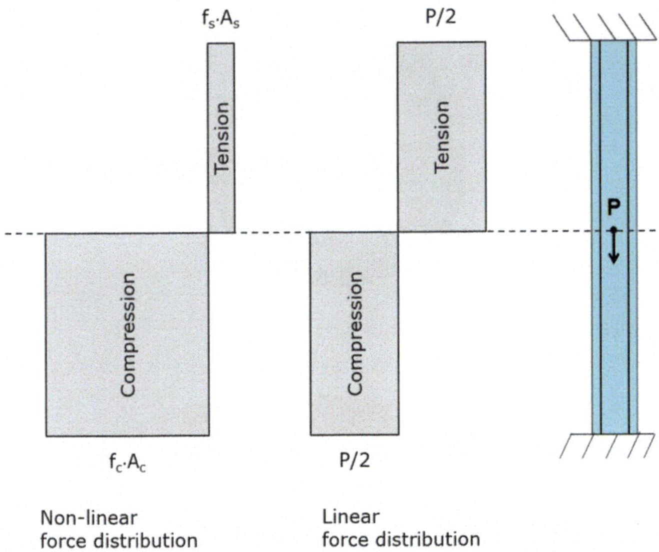

Fig. 15 Illustration of too safe/expensive and unsafe, if assuming concrete behaves linearly elastic

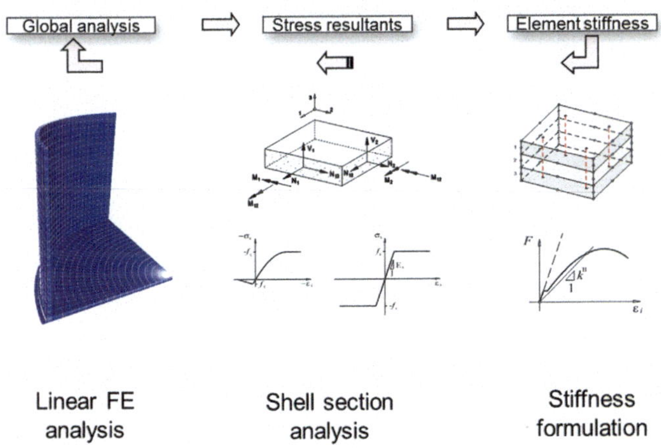

Fig. 16 ShellDesign flowchart

Many institutions have contributed significantly to design rules and design practices: ACI (Committee 357), DNV-GL, ISO, Norwegian Standards and fib.

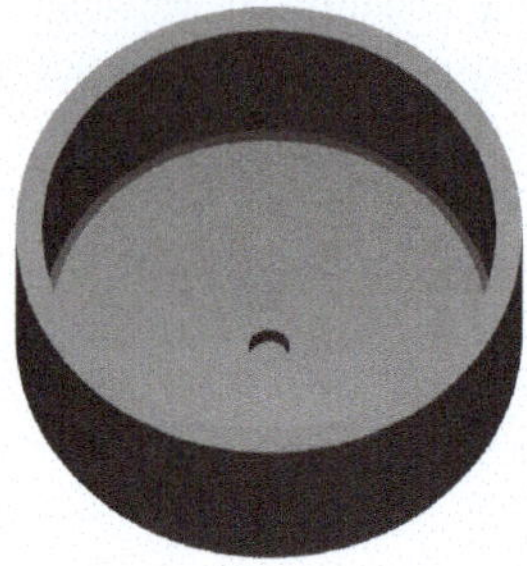

Fig. 17 The fish farming bucket

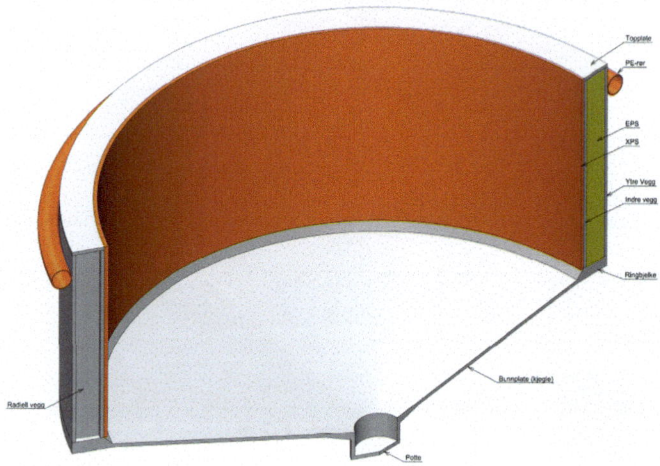

Fig. 18 Principle of fish farming bucket

6 Fishfarming Cages

The problems with fish farming with traditional open nets, in "crowded" areas, are already explained. The obvious solution is to close the cage. The closed fish farming "bucket" of Fig. 10 will be described in detail. In principle it is a simple bucket, made of concrete (see Fig. 17). It can hold 1000 m^3 of water.

It has a slender cone in the bottom, with a pot to collect feces and dead fish. The walls are double with Styrofoam in between, to provide buoyancy, Figs. 18 and 19.

The bottom cone has rebars to carry the weight while lifted, the ring beam has pretension cables to carry the thrust from the cone while floating with less water inside than outside. Otherwise the structure is reinforced with steel fibers (which are handled in ShellDesign as shown in Fig. 20).

The bucket, named Salmon Home #1 and owned by FishFarming Innovation, is the first concrete fish farming bucket built (see Figs. 21 and 22).

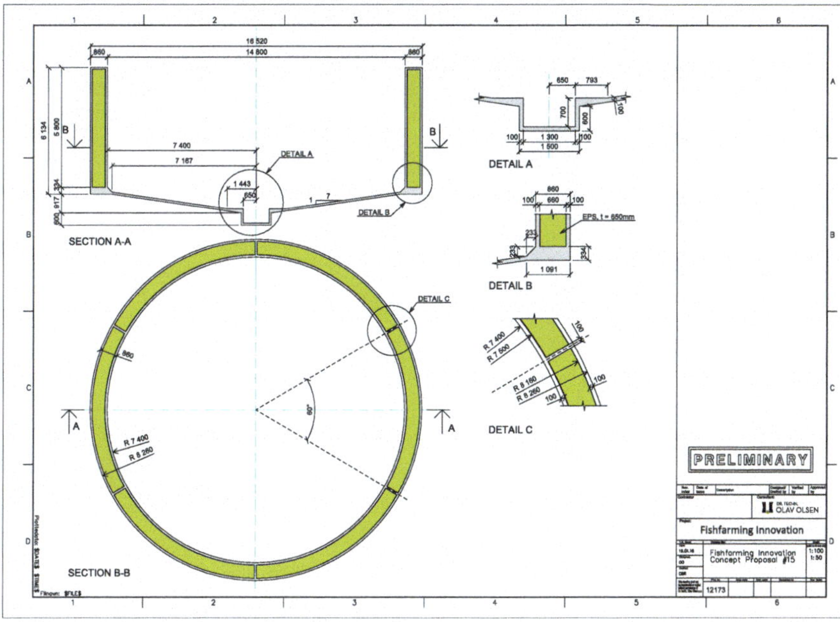

Fig. 19 Drawing of bucket

Fig. 20 FE model

The bucket now produces salmon with very good results (Fig. 23). Also working on the bucket is comfortable and safe.

Another proposed closed cage, housing 37,000 m^3 of water, is named Stadion Laks. This too is intended for "crowded" areas, and illustrated in Figs. 24 and 25. This concept is equipped with a lot of facilities, such as feed, feces, handling equipment for the fish, cranes, quarter, pumps, safe landing system for ships, and energy supply.

For "non-crowded" locations, like off shore, open nets are acceptable, if protected against ship impacts. One such concept is the Blue Farm, a tension leg offshore cage for a very large number of salmon (see Figs. 26 and 27).

Fig. 21 Completed bucket ready for lift out

Fig. 22 Lift out

There are some relatively advanced analyses that are required for this type of concept, but because of experience acquired from similar structures, it is manageable (Fig. 28). The hydrodynamic analysis of the net requires special competence.

Cost is a vital parameter. The bucket described above was a prototype, and relatively expensive because many things had to be pioneered and developed. The plan is to build several more, and a standardized construction will give lower costs. Costwise it is difficult for a concrete wall to compete with a net, per m^2 (Fig. 29).

Fig. 23 In successful operation

Fig. 24 The Stadion Laks

Fig. 25 Salmon export

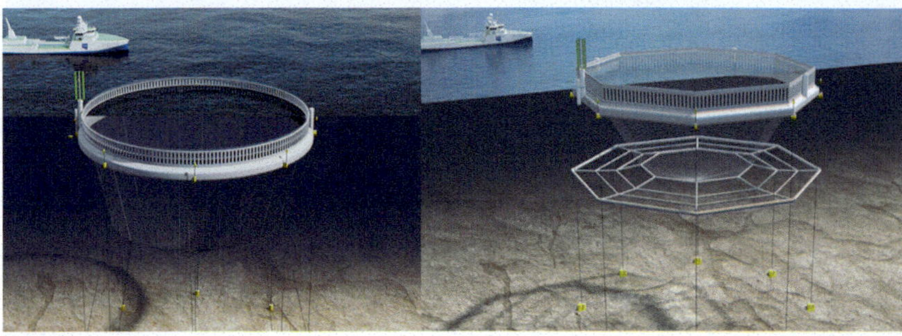

Fig. 26 Blue Farm ocean farm. Initial concept to the left, improved for construction reasons, to the right

A recent Norwegian research report [4] describes fish farming in detail, and concludes estimates of cost of farming, in Norwegian kroner NOK (1 € ~ 10 NOK) as shown in Table 1.

Expected average selling price in 2019 is 62 NOK/kg. Business is good now, it has been for some years. But it is not always good. By looking at the numbers and the assumptions behind them, it is clear that comparisons of cost should include more than what is included in the research report, such as:

- It is allowed (Norway) to have a higher density of fish in closed cages (50–75 kg/m^3) as compared to open nets (25 kg/m^3)
- Closed cages can be more closely "packed", so space is saved
- Open net farming requires location to be left idle for 2–3 months after a production cycklus, for environmental reasons

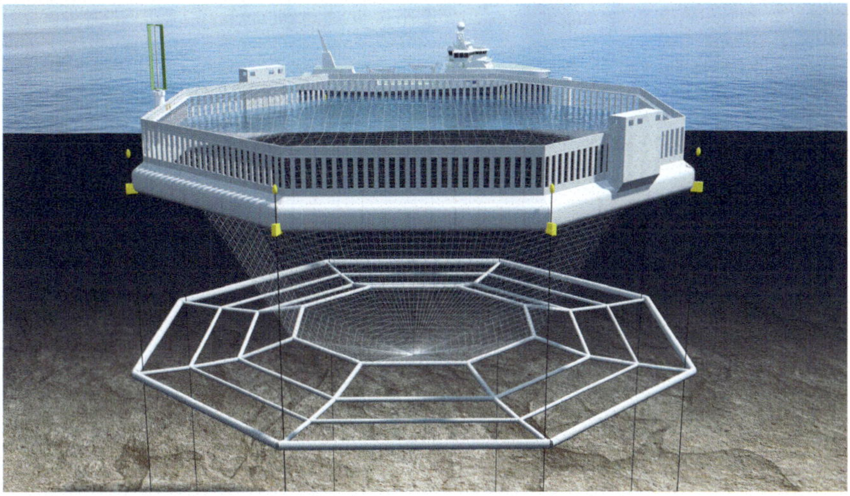

Fig. 27 Blue Farm ocean farm, details

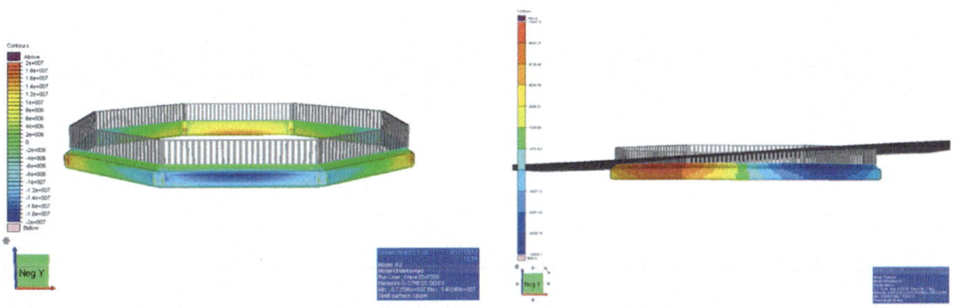

Fig. 28 Hydrodynamic and structural analyses of Blue Farm

Fig. 29 Net and concrete wall

Table 1 Reported cost of farming salmon, for different production methods

		NOK/kg
Open cages in the sea		30.60
Fish pens on land		43.60
100 g fish on land, then in open cage in the sea	0 delicing treatments	28.00
	5	31.30
	10	33.80
500 g fish on land, then in open cage in the sea	0 delicing treatments	28.90
	3	30.70
1000 g fish on land, then in open cage in the sea	0 delicing treatments	30.80
	2	32.40
Closed cage		37.90

- Closed cages can have on board silo for feed, space for handling the feces, space for oxygen production etc., and will not need an expensive work boat for effective production
- Closed cages allows control of water quality, such as salinity, temperature and oxygen
- In closed cages you can make a current in the water, for better fish welfare
- In closed cages, as mentioned, you avoid sea lice and sickness coming in with the water, can take care of the feces, avoid escaping, and overfeeding the surrounding fish
- A closed concrete cage virtually lasts for ever.

By including these elements, it is believed that fish farming in closed concrete cages is cost efficient also, not only environmentally friendly. The future will see an increased use of concrete in the marine environment; concrete will indeed be a valuable partner. We need to be creative to fully utilize its potential.

References

1. Olsen, T, O., Weider, O., & Myhr, A. (2015). Large marine concrete structures: The Norwegian design experience. In C. M. Wang & B. T. Wang (Eds.), *Large floating structures*. Singapore: Springer (2015).
2. Wang, C. M., Dai, J., Jiang, D., & Olsen, T. O. (2017). Large floating concrete structures. *Concrete in Australia, 43*(2), 29–40.
3. Pettersen, J. S., Bjønnes, S. G., Hausmann, R. E., Fiskum, K., & Nyhus, B. S. (2017). ShellDesign—innovative concrete design. In: *HPC/CIC Conference, Tromsø, Norway*, March 2017.
4. Bjørndal, T., & Tusvik, A. (2018). *Økomisk analyse av alternative produksjonsformer innan oppdrett*. SNF-rapport nr. 07/18, Samfunns-og næringsforskning AS, Bergen, Norway, Oktober 2018, ISSN 1503-2140.

Technology-Driven Sustainable Aquaculture for Eco-tourism

Ban Tat Leow and Hoon Kiang Tan

Abstract Aquaculture farming provides an avenue to grow food fish for the table in the face of rapid depletion of fish stock around the world due to over-fishing. However, traditional aquaculture farming has been largely dependent upon the environment and waste management issues have not been adequately addressed to make it sustainable in the long term. There have been increasing reports of widespread diseases and even mass fish deaths from the deterioration of water quality or pollution. This pollution may be from farming activity itself or from external sources through natural or man-made incidents. This has created concerns over food safety as companies use more vaccines or antibiotics to treat the diseases. There is also a concern over the spread of diseases from escaped farmed fish into the wild. These issues need to be addressed to ensure the continued viability of aquaculture farming as a critical food source. This paper seeks to address the issues through the application of technology across the value-chain of production for: (1) cost effective production, i.e. "more for less" to generate significant financial returns, and (2) sustainable aquaculture farming without polluting the environment to preserve ocean health. The system provides a perfect platform for the implementation of circular economies through integrated multi-trophic aquaculture for better economics and further protect the eco-system of the ocean. Faced with the current increasing negative perception towards aquaculture, it is proposed that eco-tourism opportunities be utilized to promote the adoption of these technologies as well as to further public awareness and education in this sector. It is also hoped that this will spur more research and development for further improvements.

Keywords Aquaculture · Sustainability · Food security · Food safety · Ocean health · Public perception · Eco-tourism

---

B. T. Leow · H. K. Tan (✉)
Nature Resources Aquaculture Pte Ltd., Singapore, Singapore
e-mail: hk.tan@nature-resources.com

B. T. Leow
e-mail: bt.leow@nature-resources.com

C. M. Wang et al. (eds.), *WCFS2019*, Lecture Notes in Civil Engineering 41,
https://doi.org/10.1007/978-981-13-8743-2_11

1 Introduction

It is an increasing challenge to feed the growing population around the world. The majority of the highest rates of population growth are predicted to occur in urban areas, which are highly dependent on the agriculture sector, farming crops, livestock or fisheries. Growth in this sector is seen as the most effective means for addressing this challenge. Agriculture (which includes aquaculture) has evolved over the years. The continued increase in demand has led to larger and more intensive farming methods. However, there is also a growing concern and attention on its sustainability, its impact on the environment and even the Feed Conversion Ratios (FCR) of farming different animals.

Seafood consumption has been rising steadily as fish is generally viewed as a healthy protein source. Fish has traditionally been captured from the wild, in waterways, lakes or oceans. However, the quantity of wild capture fish has declined due to large scale commercial fishing. The drastic impact from over-fishing clearly shows that this is not a sustainable option. Although there has been efforts to control the fishing, the World Wildlife Fund (WWF) reports that more than 85% of global fish stocks in our oceans are at significant risk of illegal, unreported and unregulated (IUU) fishing.

Aquaculture on the other hand, provides a means for production of fish in a farm environment to supplement the ever-growing demand. Studies have even shown the FCR of fish to be much lower than that of other farmed animals (see Fig. 1). This makes aquaculture a viable and much attractive logical option.

Aquaculture production has developed and increased over the years. The Food and Agriculture Organization of the United Nations (FAO) report indicates a growing trend in aquaculture production while captured fish production has stagnated and could potentially decline in the future if the problem of IUU is not resolved. Aquaculture production, on the other hand, has continued to increase and is projected to exceed that of capture production (see Fig. 2).

Fig. 1 Feed conversion ratios

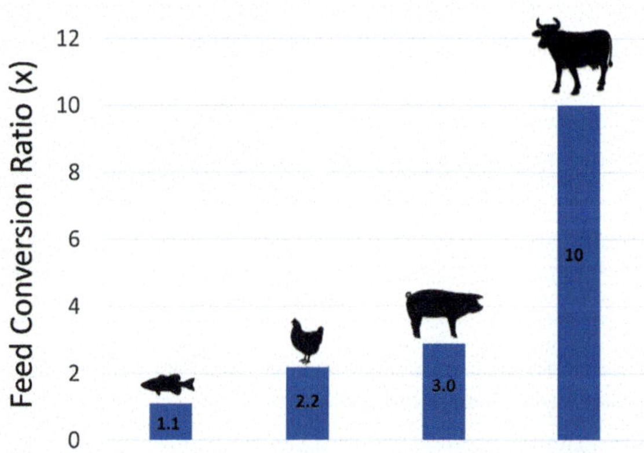

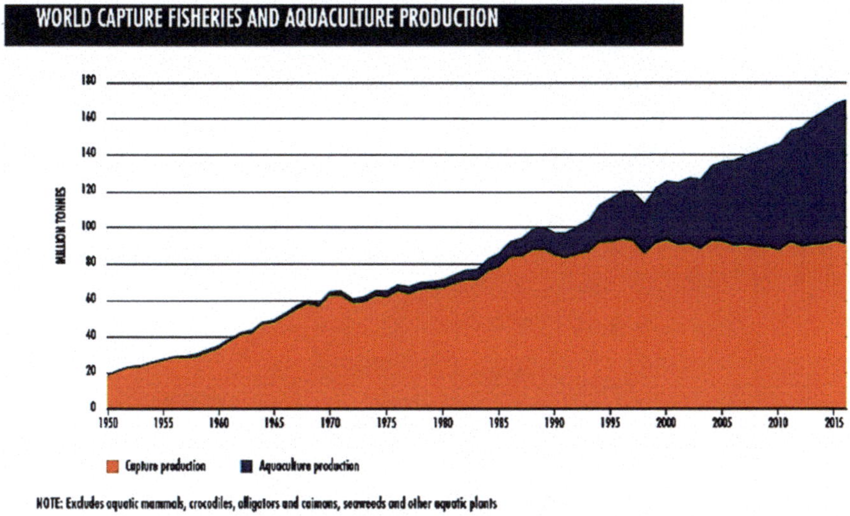

Fig. 2 Aquaculture versus capture production. *Source* FAO 2018 report

2 Aquaculture Systems

The cultivation of fish, crustaceans, algae and other aquatic animals or plants in aquaculture, has been practiced for a long time. Starting from modest small earthen ponds or plastic tanks, it has progressed to large, multiple ponds to serve growing business needs. Aquaculture has also progressed to rivers and near shore water bodies where the fishes are farmed, eventually to open net cages as the intensification of fish farming grows.

These traditional open net pens face challenges, some of which are shared by other forms of aquaculture while others are unique to themselves. Referring to Fig. 3, these challenges include:

- *Contamination by Excess Feed and Fish Waste*
  - Internal
    Excess feed and fish waste need to be removed to avoid contaminating the grow-out space in the enclosure. In fast flowing rivers or sea space with adequate flow, the excess feed and fish waste could be removed from the growing environment in the net cages. However, due to the high population density of the fishes, significant amount of excess feed and waste remain in the enclosure, causing health issues due to the deterioration of the water quality.
  - External
    The discharge of the excess feed and fish waste to the environment is also an issue that is attracting increasing concerns. While the true impact of this on the environment is still being studied, the pollution has prompted legislation in

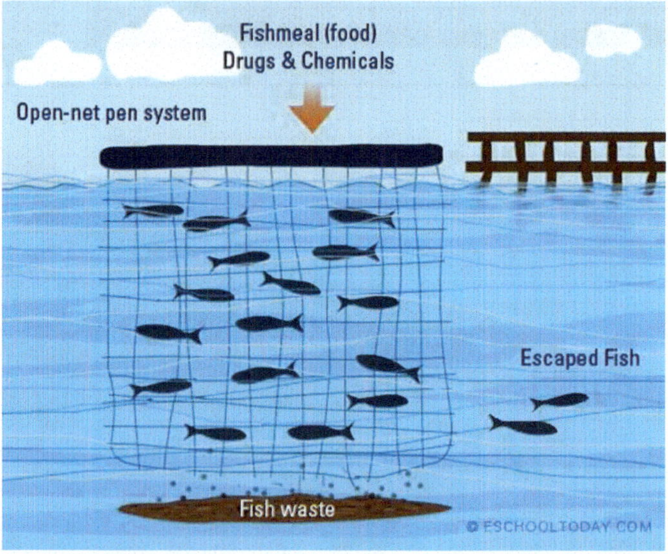

Fig. 3 Challenges faced by net pen systems

some countries to require the relocation of the farm and a clean-up of the seabed every few years.

- *Spread of Diseases due to High Density Farming*
  - Internal
    Fish health is a critical factor to the success of the farm. A high-density environment will always have a propensity for the easy spread of diseases among the fish population if not properly controlled. Fish farming in aquaculture also has a common challenge in the control of parasitic infestation from sea lice. These sea lice attach to the fishes' skin, causing ugly lesions and reducing the commercial value of the fish. In extreme cases when not properly treated it may lead to mass mortality in the pen population. Sea lice have a short, free-swimming larval phase in its life cycle where it needs to find and attach itself to a host. In the wild, its survival rate in this part of its life cycle is low but it flourishes in the high-density environment of the net pens, resulting in ease and rapid spread of the disease among the fish population.
  - External
    The presence and high concentration of the sea lice in the pens may also infect any wild fish that are attracted by the excess feeds or are swimming near or past the pens.

- *Fish Escapes*

  Open Net pens are susceptible to fish escapes due to poor maintenance or accidents. Escaped fish have the potential for spreading diseases and can also impact the

Fig. 4 Harmful algal bloom and its disastrous effects

marine ecosystem as the farmed fish in some instances are not native to the farmed areas.

- *Impact by Environment*

  Owing to their exposed nature, net pens are at the mercy of the environment. Weather events, marine accidents, pollution, shark attacks, etc. can exert a toll on the farm. Industrial pollution and run-off from agricultural farms have also adversely affected the farmed fishes. Some of these pollutions, even if not toxic on its own, can trigger rapid widespread algae growth due to the injection of nutrients into the water. This phenomenon, commonly known as Harmful Algal Bloom (HAB), often results in drastic mass mortality of the fishes in the farms (see Fig. 4).

3 Food Safety Concerns

Food safety is a universal concern. Owing to the inability to control the environment, farmers are resorting to antibiotics to treat and control the spread of diseases. Reports of contaminated food or indiscriminate use of antibiotics (real or otherwise), have a tendency to stir strong emotional responses, creating a negative perception (or not so positive perception) towards farmed fish.

4 Platform for Sustainability and Further Research

In view of the challenges faced by traditional aquaculture in the next decade to intensify aquaculture farming from traditional earthen ponds to high-tech tanks system on land, a new and innovative and cost-effective solution is critically needed. This solution needs to ensure food safety, be sustainable and also environmentally-friendly.

The comparison table (see Table 1) of different aquaculture systems shows distinctive impact on water, the environmental footprint and potential negative impact on the perception of farm-raised fish versus wild capture fish.

Figure 5 shows that the new and innovative aquaculture floating closed containment flow through (CCFT) system has added advantage over all the existing farming system. Aquaculture Centre of Excellence's (ACE) patent pending, Closed Containment Flow Through (CCFT) system on a floating structure called the Eco-Ark® attempts to address these challenges.

The Eco-Ark® is a series of tanks on a floating marine hull structure that physically isolate the growth tanks from the environment (see Fig. 6). It has a unique flow through system that controls both the water intake to the tanks as well as the effluents from the tanks before discharge. This Closed Containment Flow Through system (CCFT) detoxifies and purifies the water going to the fish grow-out tanks. Water is filtered and sterilized to remove bacteria, harmful pathogens and parasites before going into the grow-out tanks. The water from the tanks is also filtered of excess feed and fish waste upon discharge to avoid polluting the environment. This system provides a conducive environment for precision farming of the fish, enabling faster fish growth, survivability and higher productivity. The system is contained, efficient and sustainable with no waste discharge into the environment. The self-contained environment allows for opportunities for exploring Integrated Multi-Trophic Aquaculture (IMTA) where different aquatic animals are cultivated at different stages of the water system.

Table 1 Comparison of aquaculture systems

Ocean/water health index Water treatment process		Existing technology				New and innovative
		Net cage	Flow through (FT) raceways	Closed containment (CC)	Recirculating aquaculture system (RAS)	Closed containment flow through (CCFT)
Influent	Filtration	No	No	No	Yes	Yes
	O_3	No	No	No	Yes	Yes
	O_2	Partial	Partial	Partial	Yes	Yes
	Recirculation	No	No	No	Yes	Possible
Effluent	Filtration	No	No	No	Yes	Yes
	Waste treatment	No	No	No	Yes	Yes
Fallowing required		Yes	No	No	No	No
Moveable		No	No	No	No	Yes

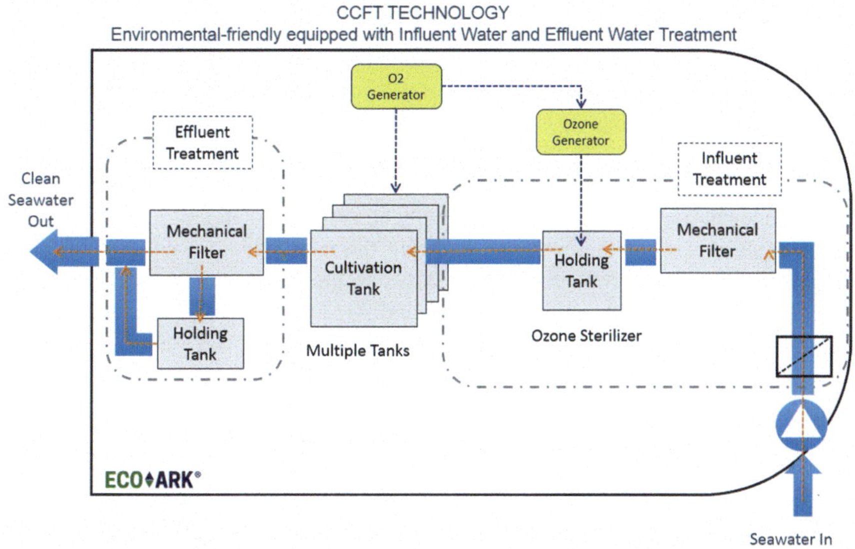

Fig. 5 Closed containment flow through (CCFT) water system

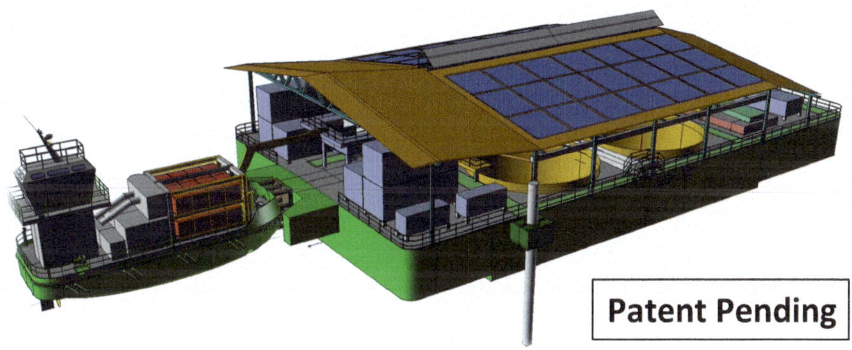

Fig. 6 Eco-Ark® CCFT

In IMTA, the byproduct from farming one species is used as input for farming another species. This is an exciting opportunity for further research and development of a Circular Economy that more closely mimic mother-nature for long term sustainability. Excess feed and fish waste could be used for cultivation of aquatic plants, crustaceans, mollusks, etc. as in the natural environment.

5 Use of Technology

Technology is a key consideration in the entire system design. It is used to address the key issues

- *System design and energy efficiency*

The physical barrier in the closed containment system isolates the tank from the environment as well as the environment from the tank. This prevents the transfer or transmission of material and diseases between the two environments. The Eco-Ark is also designed with a low freeboard and Air-Lift pumps for reduced energy consumption of the flow through water system. Solar and wave energy are also harvested wherever practical and possible.

- *Water Management*

Water quality is continuously monitored, through a series of sensors and regular testing, for O_2, pH, temperature, ORP (Oxidation-Reduction Potential), nitrates, nitrides, ammonia and pathogens for effective control of the environment in the grow-out tanks. Standard known optimum levels are maintained but can be further improved through Data Analytics. Thresholds are also set to trigger alarms for early detection and response as this is a critical part of the farm management.

- *Fish Management*

Fish feeding is automated for optimum fish growth while minimizing waste. IoT sensors monitor and provide input for automatic adjustment of the amount and frequency of feed according to the growth rate of the fish.

- *Automation*

Automation is employed in all stages of the farming process for higher efficiency and productivity. Typical areas include:

- Fish transfer system
- Automatic feeding system
- Water quality control
- Harvesting gear
- Fish processing.

- *Traceability*

Product traceability on the harvested fish provide a level of consumer assurance on food safety. The fish will be traceable by batch to its source, type of feed, growth environment, date of harvest and processing.

- *Data Analytics*

It is recognized that different aquatic animals or species may develop and grow at different rates according to different environments. Data from the water management and fish management systems can be investigated and studied through Data Analytics for a better understanding of the optimal conditions for farming of each species.

6 Eco-tourism

While the safety and sustainability concerns can be addressed by the Closed Containment Flow Through system (CCFT) on the Eco-Ark, it is proposed that Eco-tourism be also considered. Eco-tourism offers an avenue for building environmental awareness as well as education on conservation and sustainability issues, it can also be a source for business revenue. Such an awareness and understanding will help more widespread adoption of these sustainable methods and also research for further improvements.

Eco-tourism provides a platform for influencing public perception through education and understanding of the control and management of the farming process. In the Farm and Fun Concept (see Fig. 7), the integrated nature of the farming process is presented on a collection of Eco-Arks and submersible tanks. Farming of different species of aquatic animals, plants and molluscs is showcased to illustrate the benefits of the Circular Economy.

The facility is integrated with restaurants, water parks, research centers and convention halls to attract different sections of the population as well as learning institutions. This facility can also support education programs and facilitate a wider discourse on sustainable aquaculture.

Fig. 7 Eco-tourism—farm and fun concept

7 Conclusion

It is widely recognized that food production from aquaculture is essential for feeding the growing population. The intensive farming from current aquaculture methods has created significant concerns on its true impact on the environment and its sustainability to the ecosystem. Although countries have introduced legislation to mitigate some of these risks, the real impact on the environment may not be fully realized even with these measures.

While there are concerns over the risks to the environment, a more pressing concern to consumers is the risk to food safety. As farmers struggle to manage disease outbreaks, it is anticipated that they will resort to more and stronger antibiotics due to the natural increase in antimicrobial resistance in aquaculture over time.

A sustainable aquaculture model for fish farming addressing all these issues is needed. Such a system is currently available although more research may be required to investigate its application for a wider range of aquatic life, such as fish, crustaceans, molluscs, shellfish (bivalve aquaculture) and plants and worms.

It is important this new technological advanced system generates significant financial return from higher production yield coupled with high survivality rate, create employment while maintaining the ocean health.

Floating Forest: A Novel Concept of Floating Breakwater-Windbreak Structure

C. M. Wang, M. M. Han, J. Lyu, W. H. Duan, K. H. Jung and S. Kang An

Abstract The floating forest is a novel floating breakwater-windbreak structure that can be deployed in a water environment to reduce both wind speed and wave height behind it. Its purpose is to protect fragile coastlines, port terminals, marinas, and floating structures from severe storms. It can also be used to create a landing sea strip for seaplanes. The floating forest comprises several segments of breakwater hull in a lateral arch shape, caissons or mooring lines at the ends of the segments to keep the arch segments in place, and a tilted deck installed with arrays of tubes. The breakwater hull segment is typically a few hundred meters long, but the scale may be adjusted on different demands. The width is adjusted to fit the incoming wave length. A shallow draft is used since for surface waves most of the wave energy is concentrated near the mean water level. The deck of each hull segment has a gradient to create a beach run-up, and tube arrays are installed on the tilted deck. The hollow tubes provide resistance against the incoming wind, and are connected to the internal channels inside the hull that end with openings on the vertical front wall of the hull. The mooring system comprises either caissons in shallow water depths or several groups of steel mooring chains that are spread around the floating breakwater in deep waters. The primary material of the floating forest is marine prestressed concrete. As a part of the feasibility study, the structure was modelled by using linear BEM software package HydroSTAR (developed by Bureau Veritas) to study the 3D wave diffraction near the hull and its wave transmissibility with inputs of measured wave data in Gold Coast, Australia. Parametric studies were carried out to optimize the main dimensions of the structure. The results show that the arch shape floating structure has a good performance as compared with traditional rectangular breakwaters, and wave height can be reduced by half at the peak wave period. CFD analysis was also performed and it was found that the wind speed could be reduced by

C. M. Wang (✉) · M. M. Han · J. Lyu
School of Civil Engineering, University of Queensland, St Lucia, QLD 4072, Australia
e-mail: cm.wang@uq.edu.au

W. H. Duan
Department of Civil Engineering, Monash University, Victoria, Australia

K. H. Jung · S. Kang An
Hyundai Engineering and Construction R&D Division, Seoul, South Korea

C. M. Wang et al. (eds.), *WCFS2019*, Lecture Notes in Civil Engineering 41,
https://doi.org/10.1007/978-981-13-8743-2_12

20% for 1 km behind the floating forest and up to 30 m in height; thereby establishing its effectiveness as a windbreak as well.

Keywords Floating breakwater · Windbreak · Diffraction · Hydrodynamics · Boundary element method · Computational fluid dynamics

1 Introduction

Australia's coastal zone is one of the largest and most varied of any nation. Extending from hot tropical to cool temperate latitudes, the environment within this coastal zone is the result of interaction between many factors including varied geology, large-scale oceanic currents, climatic variability and diverse ecosystems. It attracts millions of tourists every year. Yet, strong wind/wave conditions, soft seabed, deep water, and stringent environmental protection pose problems for present coastal structures to protect the Australian coastline. For example, a traditional bottom-founded breakwater is not a cost-effective way for the Australian coastline. By comparing with a bottom-founded breakwater, a floating breakwater is more suitable when the seabed is soft and the water depth is large (say greater than 20 m) [1, 2]. The efficiency of a floating breakwater is measured by its ability to reduce the wave from transmitting to leeside. Numerical and experimental methods have been applied to optimize the design of floating breakwaters. The simplest form of the floating breakwater is a rectangular block-shaped structure, and many papers have been reported its efficiency in reducing wave forces. Numerous new designs of the floating breakwater have been proposed to save material costs and improved their performances. For example, a dual pontoon floating breakwater has been studied by Ji et al. [3], which focuses on increasing the virtual width of the breakwater, and bringing extra damping between the two pontoons. Kolahdoozan et al. [4] mooted the design of a π-type floating breakwater that helps block the wave by increasing the effective depth of the structure. A design with similar geometry was proposed by Neelamani and Ljubic [5] in which several porous skirt walls are attached under the breakwater and dissipating the wave energy by turbulence. Recently, Dai et al. [6] presented a comprehensive review on the recent research and developments of different types of floating breakwaters. However, traditional floating breakwaters are only effective in mild wave climates and are not designed for mitigating the wind hazards. In order to protect fragile beaches, marinas, port terminals and coastal assets/infrastructure from extreme storms, both wind and wave loads have to be reduced. A windbreak is needed as well. A windbreak is a type of barrier designed for sheltering humans and their assets from wind hazards on land. Depending on its material constituents, it can be categorized as a vegetative windbreak or a man-made windbreak. A vegetative windbreak often presents in the form of several rows of trees and bushes while a man-made windbreak usually appears as fences and walls [7]. Overtime, windbreaks on land have shown to be very effective and they have been gaining popularity in a variety of areas, including both in civil engineering and agroforestry [8]. The successes of windbreak on land assure their application in the marine environment to combat strong winds from the warm ocean. The wind speeds in the leeward area of the windbreak are believed to give an indication of the performance of the windbreak [9]. Wind speed is often divided into a mean wind speed component and a fluctuating wind speed component based on Reynolds decomposition [10]. As a time-averaged component, the mean wind speed component has been in the center of most research work to quantify the protection of the windbreak [7, 9, 11] as it forms the major part of the peak wind speed. Therefore the effectiveness of the windbreak structure has been assessed by the mean wind speed reduction in the lee.

In order to reduce both wave and wind forces, we need to have a windbreak on a floating breakwater. This prompted the authors to develop a new design concept of "floating forest" which is a mega floating breakwater cum windbreak structure which will be described herein. Parametric studies involving wave transmission analysis using linear BEM software and computational fluid dynamics analysis of wind flow through the floating forest will be conducted to establish the key dimensions of the floating forest.

2 Floating Forest Concept

The proposed floating forest (breakwater-windbreak structure) is inspired by the principle of windbreaks and salt marshes. It comprises a number of lateral arch shape concrete segments whose ends are kept in position by caissons when the water depth is shallow or by mooring chains when the water depth is large. The floating forest is installed with a heading perpendicular to the main direction of incoming waves so that the concrete arch breakwater is in a compressive state under the incident wave action. The deck is tilted towards the upstream to allow a wave run-up for wave energy dissipation. Several arrays of hollow column tubes (which form the "trees" of the floating forest) are fixed on the tilted deck area facing the wind direction. The arrays of tubes on the deck reduce wind speed in a similar way to the tree windbreak. The reduction of wind speed is contributed by the divergence of the wind flow and the energy extraction by the artificial trees. The hollow tubes are made of high density polyethylene (HDPE) and connected to the internal channels inside the hull which opens at the front wall to allow water entry, so that a curved water path is formed in the breakwater. Under the incoming wave, seawater will rush into the curved internal channels, creating turbulence near the openings on the wall, so that wave energy is dissipated. Figure 1 shows an artistic impression of the floating forest while Fig. 2 shows a possible arrangement of hollow tubes on the tilted deck.

Table 1 provides typical dimensions of a 3-segment floating forest from assumed design inputs, with the width and the draft to be determined by parametric studies.

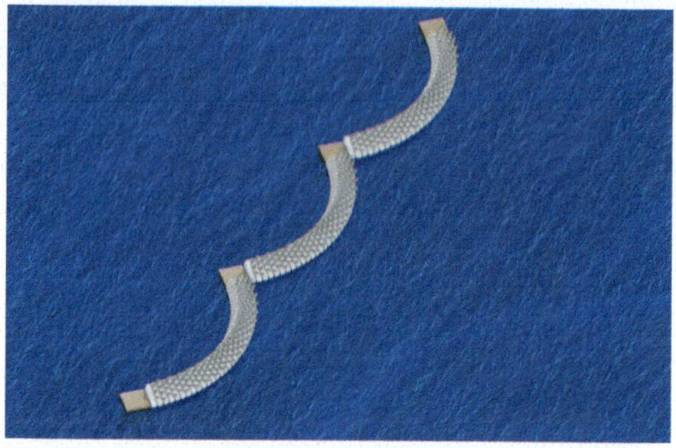

Fig. 1 Floating forest

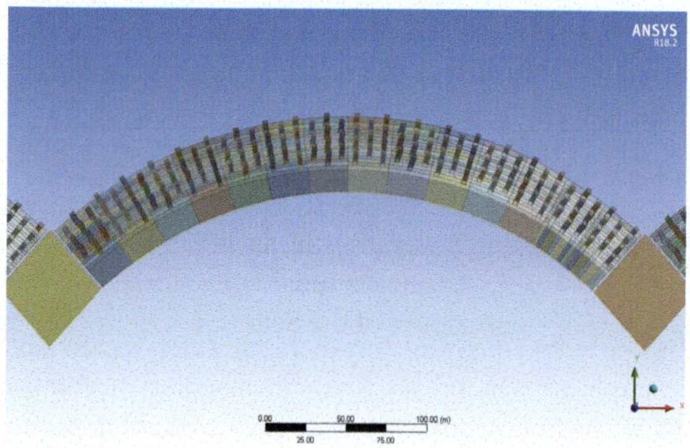

Fig. 2 Arrangement of hollow tubes on deck

Table 1 Floating forest geometry

Symbol	Quantity
Total span length, L	1080 m
Depth of Floating Segment, D	22 m
Arch Curvature, α	84°
Deck Tilting Angle, θ	25°
Deck Height, H	15 m
Column Tube Diameter, $\emptyset$	2.5 m
Maximum Tube Height, h	31 m

3 Parametric Studies on Main Dimensions of Breakwater

Parametric studies were carried out to determine the dimensions of the floating breakwater hull. Since the length of the hull is governed by the design sheltered area, only the influence of width and draft of the floating breakwater is studied. The wave transmissibility is calculated to assess the wave blocking efficiency of the breakwater. In addition, the natural frequency of structural rigid body motion is also considered so that large heave and roll motions do not appear when the mooring system is applied. The hydrodynamic analysis was carried out by using the linear BEM program HydroSTAR which was developed by Bureau Veritas.

Under linear assumption, the overall wave potential is furnished by the superposition of incoming wave potential φ_I and perturbation wave potential φ_P , i.e.

$$\varphi = \varphi_I + \varphi_P \tag{1}$$

The incoming wave potential is assumed to be given by [12]

$$\varphi_I = -\frac{gA}{\omega}\frac{cosh(k(z+h))}{coshkh}e^{ik(x\cos\beta+y\sin\beta)}, \tag{2}$$

where x and y are the Cartesian coordinates in the horizontal plane, k is the wave number determined by linear dispersion relation $gktanhkh = \omega^2/g$, g the gravitational acceleration, A the wave length, ω the wave frequency in radians/sec and β the incident wave angle.

When the body is assumed to be motionless (due to the presence of caissons), the wave perturbation will be decided by diffraction alone. The diffraction potential is obtained by solving the Laplace equation in the frequency domain [12]

$$\nabla^2\varphi = 0 \tag{3}$$

subject to the following boundary conditions

$$\frac{\partial\varphi}{\partial z} = 0 \text{ at } z = -h \text{ (on the seabed)} \tag{4}$$

$$\frac{\partial\varphi}{\partial n} = -\frac{\partial\varphi_I}{\partial n} \text{(on the undersurface of the body)} \tag{5}$$

$$-\frac{\omega^2}{g}\varphi + \frac{\partial\varphi}{\partial z} = 0 \text{ at } z = 0 \text{ (on free surface)} \tag{6}$$

where z is the vertical coordinate, h the water depth, and n the normal vector to the surface.

The hydrodynamic performance may be evaluated by the wave elevation in the sheltered area at the downstream side of the breakwater [13]. After solving the wave potential, the wave elevation η is determined by the dynamic free surface condition

$$\eta = \left(\frac{i\omega\varphi}{g}\right)_{z=0} \tag{7}$$

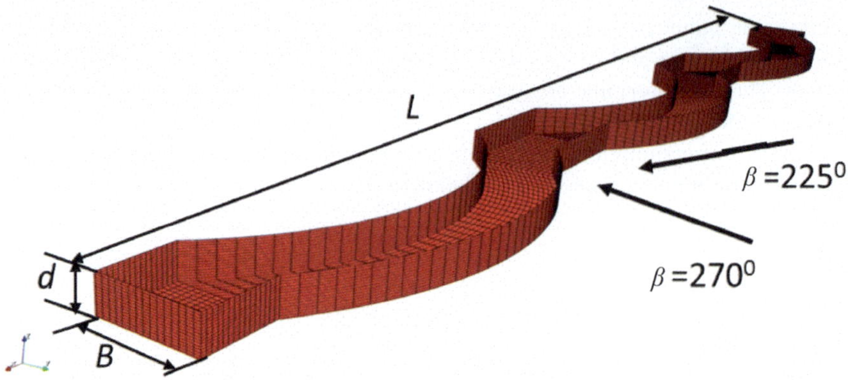

Fig. 3 Numerical model for breakwater

The 3D diffraction results in a relatively more complicated wave field when compared with 2D models. In order to evaluate the wave transmissibility in a more accurate way, a large shelter area behind the breakwater is selected for consideration, and the average wave elevation in the sheltered area is calculated to represent the global transmissibility performance of the breakwater. Thus, the wave transmissibility defined herein is expressed as

$$K_T = \frac{\overline{A_T}}{A_I} \tag{8}$$

where K_T is the average wave transmissibility ratio, $\overline{A_T}$ the average wave elevation in the sheltered area, and A_I the amplitude of the incoming wave.

By considering the total length of the floating breakwater as $L = 1080$ m and a width of 45 m and specifying the origin of the Cartesian coordinates system at the middle of the breakwater, Fig. 3 shows the numerical model. The floating breakwater, together with the caisson is considered as a single solid body. The internal channels are neglected in the wave diffraction study. Two incident wave headings at $\beta = 225°$ and $\beta = 270°$ are studied to represent the oblique sea and beam sea condition, respectively.

The sheltered area is assumed to take the form of a square covering an area of -540 m $< x <$ 540 m, -200 m $< y <$ 0 m. In the case of a 55 m width breakwater, the sheltered area is taken as -540 m $< x <$ 540 m, -210 m $< y < -$ 10 m. In this area, 1200 evenly distributed measurement points are taken as sample points, and the average of all the wave heights at these points will be considered as an indicator of global wave transmissibility (see Fig. 4).

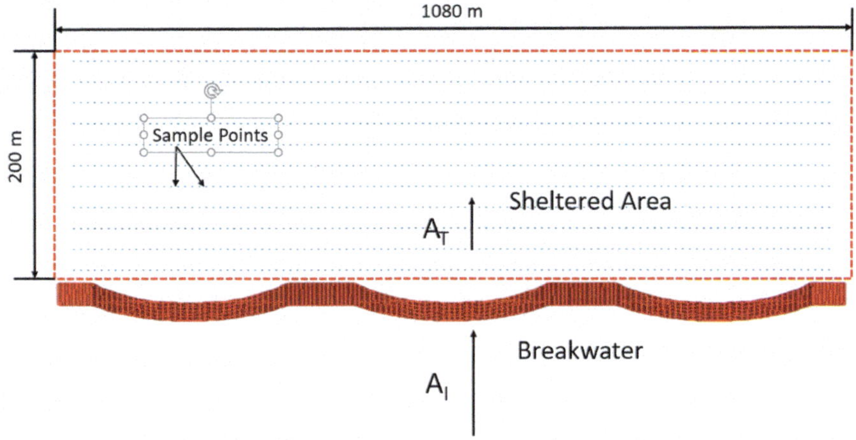

Fig. 4 Sheltered area and sample points considered for wave transmission analysis

In the following analysis, two different structural widths: $B = 45.0$ m and $B = 55.0$ m and two drafts $d = 13.5$ m and $d = 20.0$ m are considered. The water depth is assumed to be 40.0 m, which is a design input based on real site measurement in the Gold Coast, Australia. The wave conditions at site for both 1-year and 100-year wave are considered. The irregular wave is modelled by the JONSWAP spectrum with a significant wave height $H_s = 4.5$ m for a 1 year wave and $H_s = 8.4$ m for a 100 year wave. The peak wave periods are $T_p = 9.4$ s to 14.0 s for 1 year wave and $T_p = 13.9$ s for a 100 year wave. Figure 5a, b show the computed average wave transmissibility based on the various considered breakwater dimensions. It can be noticed that by increasing the width B and the draft d, the wave transmissibility can be reduced as expected. In all configurations, the wave transmissibility drops as the wave frequency increases and wave length decreases, because it is easier for long waves to pass through the barrier while short waves tend to be reflected back. All the four configurations show a similar wave transmissibility for short waves with wave period from 0.65 rad/s to 1.0 rad/s, with a larger difference in the long wave region. In terms of wave heading, the oblique wave is less blocked by the floating forest as compared with a beam sea. At the peak wave period $T_p = 14.0$ s (i.e. a frequency of 0.45 rad/s), all the four configurations could reach a wave transmissibility lower than 0.5, which proves a good wave block efficiency of the proposed floating breakwater. Figure 6a, b show the wave transmissibility distributions in the oblique sea and in the beam sea, respectively.

Spectral analysis is conducted to investigate the spectra response in random waves. For the 1-year wave with a peak period range, the upper bound of wave peak period $T_p = 14.0$ s is selected as the design wave because its wave energy is more concentrated on the long wave side; thereby providing a more conservative estimation of the floating breakwater performance. It also makes the 1-year and the 100-year peak

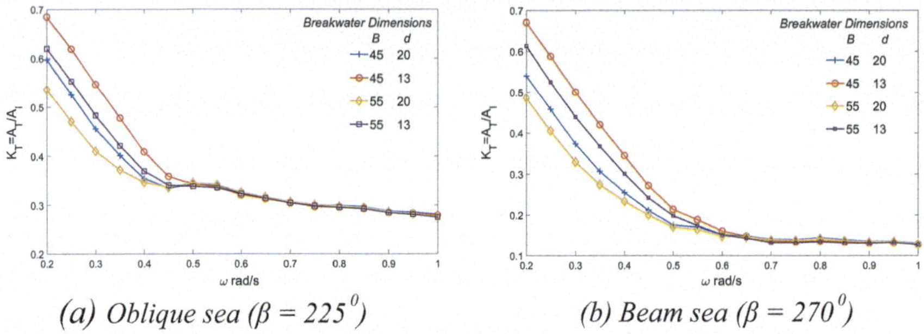

Fig. 5 Average wave transmissibility

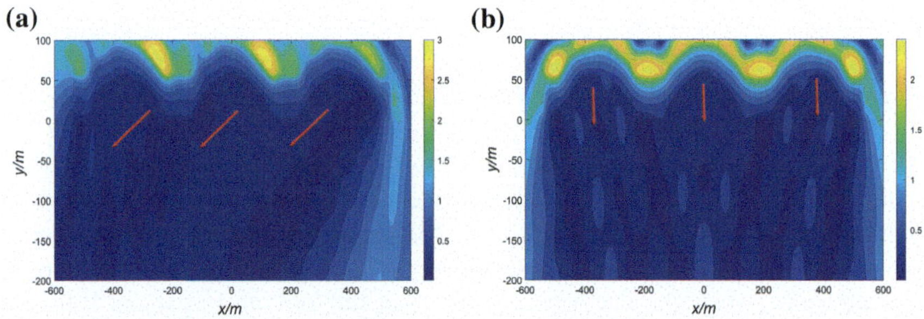

Fig. 6 Wave transmissibility distribution. **a** Oblique sea, **b** beam sea

wave period identical, so only the 100-year wave is analyzed. Referring to the DNV rule [14], the JONSWAP wave spectrum is expressed as

$$S_J(\omega) = \frac{5}{16} H_s^2 \omega_P^4 A_\gamma \omega^{-5} \exp\left(-\frac{5}{4}\left(\frac{\omega}{\omega_p}\right)^{-4}\right) \gamma^{\exp\left(-0.5\left(\frac{\omega-\omega_p}{\sigma\omega_p}\right)^2\right)} \tag{9}$$

in which γ is the wave spread parameter which varies from 1 to 7, and is taken as 3.3 based on DNV recommendation in this analysis; $A_\gamma = 1 - 0.287 \ln\gamma$ is a normalizing factor and σ is a shape factor as recommended in the DNV rule. By using the calculated wave transmissibility RAO as shown in Fig. 5a, b, the reduced wave spectrum is calculated and presented in Fig. 7. The difference between the original wave and reduced wave spectrum shows that most of the wave energy is kept outside the sheltered area. Since all the four configurations are able to achieve acceptable performance (i.e. a reduction of the peak spectrum density by 75%), the smallest main dimensions $B = 45.0$ m, $d = 13.5$ m are selected as the width and draft for the floating forest.

In order to avoid severe motion under wave excitation, the breakwater should be designed to have a natural frequency away from the frequency range where the wave energy is concentrated. Given the mass, center of gravity and moment of inertia based

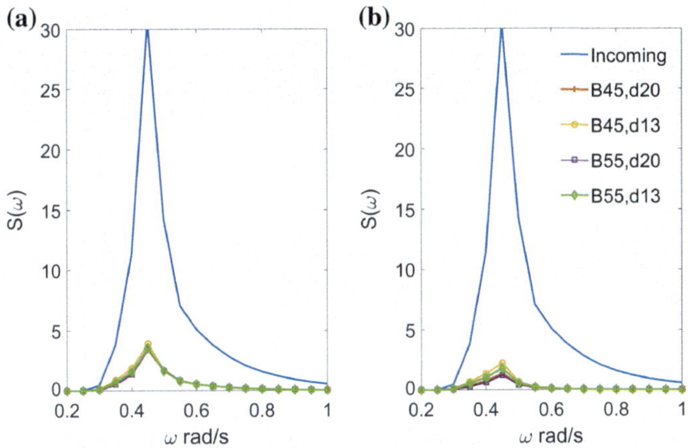

Fig. 7 Reduced wave spectrum under 100 year wave. **a** Oblique sea; **b** Beam sea

on the detailed structural design, the natural frequency of heave and roll motions and the wave frequency motion response RAO have been calculated. The vertical centre of gravity (VCG) is estimated to be 0.5 m above the mean water level, while the gyration radii $R_{xx} = 32.77$ m, $R_{yy} = 29.94$ m, $R_{zz} = 44.07$ m, $R_{xy} = 51.37$ m, $R_{yz} =$ -5.276 m, and $R_{zx} = 76.34$ m. Coupled terms are non-zero because of the asymmetry of the structure geometry along x and z axis and the resultant asymmetrical mass distribution. The sway, heave, roll and yaw motion RAO are plotted in Fig. 8. It can be seen that the floating forest has a heave and roll frequency close to the design wave peak frequency 0.45 rad/s, but the peak response is mild. The maximum heave motion RAO is 1.17 m/m while the maximum roll motion RAO is 1.60 degree/m, which may result from the large waterplane area and a shallow draft. Coupling effect between different degrees of freedom can be seen in the motion response, as fluctuation appears in the curves, especially in oblique wave condition. This coupling effect is a result of the non-zero coupled gyration radii terms.

4 CFD Analysis of Wind Speed Reduction Behind Floating Forest

Computational Fluid Dynamics (CFD) will be used to simulate the wind flow passing through the floating forest [15] with the primary focus on the mean wind speed in the leeward side of the floating forest. ANSYS Fluent will be adopted as this CFD software has proven its reliability as reported in many publications [16].

We adopted the steady Reynolds-averaged Navier-Stokes Simulation (RANS) approach in ANSYS Fluent because it consumes much less time than the Large Eddy Simulation (LES) approach and it is known to provide reasonably accurate

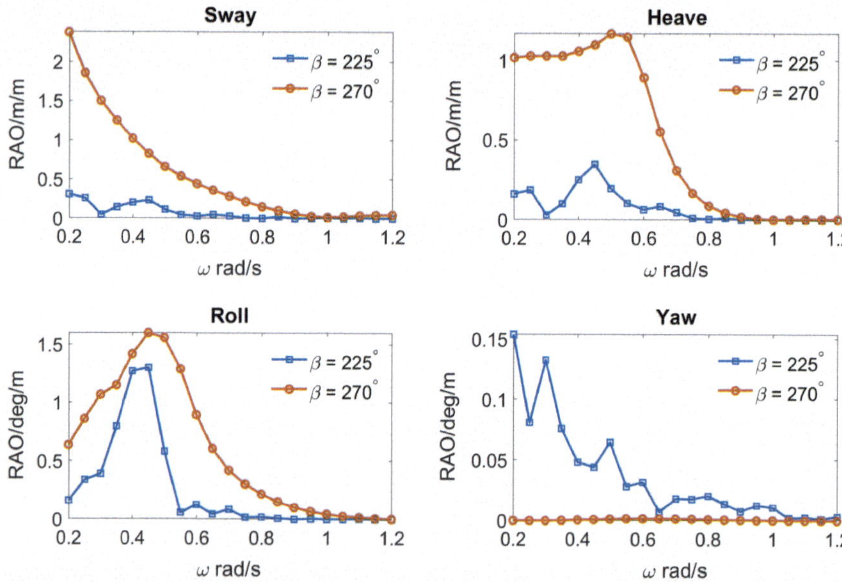

Fig. 8 Motion RAO of floating forest

results for the mean wind flow [17]. The approximate form of Navier-Stokes (N-S) equations used in RANS starts from decomposing the instantaneous variable $\overrightarrow{u(t)}$ into a mean component $\vec{U}$ that does not change with time and a fluctuating component $\overrightarrow{u'(t)}$, i.e.

$$\overrightarrow{u(t)} = \vec{U} + \overrightarrow{u'(t)} \tag{10}$$

The decomposed components are then inserted into the N-S equations, and by replacing the instantaneous variables, one obtains the following continuity equation and momentum equation:

$$\frac{\partial U_i}{\partial x_i} = 0 \tag{11}$$

$$\frac{\partial U_i}{\partial t} + U_j \frac{\partial U_i}{\partial x_j} = -\frac{1}{\rho}\frac{\partial P}{\partial x_i} + \frac{\partial}{\partial x_j}\left(2\upsilon S_{ij} - \overline{u_i' u_j'}\right), \tag{12}$$

where U_i is the mean velocity, u_i' the fluctuating velocity, P the pressure, x_i the position, t the time, ρ the air density, υ the kinematic viscosity of the air, the horizontal bar means averaging, S_{ij} is the mean strain-rate tensor given by

$$S_{ij} = \frac{1}{2}\left(\frac{\partial U_i}{\partial x_j} + \frac{\partial U_j}{\partial x_i}\right) \tag{13}$$

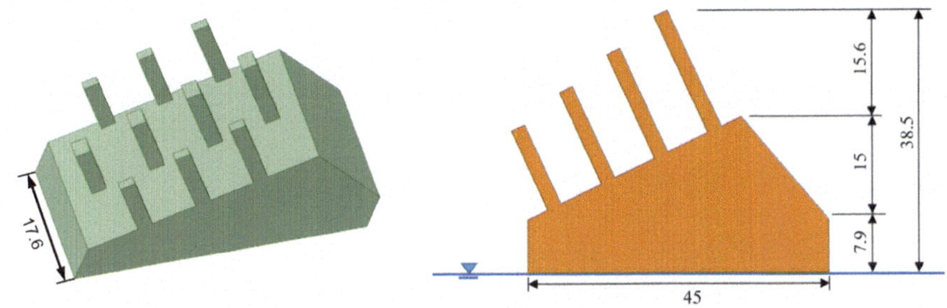

(a) *small segment of floating forest* (b) *dimensions in centimeters*

Fig. 9 Analysis model adopted for CFD analysis

As the number of the discretized volumes of the fluid space is proportional to the dimensions of the model, we chose a length scale of 1:100 to scale down the prototype in order to save computational costs and time. The flow Reynolds number based on the windbreak height, *H*, falls between 10^5 and 10^7, implying that simulation results should be representative of the reality [10]. For CFD analysis, we consider a small sample segment of the floating forest as shown in Fig. 9. The length dimensions of the analysis model are in centimeters. The part that is under the mean sea level is excluded from the model because it will not affect the wind flow. The model was constructed using ANSYS SpaceClaim.

Owing to limited computing resources, we consider a finite width of the wind flow. The width of the domain is chosen to be equal to the width of the model, considering the width needed to build representative parts of the tubes, which is 17.6 cm. In order to minimize the influence of the size of the wind flow domain on the computed results, we follow some best practice guidelines that have been summarized in Refs. [18–21]. These guidelines specify that

- the distance from the inlet to the front of the structure should be larger than $5H$ where H is the maximum height of the structure including the tubes.
- the distance from the back of the structure to the outlet of the computational domain should be larger than $15H$.
- the height of the wind flow domain should be no less than $6H$.

Starting from the above minimum wind flow domain requirements, we analyzed several larger sizes of domains. Based on the sensitivity of the results with respect to the domain size, we chose the wind flow domain to be $8H$ for the distance between the inlet and the front of the structure, $30H$ for the distance between the structure and the outlet and $8H$ for the height as shown in Fig. 10. $30H$ is chosen since the interests lie in the wind speed 1 km behind the structure.

With regard to the boundary conditions for the wind flow domain, the top, the left as well as the right side of the domain are allocated symmetry type of boundary in the ANSYS Fluent software. The bottom of the domain is modelled as a no slip wall with the equivalent roughness height k_s defined by

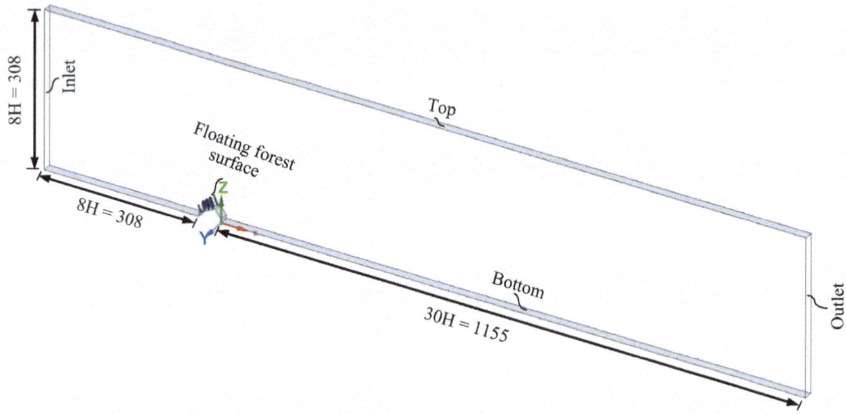

Fig. 10 Size of wind flow domain (cm)

$$k_s = \frac{9.793z_0}{C_s}, \tag{14}$$

where the roughness constant C_s is taken as 1 and the roughness length z_0 is assumed to be 2×10^{-6} m in the model scale. It represents the sea surface roughness in full scale, which is 2×10^{-4} m [10]. The surface of the floating forest is modelled as a no slip wall with $k_s = 0$. Zero-static pressure is specified at the outlet of the domain. As for the inlet, the wind characteristics are specified by the horizontal mean velocity, turbulent kinetic energy, and turbulent dissipation rate. Then the test was carried in an empty domain (i.e. without the presence of the structure). In each run, the wind flow profile in the outlet was compared with its inlet counterpart. The empty domain test ran for more than 30 times until the profile in the outlet are in close agreement with its inlet counterpart as shown in Fig. 11. This is to ensure the alignment between the input roughness height on the ground and the inlet profile as well as the alignment between the inlet profile and the profile at the location of the structure [17]. This final wind profile is then applied as the inlet condition with the structure in place for simulations.

The mesh construction is based on the principle that higher resolution grids are required in the area of interests as well as in the area where large flow gradients are expected [22]. As such, we used a finer mesh design near the structure, below $1H$ in the front and below $2H$ in the lee until the flow reattaches to the bottom of the wind flow domain. 15 layers of prisms were applied on the surface of the bottom of the wind flow domain and another 15 layers of prisms on the structural surface to capture the feature of the boundary layer. The total number of cells after meshing is 2,985,238. Poly-hexcore method was used, meaning that prism elements and hexahedron elements occupy the majority of the mesh, especially in the core region while polyhedrons are used in the transition area.

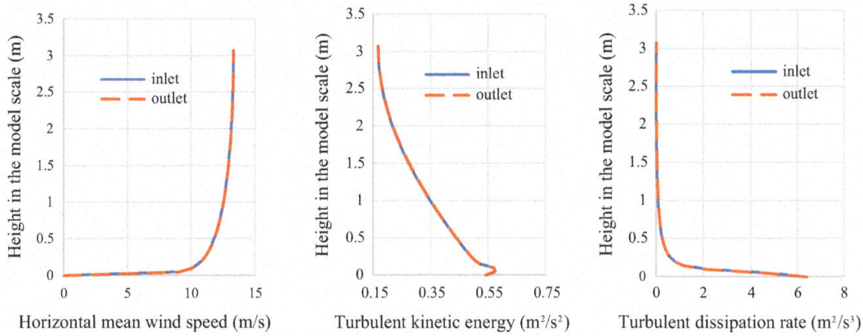

(a) Horizontal mean wind speed profile *(b) Turbulent kinetic energy profile* *(c) Turbulent dissipation rate profile*

Fig. 11 Comparison between inlet and outlet profiles in empty domain test

The turbulence model used for the 3D simulation is the k-ω SST model. Pressure-velocity coupling was achieved by the SIMPLE algorithm. Hybrid-initialization was applied. Second-order discretization schemes were used for both convection and viscous terms of the governing equations.

As the solution procedure is iterative, the residuals specified by the user determine when the calculation will be terminated. Ideally, the smaller the residuals are, the more accurate the results would be. The Fluent User Manual [16] recommends that convergence is achieved when all residuals reach a minimum value, say 10^{-6} for x, y and z velocity, 10^{-5} for turbulent kinetic energy k, and 10^{-4} for specific turbulence dissipation rate ω and for continuity [23]. The final values for all the residuals are within 10^{-6}; thereby satisfying the rigorous convergence criteria as stated.

Figure 12 shows a contour plot of the mean wind speed as the wind passes through the floating forest model. In the plot, the local horizontal mean wind speed U_z is normalized by the horizontal mean wind speed $U_{0,z}$ at the height of z in the approach flow profile. So the numbers in the contour plots actually show how much percentage of horizontal mean wind speed is left as compared to the scenario without the floating forest model. For a better interpretation, the numbers in the horizontal and vertical axes have been converted back to the real distance or height in the full scale. The height is calculated from the mean sea level. The origin of the horizontal axis is placed right behind the floating forest.

20 and 30% reduction areas (i.e. 80 and 70% contour lines) are highlighted. It can be seen that after 1 km behind the floating forest, there is still 20% reduction of the horizontal mean wind speed up to 30 m height regardless of the section of choice. The 30% reduction area extends up to 20 m in height and 800 m in distance. Before the floating forest, there is a small area that has very low absolute mean wind speed (around 20% of the incoming mean wind speed), which provides a potential for deploying wave energy converters. Above the height of floating forest, the mean wind speed is increasing due to the presence of the floating forest, meaning that the

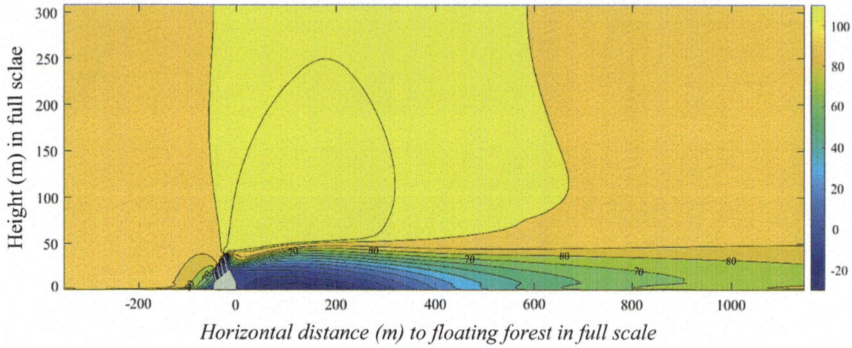

Fig. 12 Contour plot of horizontal mean wind speed for wind flow through floating forest model

floating forest has directed the wind to a higher area; thus providing a significant reduction of the mean wind speed in a broad area in the lee.

5 Concluding Remarks

Presented herein is a novel floating forest, which is a mega floating breakwater-windbreak structure, that protects fragile coastlines and coastal infrastructure/assets by reducing the wave and wind forces of extreme storms. The paper illustrates the determination of the size of the floating forest for an effective wave blockage as reflected by the desired wave transmissibility. Results show that the selected floating forest size can ensure that the wave transmissibility is under 0.5 at the peak wave period T_{p}. Owing to a large waterplane area and shallow draft, the structure has a heave natural period near T_{p}, but the peak response amplification is only 1.17. The CFD analysis on wind flow around the floating forest confirms its effectiveness in reducing wind speeds to provide a sheltered area for objects behind it against wind hazards. Normalized results shows that considerable mean wind speed reduction of 20% remains beyond a distance of 1 km from the floating forest and up to 30 m height. Given that the pressure loaded on the structure is squared to the surrounding wind speed, such a mean wind speed reduction is significant in reducing the wind forces.

A more detailed study on the effect of water channels inside the structure, the effect of large motion response to the wave transmissibility, and extreme wave response will be conducted. Furthermore, a scaled model test will begin soon in the Hydraulics Lab and the Wind Tunnel of the University of Queensland to calibrate the numerical models and further validate the floating forest concept and the numerical wave and wind results as well as to explore other configurations of the concept.

Acknowledgements The authors are grateful to the Australian Research Council for providing the Discovery Project DP170104546 grant, as well as to Hyundai Engineering and Construction and the ARC NanoCOMM Hub for additional grants to perform this study. The third author wishes to acknowledge the scholarship provided by The University of Queensland for his Ph.D. study.

References

1. Wang, C. M., & Wang, B. T. (2015). Great ideas float to the top. *Large floating structures* (pp. 1–36). Singapore: Springer.
2. McCartney, B. L. (1985). Floating breakwater design. *Journal of Waterway, Port, Coastal, and Ocean Engineering, 111*(2), 304–318.
3. Ji, C., Cheng, Y., Yang, K., & Oleg, G. (2017). Numerical and experimental investigation of hydrodynamic performance of a cylindrical dual pontoon-net floating breakwater. *Coastal Engineering, 129,* 1–16.
4. Kolahdoozan, M., Bali, M., Rezaee, M., & Moeini, M. (2017). Wave-transmission prediction of π-type floating breakwaters in intermediate waters. *Journal of Coastal Research, 33*(6), 1460–1466.
5. Neelamani, S., & Ljubic, J. (2018). Experimental study on the hydrodynamic performance of floating pontoon type breakwater with skirt walls. *Journal of Offshore Mechanics and Arctic Engineering, 140*(2), 021303.
6. Dai, J., Wang, C. M., Utsunomiya, T., & Duan, W. H. (2018). Review of recent research and developments on floating breakwaters. *Ocean Engineering, 158,* 132–151.
7. Li, B., & Sherman, D. J. (2015). Aerodynamics and morphodynamics of sand fences: A review. *Aeolian Research, 17,* 33–48.
8. *Estimation of Shelter Provided by Solid and Porous Fences* (1998). Engineering Sciences Data Unit.
9. Heisler, G. M., & Dewalle, D. R. (1988). Effects of windbreak structure on wind flow. *Agriculture, Ecosystems and Environment, 22/23,* 41–69.
10. Holmes, J. (2015). *Wind loading of structures* (3rd ed.). Boca Raton, US: CRC Press.
11. Judd, M. J., Raupach, M. R., & Finnigan, J. J. (1996). A wind tunnel study of turbulent flow around single and multiple windbreaks, part I: Velocity field. *Boundary Layer Meteorology., 80,* 127–165.
12. HydroSTAR User Manual. (2018). *Bureau Veritas*. Paris: France.
13. Rueda, A., Crespo, A. J. C., & Rodríguez, G. (2013). Evaluation of wave energy transmission through a floating breakwater using the SPH method. *Fluid Structure Interaction, 129,* 13–22.
14. DNV-RP-H103. (2011). *Modelling and analysis of marine operations*, April 2011.
15. John, D., & Anderson, J. (1995). *Computational fluid dynamics: The basics with applications*. McGraw-Hill, Sydney, Australia: Mechanical Engineering Series.
16. (2017). *Fluent 18.0 user's manual.* ANSYS Ltd., Canonsburg, US.
17. Blocken, B. (2015). Computational fluid dynamics for urban physics: Importance, scales, possibilities, limitations and ten tips and tricks towards accurate and reliable simulations. *Building and Environment, 91,* 219–245.
18. Franke, J. (2007). *Best practice guidelines for the CFD simulation of flows in the urban environment*. Hamburg, Germany: Meteorological Institute.
19. Franke, J., Hirsch, C., Jensen, A., Krüs, H., Schatzmann, M., Westbury, P., Miles, S., Wisse, J., & Wright, N. (2004). Recommendations on the use of CFD in wind engineering, *KU Leuven Project*. Available online at: https://www.kuleuven.be/bwf/projects/annex41/protected/data/Recommendations%20for%20CFD%20in%20wind%20engineering.pdf. Accessed December 19, 2018.
20. Hall, R. (1997). *Evaluation of modelling uncertainty-CFD modelling of nearfield atmospheric dispersion, EU project EV5V-CT94-0531*. Surrey, UK: Final Report. WS Atkins Consultants Ltd.

21. Tominaga, Y., Mochida, A., Yoshie, R., Kataoka, H., Nozu, T., Yoshikawa, M., et al. (2008). AIJ guidelines for practical applications of CFD to pedestrian wind environment around buildings. *Journal of Wind Engineering and Industrial Aerodynamics, 96,* 1749–1761.
22. Peric, M., & Ferziger, J. H. (1996). *Computational methods for fluid dynamics*. Berlin, Germany: Springer.
23. Ramponi, R., & Blocken, B. (2012). CFD simulation of cross-ventilation for a generic isolated building: Impact of computational parameters. *Building and Environment, 53,* 34–48.

Classification Principles for Very Large Floating Structures

ChunWee Ng and Rongrong Jiang

Abstract Very Large Floating Structures (VLFS) have been widely used in the offshore oil & gas industry for the past decades as a versatile platform for the exploration, exploitation and processing of seabed mineral resources. The classification principles of ships have been successfully applied to these Mobile Offshore Units (MOUs) by providing assurance that a set of requirements laid down in the classification rules are met during its design and construction and maintained during its operational phase with the aim to ensure that the required safety standards are met. Recently, the concept of VLFS is promulgated in other industries like power generation, residential, agriculture, etc. As an example of how classification principles can be applied to VLFS in new industries, the DNV GL's new rules on offshore fish farming units and installations (DNVGL-RU-OU-0503) is being applied to an offshore fish farm. This paper will look at how application of classification rules can help in the design of VLFS to help it rapidly gains recognition from the authorities in terms of achieving an adequate level of safety and quality.

Keywords Classification · Very large floating structures (VLFS) · Offshore fish farm

1 Introduction

Historically, ships were the biggest man-made floating structures in the seas. This concept of creating habitable conditions for humans at seas have been successfully replicated in the offshore oil and gas industry since the 1960s. The use of classification societies as independent parties to determine that an adequate level of safety had been achieved in the design, construction and operation of these floating structures is instrumental in making how the marine and offshore industries look today.

The concept of Very Large Floating Structures (VLFS) outside marine transport and offshore oil and gas industry has been studied in new applications like airport,

---

C. Ng (✉) · R. Jiang
DNV GL, Offshore Classification, Singapore, Republic of Singapore
e-mail: chun.wee.ng@dnvgl.com

C. M. Wang et al. (eds.), *WCFS2019*, Lecture Notes in Civil Engineering 41,
https://doi.org/10.1007/978-981-13-8743-2_13

bridge, fuel storage facility, entertainment centre and accommodation [1, 2]. Unlike ships and offshore units where the use of international conventions and class rules is well established and is generally taken for granted, it is still a discussion if these novel VLFS should be considered as a vessel, offshore facility or something else. The answer to this will determine the relevant design standards to be applied [3].

In some instances of existing VLFS, some elements of the classification process had been applied. The Kamigato and Shirashima oil storage bases in Japan were designed, surveyed and constructed to Japanese classification society, ClassNK's rules and regulations and class notations *KAMIGOTO NO.1 NS* (Smooth Water Service) (Oil Storage Barge, Oils Flashpoint below 61 °C)* was assigned to the storage base in Kamigoto [4].

Another example is the Marina Bay Floating Platform in Singapore which is the world's largest floating performance stage. In the structural design, the American Bureau of Shipping Rules for Building and Classing Steel Vessels for Service on Rivers and Intracoastal Waterways was applied and the Lloyd's Register Rules and Regulations for Classification of Linkspans was also used in link-bridge design and analysis. For stability of the platform, the International Maritime Organisation (IMO)'s Guidelines for the Design and Construction of Offshore Supply Vessels had been used [5].

It can be argued that classification process will be helpful in accelerating the growth of VLFS especially those with predominantly steel structures and with heightened risk levels, for example oil tanks that the classification societies have vast experiences in handling. The systematic framework in the classification process does offer a multi-disciplinary approach which covers areas like stability, structural, mechanical, electrical, instrumentation and safety which is something a traditional civil and structural engineering code will not be able to cover.

This paper will introduce the principles of classification process and how it can be applied to an example of VLFS like offshore fish farm using the latest DNV GL's new rules on offshore fish farming units and installations. Issues affecting VLFS design will also be discussed in the context of classification and related processes and conclusions drawn regarding the future application of class concepts in VLFS.

2 Classification Principles

2.1 History of Marine Classification Societies

Classification societies in the marine industry has its roots in mid-18th century at the time where the international trade was entirely carried by ships. Marine technology was still in its infancy and there was no control on the condition of the merchant ships. Long sea voyages were risky endeavours and there were many losses of ships, lives and precious cargoes. That prompted those in the ship financing, especially the ship insurers to set up the first classification society, Lloyd's Register to establish objective safety criteria and regular surveys of ships to reduce ship losses.

Subsequently, more classification societies were set up in several other countries and continued to play a major role as a qualified third-party in maintaining maritime safety and environmental protection for the world's fleet of ships by creating their unique set of rules that defined the minimum technical standards and requirements to follow in the ship's lifecycle of design, construction and operations.

2.2 *The Classification Process*

Classification provides assurance that a set of requirements laid down in rules established by classification societies are met during design, construction and maintained during operation of a ship or offshore unit. This has gained worldwide recognition to ensure an adequate level of safety and quality for ships and offshore units.

Before a ship can "enter" class, relevant engineering documents of the ship will be submitted to the class and reviewed with respect to compliance to the class rules. Designs deemed to have met the technical requirements will be stamped approved and returned to the designers. Shipbuilders will then construct the ship according to approved plans witnessed by the class surveyor who will issue the final classification certificate to attest that the design and construction of the new ship has met the class requirements. This class certificate is often part of the deliverables that the shipbuilder will need in order to obtain payments from the shipowner. During the operational phase, in-service class surveyors will also board the ship in annual surveys as well as intermediate and renewal surveys in 2.5 and 5 years mark to confirm that hull, machinery, equipment and systems remain in satisfactory condition and in compliance with approval according to accepted standards. An overview of this process is provided in Fig. 1.

2.3 *Classification Societies as Recognised Organisations*

As most ships are operating in an international setting, they are subjected to the maritime regime of international conventions and the overarching United Nations Convention on the Laws of the Seas (UNCLOS). The International Maritime Organisation (IMO) is an agency of United Nations in developing and updating international maritime conventions like Safety of Life at Sea (SOLAS) and International Convention for the Prevention of Pollution from Ships (MARPOL).

Maritime states, either in their capacity as Flag States that register ships and enforce its law or in another role as Coastal States that enforce their laws on ships in their coastal waters, are Member States of the IMO. These Member States participate and vote in IMO meetings which results in international regulations when ratified. The responsibility of the implementation of these regulations falls upon the Member

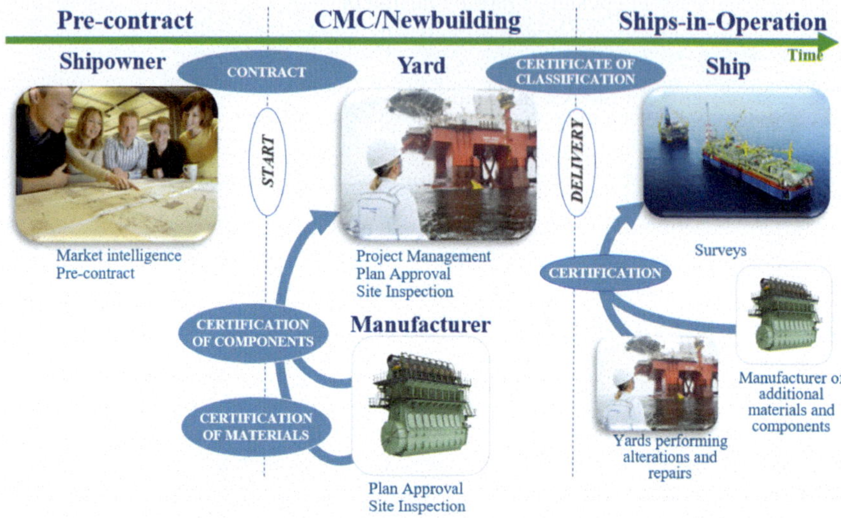

Fig. 1 Overview of the classification process

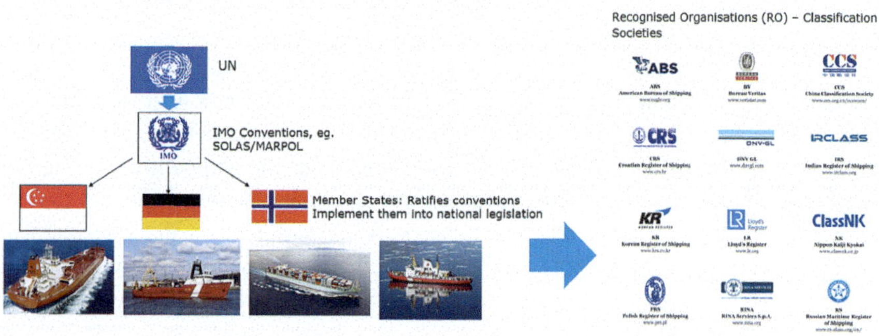

Fig. 2 Overview of the maritime regime

States who have to adopt national legislations so that they come into effect for ships under their jurisdiction.

As the lifecycle of a ship spans many countries, the job of verifying compliance of the ship to these regulations is delegated to multi-national organisations like classification societies who will issue statutory certificates documenting compliance after successful design approval and site survey. Under this capacity, the classification societies are acting as Recognised Organisation (RO) acting on behalf of the flag state of the ship. Attention is to be drawn to the difference between the "class certificate" described earlier and the "statutory certificate" issued by RO on behalf of flag states. An overview of this is shown below in Fig. 2.

2.4 Benefits of Classification

Owing to the international nature of the marine industry, the classification process instils trust and confidence in the level of safety and quality of the ship by engaging independent third party systematic of engineering review and periodic surveys.

At the newbuilding stage of a ship, the issue of a class or statutory certificate by the surveyor indicates that the completion is such that a satisfactory level of safety is achieved and that is used by shipyards to obtain payments. The same is that for manufacturers of components when a product certificate is issued.

In the operational phase, certain capabilities of a ship need to be demonstrated to the charterer, for example Dynamic Positioning which can be added optionally on top of the basic class verification. This is done by the addition of an optional notation like *DYNPOS* to the main class notation on the class certificate.

Lastly, when a ship is sent for scrapping at the end of its useful life, the scrapyard would want to know the hazard materials being used or stored onboard to ship so that precautions can be taken to protect humans and environment. This is another optional notation like *Recyclable* that can be verified by class societies.

From this, one can see how closely knitted the classification process is integrated into the lifecycle of a ship or offshore unit that has an impact in the rapid development of marine and offshore technology for worldwide usage.

3 Case Study on Classification of an Offshore Fish Farm

Among the various new VLFS concepts being studied, the most advance can be said to be the offshore fish farms. The world's first offshore fish farm with a semi-submersible design, Ocean Farm 1 is already operating off Frohavet in Norway since end 2017. Ocean Farm 1 is owned by Norwegian owners, SalMar and was built in CSIC Shipyard in China.

Another project called Havfarm 1 which is based on ship hull design was ordered in early 2018 and will become one of the longest ship of the world at 375 m. The Norwegian owner, Nordlaks will be building the vessel at CIMC Raffles Shipyard in China and has planned for the operations to start in Hadseløya in Norway by mid-2020.

Since these are novel VLFS concepts applied for the first time to offshore fish farms, DNV GL was involved in the following activities of these projects:

- Advisory services for structural design and to establish the regulatory framework for offshore fish farm
- Design verification and independent analyses
- Fabrication follow-up in shipyards
- Certification of Material and Components

Fig. 3 Ocean Farm 1 (left) and Havfarm 1 (right)

- NYTEK certification for Norwegian fish farming regulations
- Marine warranty services for heavy lift transport from shipyard to site
- Preparation of risk-based inspection plans for vessel's structure (Fig. 3).

Although full classification process had not been performed in these projects, relevant offshore rules and standards had been applied, pertaining to structures, main systems and safety systems design. The experience gained from these rapidly maturing projects led to the world's first issue of the *DNV GL Rules for Offshore Fish Farming Units and Installations DNVGL-RU-OU-0503* in July 2017 (referred to as the Rules from now onwards) [6].

This section will present how classification rules can be applied to a new offshore fish farm operating in Norwegian waters and provide insights on how similar principles can be used in other emerging VLFS designs as well.

3.1 General Regulations and Principles

Classification is a process that comprises the development and maintenance of independent technical standards known as class rules for ships and offshore units, and to verify compliance with these rules and standards throughout their lifetime.

Class rules are not "cast in stone" but they are updated occasionally with interval of usually one year with advances in state of technology and experiences from operations. Once rules are published, there is a coming into force date, usually six months later to help the industries to ease into new requirements. In a newbuilding project, a set of applicable rule edition, for example the Rules published in July 2018 will be used as the applicable rules to help the different stakeholders, shipyard and owner to have a common technical basis of design and construction for units contracted after January 2019.

The Rules are intended for offshore fish farms under a maritime regime, where the classification concept is used as part of the certification or license regimes for obtaining compliance with applicable national requirements (e.g. aquaculture acts and regulations of the flag state administration. DNV GL will not verify compliance with statutory requirements unless authorised by national authority/flag state administration.

It is often a question whether an offshore fish farm can be categorised to follow the maritime regime and which conventions, regulations and requirements will be applicable. The Rules can also be used if non-maritime regime is followed. In any case, the relevant national authorities should be contacted for clarification at early stage.

The Rules are predominantly concerned with offshore fish farms of conventional steel or other metallic designs as follows:

- ship-shaped type
- column-stabilised type
- self-elevating type
- cylindrical type
- deep draught type.

There are several items not covered by the Rules in the following paragraph.

- Offshore fish farms where floating collars are made of polymers or concrete
- Net pens and associated equipment
- Fish escape
- Fish feeding and production facilities
- Feedstock, its processes and operations.

The first two are excluded as these are not steel structures that class has deep expertise in. The last three are dealing with aquaculture and associated processes that are already covered in existing standards like the Norwegian standard *NS 9415* on *Marine Fish Farms Requirements for Site Survey, Risk Analyses, Design, Dimensioning, Production, Installation and Operation* and NYTEK's regulations for technical standards for floating aquaculture facilities (Fig. 4).

Under the class systematics, each classed fish farm will be given a class designation comprising mainly the main function, structural design and additional optional class notations that can be used to describe the unit.

Table 1 list the possible class notations of an offshore fish farm: ✠*1A Column-stabilised Offshore Fish Farming Unit DYNPOS POSMOOR Recyclable* that can be assigned.

3.2 Design and Construction Requirements

Under the Rules, the following disciplines are covered under the main class notations:

- safety principles and arrangement
- materials
- hull design and fabrication
- temporary mooring and towing

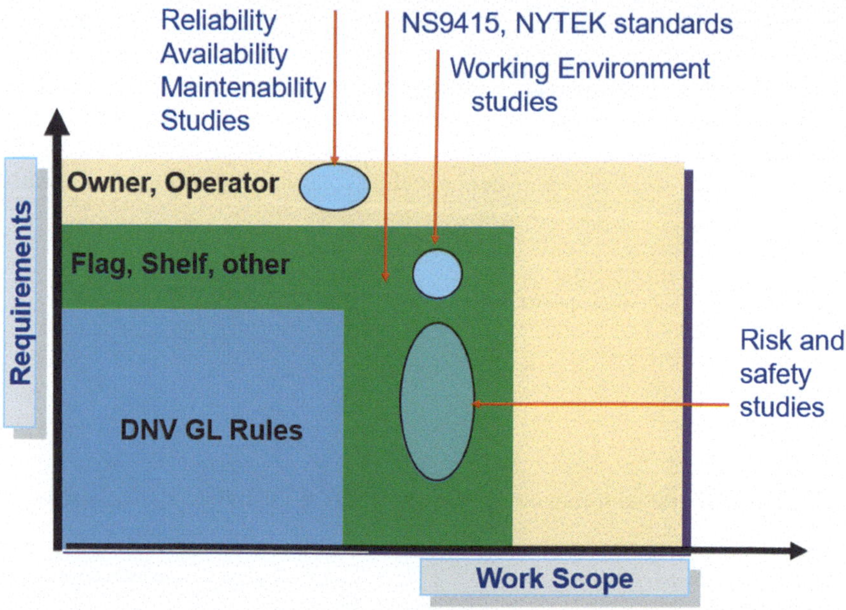

Fig. 4 Summary of classification scope for offshore fish farm

Table 1 Class notations summary

Class notation	Type	Description
✠	Construction symbol	Units built under supervision of DNV GL
1A Offshore Fish Farming Unit	Main character of class	Fish farm with hull and marine machinery and equipment built to DNV GL rules
Column-stabilised	Structural design	Structure dependent on buoyancy of widely spaced columns for floatation and stability
DYNPOS	Additional class notation	Dynamic positioning system
POSMOOR	Additional class notation	Position mooring
Recyclable	Additional class notation	Inventory of hazardous materials

- stability, watertight and weathertight integrity
- marine and machinery systems and equipment
- electrical systems and equipment
- instrumentation and telecommunication systems
- fire protection.

The Rules make references to other DNV GL offshore standards, guidelines, recommended practices and other internationally recognised codes and standards. For instance, safety principles and arrangement should comply with the generic safety principles given in the *DNV GL Offshore Standards for Safety Principles and Arrangements DNVGL-OS-A101*.

A hierarchical approach of how the Offshore Rules (RU-OU), Offshore Standards (OS) and Recommended Practices (RP) are shown in Fig. 5.

The DNV GL Offshore Standards contain detail technical requirements, principles, and acceptance criteria for offshore units in the oil and gas industries like drilling and production units. The *DNVGL-OS-A101* covers design principles and accidental loads, area arrangement, hazardous area classification, emergency shutdown (ESD) principles and requirements, escape and communications.

These offshore standards assume that the unit will be designed and constructed with sufficient integrity to withstand operational and environmental loading throughout its lifecycle. The underlying safety principles behind the design is that it shall be sufficiently robust to tolerate at least one failure or operator error without resulting in a major hazard or damage to the unit. Suitable measures shall be provided to enable timely detection, control and mitigation of hazards. Also, escalation to plant and areas that are not affected by the initiating event shall be avoided. This concept is often explained in a bow-tie diagram shown in Fig. 6.

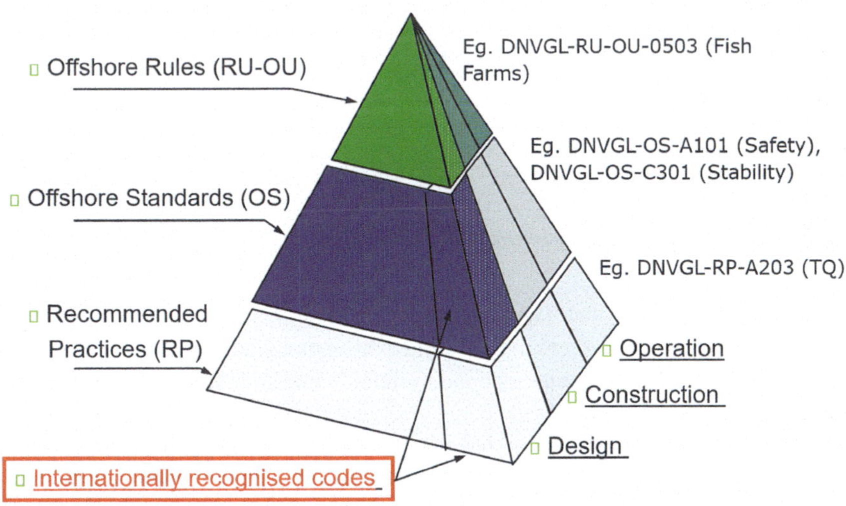

Fig. 5 Hierarchy to DNV GL classification approach

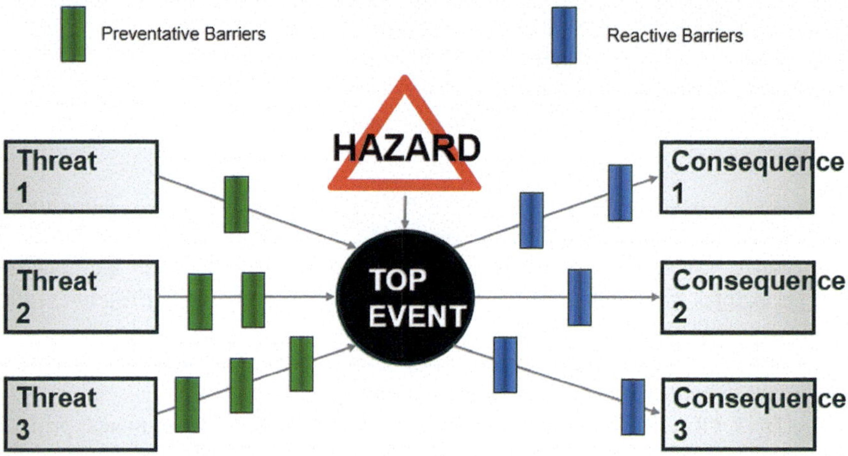

Fig. 6 Generic bow-tie diagram

Indeed, it can be argued that the application of very stringent requirements originating from the oil and gas industries can be an overkill for the offshore aquaculture industry. The process of ensuring an adequate level of safety will nonetheless be very useful to identify any hazards onboard and manage them. For example, the Rules require the customer to prepare a design brief at the beginning of the project to describe the design and operations philosophies and limitations and justify the use of any alternative design standards with risk mitigation in order to ascertain that the design will be carried out within an overall safety standard equivalent with the intention of the rules. This allows equivalent solutions and rule deviations to be accepted as long as they meet the above requirement and are documented and agreed between all parties involved.

Hence, the Rules are written to be flexible to accept non-conventional designs like offshore fish farms with the final aim of establishing an acceptable level of safety whilst promoting safety improvement through experience and available technology.

3.3 In-Operation Requirements

In order for the offshore fish farm to retain class with a classification society, it is necessary for periodical surveys to be arranged throughout the operating life to ascertain that hull, machinery, equipment and systems remains in satisfactory condition in accordance to the Rules.

Periodic surveys can be categorized as either an annual survey, intermediate survey, complete survey or a renewal survey that leads to the issuance of a new certificate. Annual survey is due on the anniversary date of class assignment, intermediate survey is due about 2.5 years after the expiry of previous class certificate and the renewal

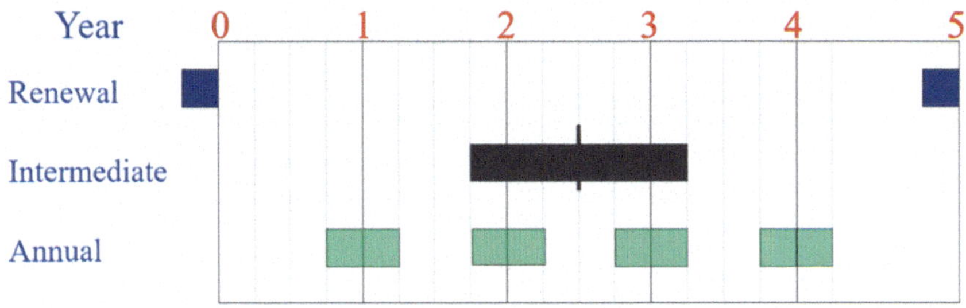

Fig. 7 Survey windows

survey is due five years from the expiry date of the classification certificate. Various specific surveys like bottom survey and boiler survey are scheduled together with one of the above surveys (Fig. 7).

An In-service Inspection Plan (IIP) is developed that contains the structural items to be surveyed on the basis of a general experience-based scope in combination with design and fabrication particulars for the actual unit as well as experience from in-service surveys of units of similar type.

Most periodic surveys will allow surveys to be done on location with minimum interruption to operations. The survey that may require dry-docking are the bottom surveys as required by statutory and class regulations to be done twice in a five-year period to inspect the integrity and watertightness of the hull. The Rules allow what is known as *Under Water Inspections In Lieu of Drydock (UWILD)* survey to replace the traditional drydocking related to these bottom surveys. This is achieved by using divers or Remote Operated Vehicles (ROV) who will relay information on underwater condition of the hull structures with pictorial and video equipment. The class surveyor will witness this and will need to be satisfied that with the quality of the information relayed.

Finally, the class survey also allows the use of alternative survey arrangements for the machinery and structural surveys. For machinery surveys, other arrangements like continuous, planned or condition maintenance system can be used to substitute the time-based renewal surveys. For structural surveys, besides the default renewal surveys, continuous or structural integrity management (SIM) can be selected to create greater flexibility for survey of offshore fish farms.

4 Discussions

There are several issues which are important in the application of classification process to VLFS and these will be discussed in this section with potential benefits that can be reaped.

4.1 Hydroelastic Response and Structural Design

A conventional offshore platform, such as a semi-submersible unit, is normally considered as a non-deformable body when determining the motion of the hull through seawater. In this way, the hydrodynamic response of the hull can be estimated in a relatively simplified way, yet with sufficient accuracy.

However, the situation is different for a VLFS. One example is the floating airport, the size in one dimension (vertical) may be very small comparing to other. This results in very flexible character of the hull structure that its deformation cannot be neglected when calculating its hydrodynamic response. Another example is the offshore fish farm which consists of very slender members and non-steel components like the fish net. The behaviour of these individual members will influence the behaviour of the fish farm structure as a whole. It takes considerable effort reiterating the interactions between a VLFS and the water, to determine the correct hydrodynamic response.

In either way, the purpose is to obtain hydrodynamic response with sufficient accuracy for subsequent structural design of strength, fatigue, and serviceability. The class rules recognise this difference and accept direct analysis methods using finite element analysis or model tests to demonstrate the structural strength is ensured in relations to the hydrodynamic forces and hence will only need minor adjustment to cater for the unique characteristics of the VLFS.

4.2 Drift Forces and Mooring Design

A conventional drilling platform needs to stay on position in the designated location with minimal drift off and very often with an orientation towards undesired environmental loads. The purpose is to ensure the integrity of the steel drilling pipe connecting the platform and the wells at the bottom of the sea, and to ensure smooth flow of pressured mud and hydrocarbon through the pipes without disturbance or leakage. To achieve this target in severe weather and in water depths up to 3000 meters, the platform needs a highly reliable mooring system that is designed to withstand environmental loads from all directions, or a dynamic positioning system that is designed and equipped with instant response and high manoeuvrability, or a combined system that achieves both. The position mooring and dynamic positioning capabilities are characterised by the DNV GL class notation *POSMOOR* and *DYNPOS* respectively.

However, these sophisticated systems may not always be technically appropriate or economically viable. DNV GL recognises this problem and is developing a set of offshore standards more suited for near-shore mooring systems that is fit for purpose and is economically viable for the owners. The new class notation for near shore mooring will account for shallow water effects and a relook at the safety factor for mooring systems.

4.3 Relevant Optional Class Notations for VLFS

Over the past few years several class notations have also been developed by DNV GL to help the development of new VLFS. An example is the world's first Floating Liquefied Natural Gas (FLNG) unit, PFLNG SATU which is owned by Malaysian state-owned oil company Petronas which started operations in end 2016 to exploit the remote and stranded offshore gas field of Kanowit in East Malaysia. Before this, offshore gas is piped to onshore facilities for treatment and export to LNG carriers. The success of this unit shows how the classification societies helped to ensure the technical and safety challenges of a novel floating LNG installation can be addressed and accepted by regulatory bodies and investors.

Another popular application related to LNG is the Floating Storage and Regasification Unit (FSRU) which is an LNG import facility consisting essentially a LNG carrier with LNG tanks and regasification facility to vaporise the LNG to gaseous state so that it can be supplied to consumers like power stations. DNV GL assigns the class notation *REGAS* to gas carriers meeting its requirements that covers vessel arrangement, structures, regasification piping and process safety, fire safety and electrical systems [7].

A recent addition is the class notation *Gas Power Plant* that can be used for a floating power station that is arranged with a gas fired power plant for the production and export of electrical energy. In this DNV GL rule, safety requirements cover total gas system from gas bunkering to end gas user onboard as well as related electrical system from electrical power generation to power export. If the additional notation qualifier, *Performance* is included, there will be additional requirements to ensure the quality of the rated electrical power delivered from the floating power plant [8].

The above section gives some examples how DNV GL has devised rules to handle the functional aspects of the VLFS. Other classification societies may have similar approaches or more examples (Fig. 8).

4.4 Design Verification at Different Project Stages

Offshore projects due to its scale inevitably carries substantial technical and commercial risks should the final product do not perform as expected or require major modifications that will take time. Having a neutral third party to perform design reviews and verifications at critical milestones in major projects can yield tremendous benefits by providing the following:

- Ensure the feasibility of a concept by having an independent third party to verify compliance with relevant rules and regulations or recognized codes and standards.
- Ensure the designs are in compliance with rules and regulations thus allowing purchase of long lead time material and components.
- Provide a statement from an independent party to market new concepts towards potential investors and other stakeholders.

Fig. 8 Example of gas power ship (*Sumbebekos, Wikimedia*)

These kinds of design review and verification can be performed by classification societies like DNV GL at different stages of design development from initial concept development, pre-Front End Engineering Design (FEED), FEED and Basic Design with maturing design information.

Where there is a high level of novelty in the design, Approval in Principle (AIP) can be offered as an independent assessment of a novel concept within an agreed requirement framework confirming that a design is feasible and that no insurmountable obstacles or showstoppers would prevent the concept from being realised. For example, a novel crane vessel will require more attention into the stability aspects (Fig. 9).

Other services that can be requested include Main Scantling Approval (MSA) and Basic Design Approval (BDA) when more detail engineering information like main scantling drawings showing plate thicknesses, stiffeners and girders dimensions is available and will help in the purchase of long lead time materials and components.

4.5 Technology Qualification of Novel Connector Design

Design of novel offshore projects may typically identify a number of critical systems which have no proven service history, for example novel connector design. Existing codes and standards are not likely to describe these systems adequately. Similar to AIP and related services described above which are applied to unit designs and con-

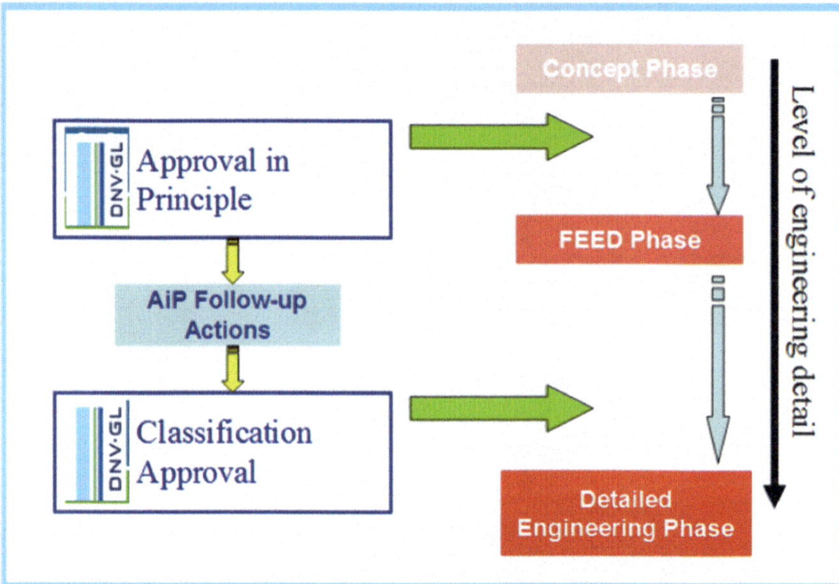

Fig. 9 Approval in principle with design phase

cepts. The Technology Qualification (TQ) process can be used to provide a systematic approach to qualification of new technology, ensuring that it functions reliably within the specified limits. The approach developed by DNV GL is described in the *Recommended Practice for Technology Qualification (DNVGL-RP-A203)* as shown in Fig. 10 [9]. TQ benefits all major players by providing a structured process to ascertain the technology readiness level of a product.

5 Conclusions

The principles for classification of Very Large Floating Structures (VLFS) have been presented in light of how it has been successfully deployed in the maritime and offshore oil & gas industries. As more novel applications of VLFS are proposed in other industries like power generation, residential, agriculture, etc., the same classification process can be used to reap the benefits.

A case study of applying a new classification rule for offshore fish farm *DNV GL Rules for Offshore Fish Farming Units and Installations DNVGL-RU-OU-0503* is presented and discussed with regards to hydroelasticity, structural, mooring and additional class notations for different VLFS. Additional services that is relevant include design verification at different stages of project and technology qualification for novel solutions.

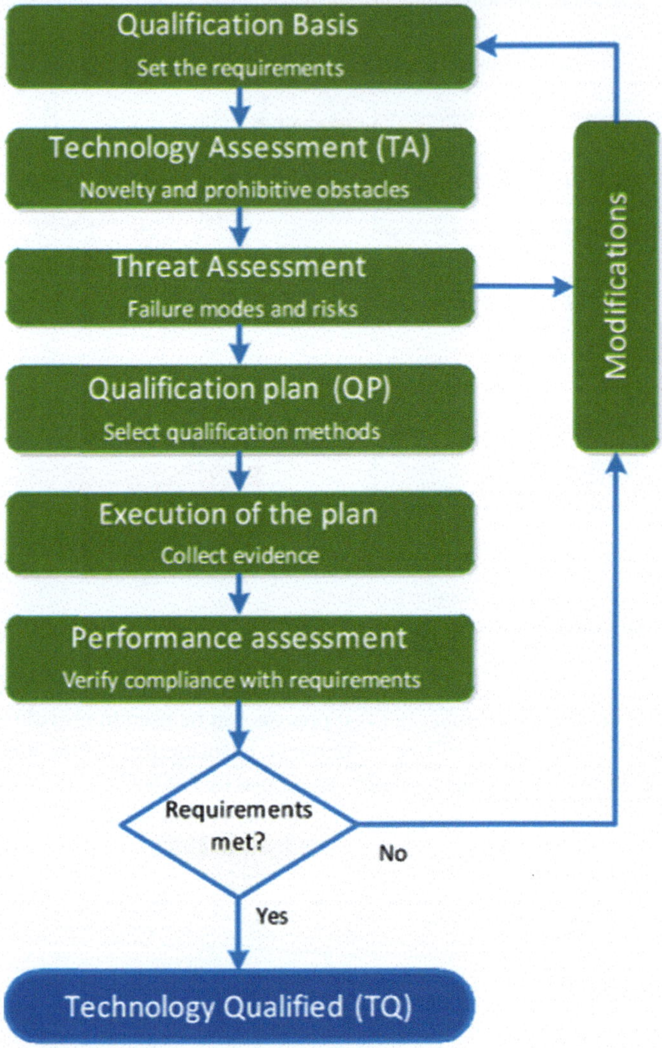

Fig. 10 Steps in basic technology qualification process

In conclusion, the classification process offers a flexible multi-disciplinary safety assurance scheme that can help to accelerate acceptance from authorities in achieving an adequate level of safety and quality.

References

1. Wang, C. M., & Tay, Z. Y. (2011). Very large floating structures: Applications, research and development. *Procedia Eng., 14,* 62–72.
2. Watanabe, E., Wang, C. M., Utsunomiya T., & Moan, T. (2004). *Very Large Floating Structures: Applications, Analysis and Design*. Singapore.
3. Lamas-Pardo, M., Iglesias, G., & Carral, L. (2015). A review of very large floating structures (VLFS) for coastal and offshore uses. *Ocean Engineering, 109,* 677–690.
4. Arai, H. (2001). Consideration on safety assessment of very large floating structures—classification surveys for large-scale oil storage system. In *24th Meeting of the UJNR Marine Facilities Panel*.
5. Koh, H. S., & Lim, Y. B. (2015). Floating performance stage at the Marina Bay, Singapore. In Wang, C. M., & Wang, B. T. (Eds.) *Large floating structures technological advances* (pp. 37–60). Springer.
6. DNV GL. (2018). *DNV GL Rules Offshore fish farming units and installations*. Available: http://rules.dnvgl.com/docs/pdf/DNVGL/RU-OU/2018-07/DNVGL-RU-OU-0503.pdf. Accessed November 20, 2018.
7. DNV GL. (2018). *DNV GL Rules for Regasification Units DNVGL-RU-SHIP Pt.6 Ch.4 Sec.7*. Available: http://rules.dnvgl.com/docs/pdf/DNVGL/RU-SHIP/2018-07/DNVGL-RU-SHIP-Pt6Ch4.pdf. Accessed November 16, 2018.
8. DNV GL. (2018). *DNV GL Rules for Gas Power Plant DNVGL-RU-SHIP Pt.6 Ch.5 Sec.20*. Available: http://rules.dnvgl.com/docs/pdf/DNVGL/RU-SHIP/2018-07/DNVGL-RU-SHIP-Pt6Ch5.pdf. Accessed November 16, 2018.
9. DNV GL. (2017). *DNV GL Recommended Practice Technology Qualification DNVGL-RP-A203*. Available: https://rules.dnvgl.com/docs/pdf/DNVGL/RP/2017-06/DNVGL-RP-A203.pdf. Accessed November 19, 2018.

Mooring Systems for Very Large Floating Structures

Aditya Sankalp and Yves De Leeneer

Abstract Owing to scarcity of land, very large floating structures (VLFS) are now being designed to cater for the increase in population and growth of coastal areas. The applications of VLFS include floating piers, floating airports, floating bridges, floating fuel storage facilities and even floating cities. One of the key design aspects of VLFS is the mooring design. Mooring design of VLFS is a challenge due to huge size of the structures, environmental loads, shallow water depths, space constraint for mooring lines and anchor installation. There are additional challenges pertaining to transportation of blocks, integration onsite and design allowance for possible future expansion of the VLFS. This paper examines the hydrodynamic and mooring design of a typical VLFS. The relevant concepts, motion response, mooring design and design criteria will be presented. The mooring design will incorporate sensitivity studies on different material choices for mooring lines. Chains, wire ropes and polyester (Dyneema) will be considered for the mooring design. The chain mooring system is compared with piles mooring system. Additional issues pertaining to installation and future expansion of VLFS will be discussed.

Keywords Very large floating structures · VLFS · Mooring · Hydrodynamics · Chains · Piles

1 Introduction

There are multiple ways for station keeping of VLFS. Mooring chains, tension legs, caisson dolphins and piles are some of the options for mooring system design of VLFS. The mooring options may be integrated with wave energy dissipation devices to minimize the complexity of the mooring system design. This paper examines two designs, namely, mooring chains and piles for station keeping of VLFS. For mooring chains, a time domain mooring analysis was carried out. The calculations are based on software calculations and can be computed in normal computers. Linear

A. Sankalp (✉) · Y. De Leeneer
DEEPBLUE Pte Ltd., 14 Benoi Sector, Singapore 629847, Singapore
e-mail: aditya.s@deepblue.sg

C. M. Wang et al. (eds.), *WCFS2019*, Lecture Notes in Civil Engineering 41,
https://doi.org/10.1007/978-981-13-8743-2_14

diffraction method was used to calculate the response amplitude operator's (RAO) and quadratic transfer function (QTF) for the VLFS. The hydrodynamic analysis was carried out using Hydrostar software. The mooring analysis was carried out by using Ariane and Orcaflex software. Pile sizing was carried out based on total environment loads (from mooring analysis). The two mooring options were then compared based on weight, location, seabed and water depth.

2 Mooring System with Chains

2.1 VLFS Hydrodynamics

The sections below provide the hydrodynamic analysis methodology, input data used for analysis and the results.

2.1.1 Analysis Methodology

The purpose of the hydrodynamic analysis is to evaluate motion and QTF RAOs for mooring analysis. The calculations are carried out by means of the Bureau Veritas computer program HYDROSTAR FOR EXPERT [1]. Based on the potential theory, this software predicts the three-dimensional flow of wave diffraction and radiation around floating or fixed bodies in deep water and water of finite depth. The singularity method of Kelvin's sources is used to solve the first order problems while Molin's method [1] is applied to evaluate the complete second order low-frequency and high-frequency wave loads.

Numerical results have shown that the Newman approximation [1] overestimates the drift loads for the extreme values. These values of drift loads are very important for the shallow water response. The Newman approximation is not efficient for this kind of study. Thus, full QTFs using near field formulation are calculated [1]. The hydrodynamic analysis follows the process represented by the scheme on Fig. 1. The hydrodynamic model is composed of flat panels representing the geometry of the submerged part of the hull.

2.1.2 MODEL Characteristics

The main characteristics of the VLFS used for analysis are given in Table 1. The water depth at the VLFS location is 20 m [1].

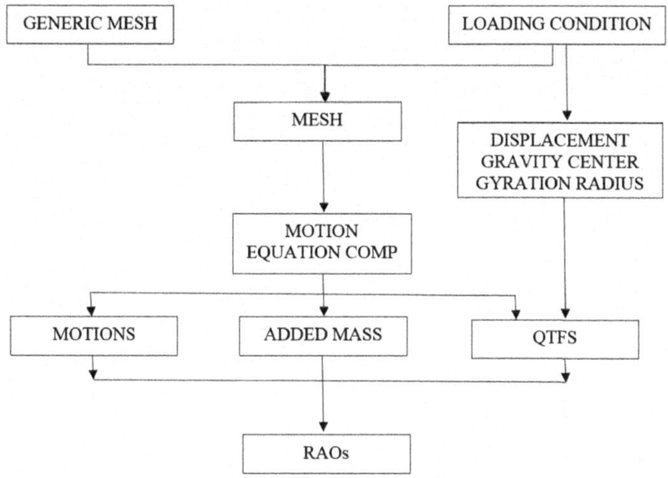

Fig. 1 Approach for hydrodynamic analysis

Table 1 VLFS particulars

Vessel type	VLFS
Length (L)	500.0 m
Breadth (B)	100.0 m
Depth (D)	5.0 m
Mean draft (T)	2 m
Displacement (Δ)	102,500 MT
Buoyant volume	100,000 m^3
LCG from aft perpendicular	250 m
VCG from baseline	1.32 m
Yaw radius of gyration	125.0 m
Pitch radius of gyration	125.0 m

2.1.3 Hydrodynamic Mesh

The mesh model is shown in Fig. 2.

2.1.4 Roll Damping

The damping due to radiation is computed by HydroStar. However, in addition to the radiation damping, there are other sources of damping acting on the floating bodies such as the fluid viscosity and the mooring systems damping. The effects of viscosity on the hull and on the appendages on roll damping are generally higher than the radiation damping.

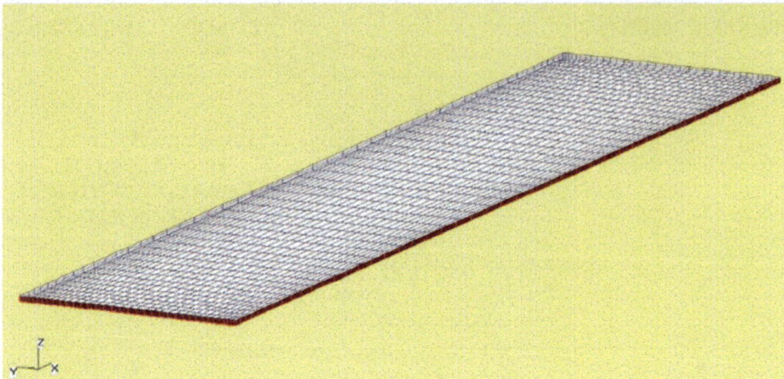

Fig. 2 VLFS mesh

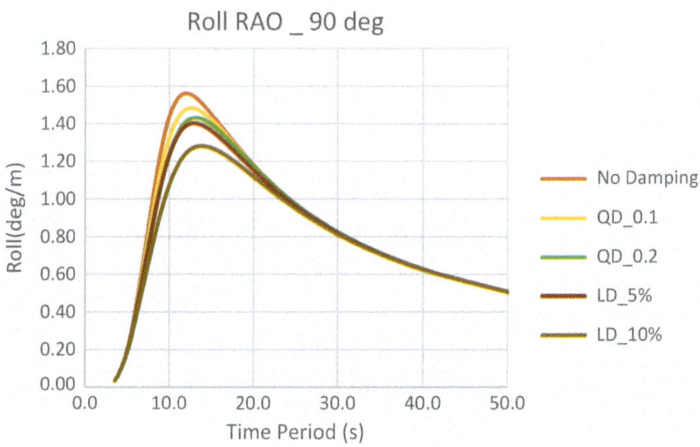

Fig. 3 Comparison between different roll damping values

Hydrostar recommends Quadratic damping for barge shaped vessels [1]. In absence of model test results of VLFS, Quadratic damping [1] was considered for analysis. The equation for quadratic damping is given by

$$B_Q = \frac{1}{2}\rho C_D B^4 L \tag{1}$$

where B_Q is the quadratic damping, ρ the fluid density, B the breadth of VLFS, L the VLFS length and $C_D = 0.1$ is coefficient [1].

Sensitivity analysis was carried out for different types of damping (linear & quadratic) inorder to assess the effect of damping on ROLL RAOs. The effect of different damping values (no damping, 5–10% linear damping, quadratic roll damping) on Roll RAO amplitude is shown in Fig. 3). The damping has a minor effect on the Roll RAO Amplitude.

2.1.5 Hydrodynamic Analysis Results

The plots of Heave, Roll & Pitch RAOs are shown in Figs. 4, 5 and 6, respectively.

Added Mass

Added mass under the specified conditions used for mooring analysis (horizontal components) are summarized in Table 2. They are evaluated at the vessel centre of gravity.

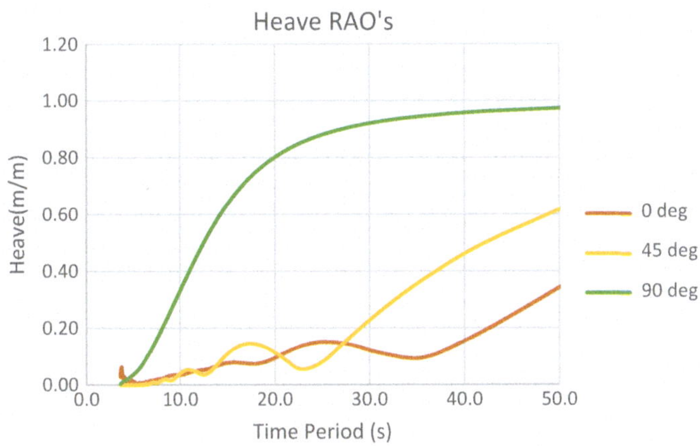

Fig. 4 Heave RAO's for head, quarter & beam seas

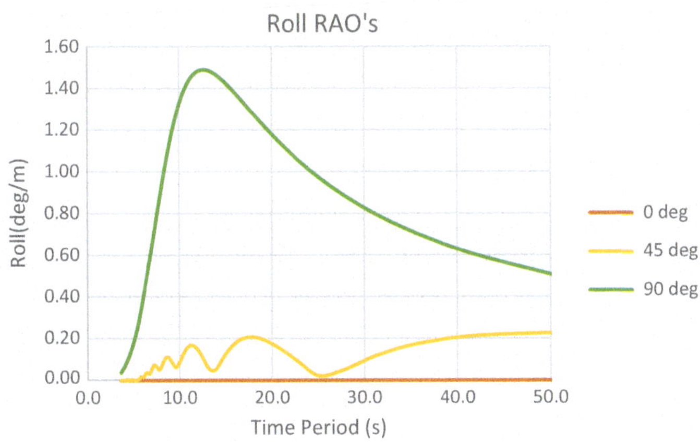

Fig. 5 Roll RAO's for head, quarter & beam seas

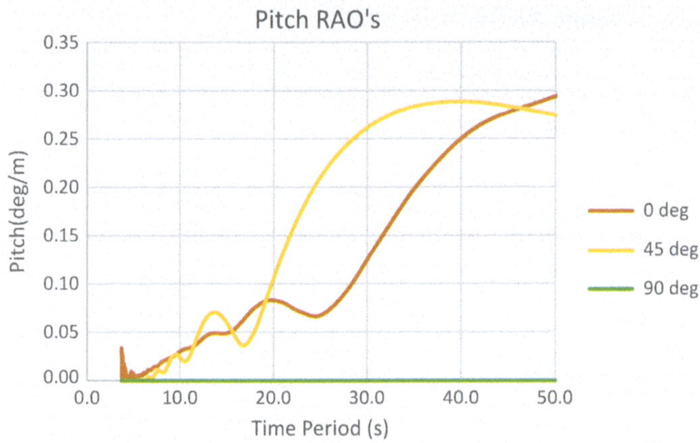

Fig. 6 Pitch RAO's for head, quarter & beam seas

Table 2 Added masses of VLFS

Component	Quantity
Surge	3.76E+06 kg
Sway	1.43E+07 kg
Sway/yaw	3.08E+03 kg m
Yaw	2.32E+11 kg m^2

2.2 *Mooring Analysis*

The sections below provide the mooring analysis methodology, input data used for analysis and the results.

2.2.1 Mooring Analysis Methodology

For Mooring Analysis VLFS is considered as rigid body. Hydroelastic motion for VLFS was not considered. There are multiple ways to minimize the hydroelastic effect [2] and thus reduce the impact of hydroelastic motion on mooring loads. The methodology for the extreme strength analysis of the spread moored system is described below.

Quasi Dynamic Analysis

Quasi dynamic analysis is performed using the Ariane software. This is to evaluate the low frequency and wave frequency loads on the mooring system. The low frequency response of the vessel is evaluated by numerical resolution in time domain and at the end of each time step of this numerical integration, the wave frequency motions are added. Then the instantaneous tensions are derived from the tension-offset curves (characteristics). In order to achieve statistical significance, for each weather combination, 10,800 s (3 h) simulations were ran and the attached maximum tensions in the mooring system were determined. A ramp up period of 2000s was used. For a given metocean state, 20 simulations were performed for randomly chosen seeds by software. These seeds correspond to sets of elementary wave components. The value of the design tension for the considered metocean state takes into account the dispersion of the maximum tensions in the lines due to the seed. It is based on the formula given by

$$T_D = T_{mean} + aT_\sigma \tag{2}$$

where T_{mean} is mean of the maximum tensions for the 20 simulations, T_σ is the standard deviation of the maximum tensions for the 20 simulations, a is a factor based on type of analysis and number of simulations (i.e. $a = 0.5$ for 20 simulations for quasi dynamic analysis [3] and $a = 0.6$ for 5 simulations for dynamic analysis [3]).

Dynamic Analysis

Dynamic Analysis is performed for the worst load cases to evaluate the maximum tension in the most loaded Line (Quasi Dynamic Analysis). The Fairlead motion of the most loaded line is used as input in Orcaflex which then uses the time series fairlead motion for the analysis. Five simulations (for 5 random seed numbers) were ran for dynamic analysis. The value of the design tension for the considered metocean state takes into account the dispersion of the maximum tensions in the lines due to the seed. It is based on the formula given by Eq. (2).

The following steps are done to post process the results from the various runs for a particular load case.

- From all the maximum line tensions standard deviation is calculated $\rightarrow T_\sigma$.
- The mean of all the maximum mooring line tensions are calculated $\rightarrow T_{mean}$.
- Design tension is computed from $T_D = T_{mean} + a * T_\sigma$.
- T_D is used for Factors of Safety (FOS) calculation.
- Offset is obtained in the similar way as Tension.

SLF (Single Line Failure) Case

For SLF (Single Line Failure) the second most loaded line in the Intact Case will be broken to find the maximum tension. The most loaded line in the Intact Case will be broken to find the maximum offset. The environmental cases to be used in analysis will be the worst loading cases from Intact Analysis. The tension and offset will be calculated the same way as for the intact case.

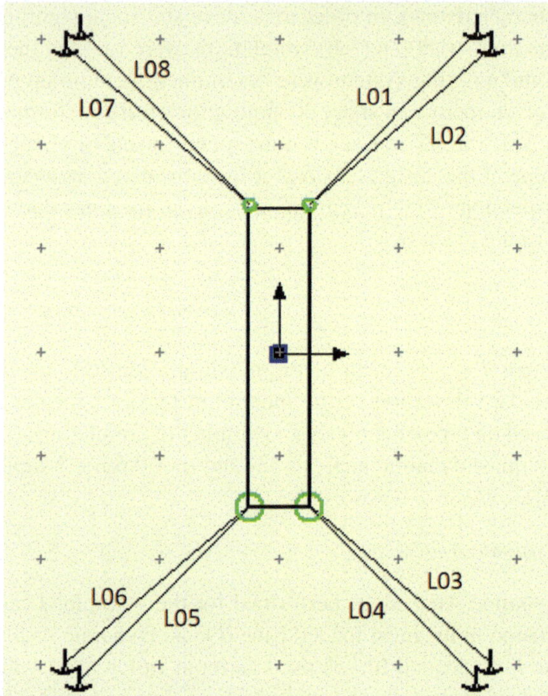

Fig. 7 Mooring layout

2.2.2 Mooring Layout

The system is composed of VLFS and 8 mooring lines. The mooring layout is as shown in Fig. 7.

2.2.3 Vessel Damping

Additional linear damping was considered for the mooring analysis. The equations below provide the formula for the calculation of low frequency damping [3].

$$B_{xx} = 0.06\sqrt{K_{Oxx}(m + Ma_{xx})} \tag{3}$$

$$B_{yy} = 0.06\sqrt{K_{Oyy}(m + Ma_{yy})} \tag{4}$$

$$\mathrm{B}\psi\psi = 0.10\sqrt{K_{O\varphi\varphi}\left[I_{\varphi\varphi} + Ma_{\varphi\varphi} + (m + Ma_{yy})x_G^2\right]} \tag{5}$$

where B_{xx} is the linear damping coefficient in surge, B_{yy} is the linear damping coefficient in sway, $\mathrm{B}\psi\psi$ is the linear damping coefficient in yaw, m is the mass, L is the length, B is the breadth, Ma_{xx}, Ma_{yy}, $Ma_{\varphi\varphi}$ refers to added mass of the vessel. *Koxx*,

Table 3 Mooring system stiffness

	K_{oxx} (kg/s^2)	K_{oyy} (kg/s^2)	$K_o\varphi\varphi$ (kg/s^2)
Mooring system stiffness	2.16E+04	1.82E+04	9.43E+08

Table 4 Linear damping coefficients

	B_{xx} (kg/s)	B_{yy} (kg/s)	$B_{\psi\psi}$ (kg m^2/s)
Linear damping coefficients	9.08E+04	8.75E+04	4.16E+09

$Koyy$ and $Ko\varphi\varphi$ are the mooring system stiffness as calculated for average position of vessel for the most probable environment, x_G refers to vessel Longitudinal centre of gravity measured from centre of vessel and $I_{\varphi\varphi}$ refers to moment of inertia in yaw calculated at centre of gravity of the vessel.

The mooring system stiffness ($Koxx$, $Koyy$ and $Ko\varphi\varphi$) have been calculated for equilibrium position of vessel and provided in Table 3. Table 4 indicates the damping coefficients calculated based on Eqs. (3), (4) and (5).

2.2.4 Mooring Lines Characteristics

Anchor radius is 350 m and the total paid out length is 355 m. Table 5 shows the line segments used for mooring analysis and sensitivity studies. The chain properties are shown in Table 6. Table 7 provides the line segments minimum breaking loads (MBL).

Table 5 Line segments

Line type	Top segment	Bottom segment
Chain only	Chain—355 m	
Wire + chain	6-strand wire—50 m	Chain—305 m
Dyneema rope + chain	Superline Dyneema—50 m	Chain—305 m

Table 6 Line properties

Line type	Value
Chain	76 mm R3 studlink
Wire	76 mm 6-strand
Dyneema	76 mm superline

Table 7 Line MBL

Line type	Value
Chain MBL (MT)	405.0
Wire MBL (MT)	378.0
Dyneema MBL (MT)	340.0

Note Corrosion allowance is 0.4 mm/year [4]. Based on field life of 20 years chain MBL is calculated for 76 mm chain

Table 8 Mooring line safety factor

Type of analysis	INTACT	SLF
Quasi dynamic	2.00	1.43
Dynamic	1.67	1.25

Table 9 Environment data

Direction (from)	Wave			Wind	Current
	γ	Hs (m)	Tp (s)	(m/s)	(m/s)
Head	5	1	3.6	25	2.2
Quarter	5	1	3.6	25	2.2
Beam	3.3	0.6	3.1	25	1

2.2.5 Design Criteria

Safety factors as per API guidelines [4] is shown in Table 8. The minimum ground length is positive to prevent anchor uplift.

2.2.6 Environmental Conditions

Table 9 shows the environmental data that is used for the mooring strength checks [5].

The waves are modeled using JONSWAP spectrum. Note: (1) directional variation of wind was done for $\pm 45^\circ$ and (2) current was considered one by one for all 3 directions.

Table 10 Load cases for mooring analysis

Wave	Wind	Current	Total load cases
0	0, −45, 45	0, 45, 90	9
45	45, 0, 90	0, 45, 90	9
90	90, 45, 135	0, 45, 90	9
Total load cases			27

2.2.7 Load Cases

Intact Load Cases

In order to find out the maximum loads and offsets, different environmental conditions were used in the analysis as shown in Table 10.

SLF Load Cases

For the worst Intact case SLF was run to find the maximum tension by breaking the second most loaded line and maximum offset by breaking the most loaded line.

2.2.8 Results

The results for the mooring analysis are presented in the sections below. Detailed results are presented in Appendix 1. The results are presented in Tables 11 and 12.

The line safety factors as shown in Table 13 are within the limits. The minimum ground length of chain for Intact case is 85 m. The minimum ground length of chain for SLF case is 20 m.

2.2.9 Sensitivity Analysis

Sensitivity Analysis was carried out with top section of line replaced with wire and then with Dyneema rope. The analysis was carried out for the beam direction for

Table 11 Intact case results

Env	Max line tension (MT)	Offset (m)
0	110.6	5.9
45	177.0	7.3
90	180.0	5.9

Table 12 SLF case results

Max line tension (MT)	Offset (m)
285.0	8.7

Table 13 Safety factors

Analysis	Max line tension (MT)	Line MBL (MT)	Safety factor
Intact	180.0	405.0	2.26 (>1.67)
SLF	285.0	405.0	1.42 (>1.25)

Table 14 Intact case results

Line type	Intact line tension (MT)	VLFS offset (m)
Wire rope	169.7	5.2
Dyneema	161.2	5.0

Table 15 SLF case results

Line type	SLF line tension (MT)	VLFS offset (m)
Wire rope	280.3	6.4
Dyneema	274.5	6.4

which maximum line tension occurred. The results are shown in Tables 14 and 15. The mooring loads are similar to loads obtained in Table 13 for chain segments.

3 Pile Mooring System

An alternative solution as compared to mooring lines would be to use steel piles to support the VLFS.

3.1 Methodology

The environmental loads (wind, wave and current) acting on the VLFS were calculated in Sect. 3. The piles were designed to withstand the environmental forces acting on VLFS and the piles structure. The piles are fixed on the seabed. Further analysis will be required based on detailed seabed profile for seabed reaction forces on piles. The VLFS will slide up and down the piles. There will be no axial compression force on Piles due to VLFS weight. Thus, bending stress will be the design criteria for sizing of Piles. The maximum allowable bending stress [6] is given in (4).

$$F_b = 0.75F_y \quad for\frac{D}{t} \leq \frac{10,340}{F_y} = 47 \tag{4}$$

where the yield stress $F_y = 220$ MPA (mild steel) and F_b is the allowable bending stress on the pile. The total bending moment acting on the piles was calculated and used to determine the diameter, thickness and number of piles required for station keeping of VLFS.

3.2 Pile Parameters

The dimensions of the pile are provided in Table 16.

3.3 Environmental Forces

The environmental forces acting on VLFS were calculated in Sect. 2 (mooring analysis) and the maximum wind, wave and current loads for head and beam seas are presented in Table 17. The wave and current forces acting on pile have been calculated in Appendix 2.

The environmental forces acting on VLFS and piles are provided in Table 17.

3.4 Calculation & Results

The bending Stress [6] is given by

$$F_b = \frac{BM}{S} \tag{5}$$

Table 16 Pile particulars

Parameters	Value
Length	20 m
Diameter (D1)	1.25 m
Thickness (t)	0.03 m
Section modulus (S)	0.034 m^3
L/r	32
D/t	41.7
Yield stress (Fy)	220 MPa

Table 17 Environmental forces on VLFS and single pile

Environmental forces	Head	Beam
VLFS (MT)	78.9	288.0
Single pile (MT)	6.0	1.7

Table 18 Bending stress on piles

Direction	Number of piles	Total force (MT)	Force arm (m)	Bending moment (kN m)	Bending stress per pile (MPa)	Allowable stress (MPa)
Head	4	102.5	20	20,147	147.1	165
Beam	12	304.4	20	60,395	147.0	165

Fig. 8 VLFS station keeping using PILES

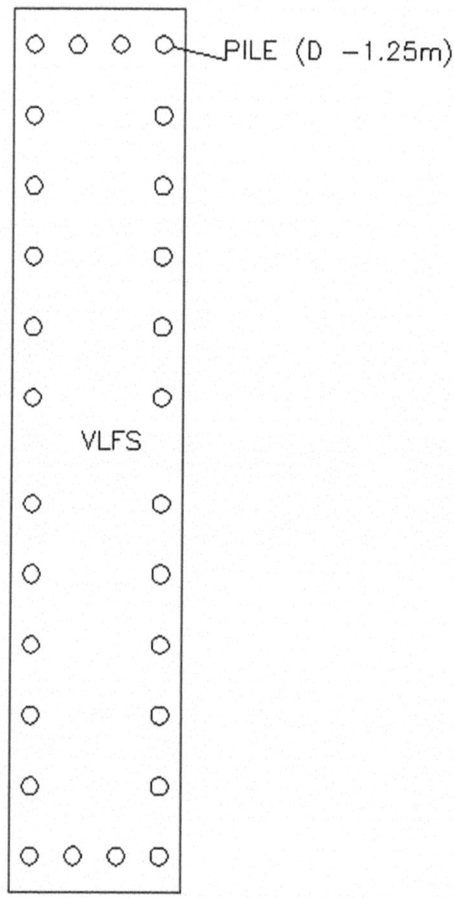

where *BM* is the bending moment (Force x Force arm) and *S* is the section modulus. The bending stress calculations for the piles are provided in Table 18. VLFS Layout with Piles is shown in Fig. 8.

Table 19 Piles versus mooring chain weight

Mooring component	Length	Number	Unit weight (MT)	Total weight (MT)
Piles	20	28	18.1	505.5
Chain system	355	8	56.9	455.3

3.5 Comparison with Mooring Chains

The total number of piles required for station keeping of VLFS is 28. Table 19 provides a comparison of total weight of piles versus chains. The chains weight is much less than Piles weight.

Note that the length of piles excludes the length of pile below seabed. The additional length will lead to further increase in pile weight. Also, the chain system includes weight of chain and anchors.

4 Installation & Future Expansion

4.1 Installation

The VLFS can be towed in multiple blocks to the field and then assembled together. The mooring anchors and chain can be installed using Anchor Handling Tug (AHT). The lines pretension can be carried out using portable winches or Harbor tugs.

4.2 Future Expansion

The mooring lines are connected in bow and stern of VLFS. Future expansion can be carried out by adding additional blocks in front or side (Fig. 9).

Additional mooring lines will be required to handle environment due to increase in size. The existing mooring lines may be required to be relaid and re-connected depending on if additional blocks are added from side or from front.

5 Conclusion

The paper provides two mooring options for VLFS station keeping. The first option is to use mooring chains and anchors. The calculations are based on software calculations and can be computed in normal computers. This method allows a quick estimate of the mooring loads on the VLFS. Hydrodynamic Analysis was carried out

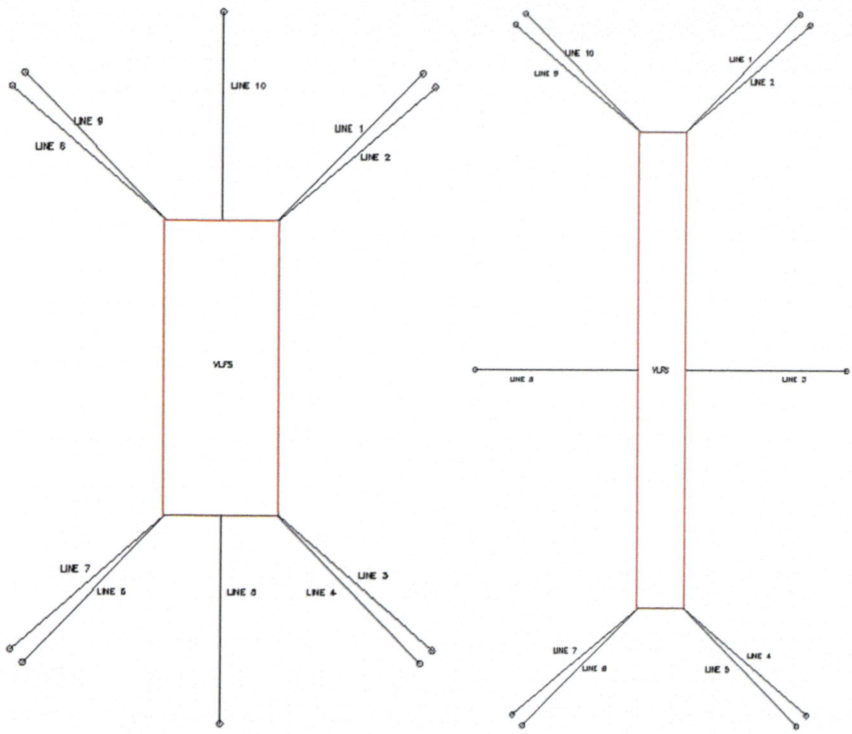

(a) 500m x 200m VLFS

(b) 1000m x 100m VLFS

Fig. 9 Mooring layout for VLFS (future expansion)

for VLFS. Hydroelastic motion for VLFS was not considered. There are multiple ways to minimize the hydroelastic effect [2]. This allows diffraction software programs to analysze the hydrodynamic characteristics of VLFS. The VLFS mooring analysis was carried out for 8 point spread moored system. The mooring loads were within safety limits as shown in Table 13. Additional sensitivity analysis was carried out for different top sections of mooring lines (Wire & Dyneema). The mooring loads were close to chain only lines loads.

The second option was use of piles for station keeping of VLFS. A total of 28 piles were required for mooring system design of VLFS. The weight comparison of mooring chains and piles showed that chains are more effective for mooring design. Piles are quite sensitive to the soil characteristics and seabed profile. Drag Anchors are less sensitive and by adjusting the fluke angles they can be deployed in different types of soil (clay, sand). However, for very shallow water depths (<10 m), piles can be more advantageous as compared to mooring chains due to reduction in pile weight. Moreover, while chains require a mooring zone of 350 m from VLFS edges, the piles can be placed within the VLFS effectively limiting the mooring zone to dimensions of VLFS.

Appendix 1: Detailed Results for Mooring Analysis

Appendix 2: Environmental Forces on Pile

Here a theoretical method for calculation of wave and current forces on pile have been presented. The linear wave theory was used for calculation of wave components. In the end, comparison was made with Orcaflex results where irregular wave (as per Table 9) was used for analysis for head and beam directions.

The total force (wave and current) exerted on a vertical cylindrical pile [7] is given by

$$dF = \int_{-h}^{0} \frac{1}{2} C_D \rho D u|u| dz + \int_{-h}^{0} C_M \rho \pi \frac{D^2}{4} \frac{Du}{Dt} dz \quad (6)$$

The wave induced velocity and current velocity are combined together [8] and the total force acting on pile is given by

$$dF = \int_{-h}^{0} \frac{1}{2} C_D \rho D \left(v_c + \frac{H\omega}{2} \frac{\cosh(k(h+z))}{\sinh(kh)} \right)^2 \cos(kx - \omega t)|\cos(kx - \omega t| dz$$
$$+ \int_{-h}^{0} C_M \rho \pi \frac{D^2}{4} \frac{H\omega^2}{2} \frac{\cosh(k(h+z))}{\sinh(kh)} \sin(kx - \omega t) dz \quad (7)$$

The drag force F_{DC} and inertia force F_{IC} constant terms are defined by

$$Let, \quad F_{DC} = \int_{-h}^{0} \frac{1}{2} C_D \rho D \left(v_c + \frac{H\omega}{2} \frac{\cosh(k(h+z))}{\sinh(kh)} \right)^2 dz \quad (8)$$

$$Let, \quad F_{IC} = \int_{-h}^{0} C_M \rho \pi \frac{D^2}{4} \frac{H\omega^2}{2} \frac{\cosh(k(h+z))}{\sinh(kh)} dz \quad (9)$$

F_{DC} is calculated from

$$dF_{DC} = \int_{-h}^{0} \frac{1}{2} C_D \rho D \left(v_c^2 + 2v_c \frac{H\omega}{2} \frac{\cosh(k(h+z))}{\sinh(kh)} + \frac{H^2\omega^2}{4} \frac{\cosh^2(k(h+z))}{\sinh^2(kh)} \right) dz$$
$$= \frac{1}{2} C_D \rho D \left(v_c^2 h + v_c H\omega \frac{\sinh(kh)}{k \sinh(kh)} + \frac{H^2\omega^2}{4\sinh^2(kh)} \frac{(2kh + \sinh(2kh))}{4k} \right)$$

Case	Wave (deg)	Wind (deg)	Current (deg)	L01 (MT)	L02 (MT)	L03 (MT)	L04 (MT)	L05 (MT)	L06 (MT)	L07 (MT)	L08 (MT)	Offset (m)
1	0	0	0	51.7	43.8	17.5	18.1	18.1	17.5	43.8	51.7	3.9
2	0	45	0	46.9	40.0	42.0	31.5	8.1	7.9	32.1	35.4	4.3
3	0	−45	0	38.5	33.9	23.5	25.3	25.3	23.5	33.9	38.5	3.2
4	0	0	45	67.3	96.4	58.8	69.4	25.6	23.1	6.1	6.4	5.4
5	0	45	45	78.3	110.6	77.4	89.2	22.2	20.0	5.9	6.3	5.7
6	0	−45	45	61.7	102.1	68.3	85.4	38.1	33.6	4.8	4.9	5.9
7	0	0	90	47.7	42.0	37.2	31.1	13.6	13.1	31.0	34.0	4.1
8	0	45	90	63.2	54.0	66.9	48.7	9.1	8.6	28.5	31.6	5.0
9	0	−45	90	33.7	33.8	37.7	40.0	23.3	21.8	21.0	22.2	3.4
10	45	45	0	117.1	91.3	136.8	73.3	7.8	7.5	74.6	84.6	6.5
11	45	90	0	125.1	94.4	144.3	77.6	7.5	7.1	78.3	90.7	6.7
12	45	0	0	86.0	70.7	85.9	56.7	14.3	15.0	74.9	81.0	5.4
13	45	45	45	134.5	171.0	137.6	132.9	35.0	30.4	36.4	42.8	6.7

(continued)

(continued)

Case	Wave (deg)	Wind (deg)	Current (deg)	L01 (MT)	L02 (MT)	L03 (MT)	L04 (MT)	L05 (MT)	L06 (MT)	L07 (MT)	L08 (MT)	Offset (m)
14	45	90	45	142.1	177.4	158.9	154.0	39.1	33.4	34.3	40.0	6.8
15	45	0	45	111.0	150.4	110.9	118.0	42.0	36.5	28.6	32.9	6.4
16	45	45	90	166.3	124.9	164.4	100.1	16.5	14.9	101.3	121.1	7.3
17	45	90	90	158.0	117.7	177.2	107.7	15.1	13.8	102.6	121.9	7.3
18	45	0	90	150.3	113.6	140.3	82.2	20.1	18.9	104.0	121.5	6.9
19	90	90	0	94.0	93.7	115.2	75.0	6.6	6.6	35.9	37.1	5.3
20	90	135	0	56.2	58.9	70.6	53.3	43.1	56.6	33.3	33.7	4.1
21	90	45	0	87.4	88.3	108.7	72.6	10.0	10.3	45.7	48.2	5.1
22	90	90	45	118.4	180.0	151.2	138.9	10.9	10.3	4.6	4.7	5.8
23	90	135	45	95.4	159.5	115.9	121.2	27.3	25.8	4.2	4.3	5.9
24	90	45	45	117.7	176.3	142.4	130.7	11.8	11.2	4.7	4.8	5.8
25	90	90	90	118.9	127.2	144.2	95.8	5.6	5.5	19.8	20.4	5.5
26	90	135	90	73.6	89.5	90.6	73.0	13.1	13.1	15.9	15.6	4.6
27	90	45	90	101.7	106.9	126.1	84.3	6.0	5.9	20.7	21.4	5.3

Table 20 Environment parameters

Direction	Hmax	T	k	Current velocity
Head sea	1.9	3.6	0.31	2.2
Beam sea	1.14	3.1	0.42	1.0

$$= \frac{1}{2} C_D \rho D \left(v_c^2 h + \frac{v_c H \omega}{k} + \frac{H^2 g k tanh(kh)}{4 \sinh^2(kh)} \frac{(2kh + \sinh(2kh))}{4k} \right)$$
$$= \frac{1}{2} C_D \rho D \left(v_c^2 h + \frac{v_c H \omega}{k} + \frac{H^2 g}{2 \sinh(2kh)} \frac{(2kh + \sinh(2kh))}{4} \right) \quad (10)$$

F_{IC} is calculated from

$$F_{IC} = \int_{-h}^{0} C_M \rho \pi \frac{D^2}{4} \frac{H\omega^2}{2} \frac{\cosh(k(h+z))}{\sinh(kh)} dz = C_M \rho \pi \frac{D^2}{4} \frac{H\omega^2}{2k} \quad (11)$$

The total force F as a function of F_{DC} and F_{IC} is given by

$$F = F_{DC} \cos(kx - \omega t)|\cos(kx - \omega t)| + F_{IC} \sin(kx - \omega t) \quad (12)$$

The maxima of F is calculated from the following equations

$$\frac{dF}{dt} = 0 \quad (13)$$

$$F_{DC}(-2\omega) \cos(kx - \omega t)(-\sin(kx - \omega t)) + F_{IC}(-\omega) \cos(kx - \omega t) = 0 \quad (14)$$

$$2F_{DC} \sin(kx - \omega t) - F_{IC} = 0 \quad (15)$$

$$\sin(kx - \omega t) = \frac{F_{IC}}{2F_{DC}} \quad (16)$$

The substitution of the value of sin(kx − ωt) in Eq. (12) furnishes the maximum value of F, i.e.

$$F = F_{DC} + \frac{F_{IC}^2}{4F_{DC}} \quad (17)$$

The environment parameters used for pile force calculation are shown in Table 20. The wave parameters were derived from Hs and Tp (Table 9) as per formulations in [9].

Table 21 Pile forces

Direction	Fd (MT)	FI (MT)	Fmax (MT)	Fmax_Orcaflex (MT)
Head sea	5.7	2.4	6.0	5.2
Beam sea	1.2	1.4	1.7	1.1

The summary of environment forces on the Pile is shown in Table 21. Comparison was also made with irregular wave analysis on a pile model in Orcaflex software. The Linear theory provided higher results and have been considered in the pile design.

References

1. Hydrostar For Experts User Manual. (2016). Research Department BV. June 2016.
2. Wang, C. M., & Wang, C. D. (2005). *Hydroelastic analysis of a super-large floating container terminal*. Floating Container Terminal Research Report, submitted to MPA and JCPL, Centre for Offshore Research and Engineering, National University of Singapore, Singapore.
3. BV, NR 493. (2012). *Classification of mooring systems for permanent offshore units*. April 2012.
4. API RP-2SK. (2005). *Recommended practice for design and analysis for station keeping systems for floating structures* (3rd ed.). October 2005.
5. FUGRO—Metocean Criteria off Changi C54120/9040. July 1, 2016.
6. API Recommended Practice 2A-WSD (RP 2A-WSD) (21st ed.). December 2000.
7. Dean, R. G., Dalrymple, R. A. (1984). *Advanced series on ocean engineering—volume 2, water wave mechanics for engineers and scientists*. Singapore: World Scientific Publishing Co Pte Ltd.
8. Faltinsen, O. M. (1990). *Sea loads on ships and offshore structures*. Cambridge, UK: Cambridge University Press.
9. DNV-RP- C205. (2014). *Environmental conditions and environmental loads*. April 2014.

Durability of Floating Concrete Platforms

Bahador Sabet Divsholi

Abstract Concrete is the cheapest construction material and the second most consumed man-maid product after drinking water, which is easily cast to any shape. Concrete technology has evolved very rapidly in recent years and continues to improve with new advancement in construction and materials technology. Early concrete platforms dated to more than 100 years ago with primitive design mixes, materials and construction technique. Since then, many concrete platform have been constructed; some with poor quality and some with very reasonable performance. However, generally the concrete platforms were more durable when compared to steel structures but they are slower to construct. In the past 30 years, there is a new wave of concrete platforms riding on advancement of concrete and construction technology. The owners of concrete platforms now demand for durability design of 100 years or more. The construction time is also significantly reduced. In this paper, durability design of floating concrete platforms is discussed, among which is the recent construction of 138 m × 46 m concrete drydock built for *Marisco* Ltd. with a lifting capacity of 9500 tons. As stationary platforms, concrete platforms are more durable, less expensive, safer, and more stable and they require less maintenance as compared to steel platforms. However, structural and durability design as well as construction experience are critical for successful execution.

Keywords Floating concrete platforms · Concrete technology · Durability

1 Introduction

Concrete is the cheapest construction material and the second most consumed man-maid product after drinking water, which is easily cast to any shape. Concrete technology has evolved very rapidly in recent years and continues to improve with new advancement in construction and materials technology.

B. Sabet Divsholi (✉)
Global Engineering Creators Pte Ltd., 53 Ubi Avenue 1 #06-05 Paya Ubi Industrial Park, Singapore, Singapore
e-mail: bsabet@ecreators-global.com

C. M. Wang et al. (eds.), *WCFS2019*, Lecture Notes in Civil Engineering 41,
https://doi.org/10.1007/978-981-13-8743-2_15

The corrosion of reinforcement in concrete is the main cause of deterioration and progressive collapse of concrete structures [1]. The parameters affecting corrosion rate are (1) the critical chloride content which depends on type of reinforcement materials and alkalinity of concrete, (2) concrete cover and permeability of concrete, (3) presence of water, (4) presence of oxygen, (5) environmental factors such as temperature, (6) cracks in the concrete and (7) corrosion cell formation [2]. Construction quality has important influence on some of these parameters and needs to be addressed separately from the intended design.

Concrete has been widely used in construction of ports, jetties and coastal structures due to its proven performance against rapid deterioration. Early concrete platforms dated to more than 100 years ago with primitive design mixes, materials and construction technique. Since then, many concrete platforms have been constructed; some with poor quality and some with very reasonable performance. However, generally the concrete platforms were more durable when compared to steel structures but they are slower to construct. In the past 30 years, there is a new wave of concrete platforms riding on advancement of concrete and construction technology. The owners of concrete platforms now demand for durability design of 100 years or more. The construction time is also significantly reduced. As stationary platforms, concrete platforms are more durable, less expensive, safer, and more stable and require less maintenance as compared to steel platforms. However, structural and durability design as well as construction experience are critical for successful execution.

Despite the large number of concrete platforms and concrete structures in marine environment, there is still a large variation in quality and design requirements of newly built structures. Some are designed and constructed with careful consideration of important parameters while others are constructed with no or little consideration about durability in a marine environment. Beside the parameters affecting corrosion rate, construction joints, construction technique, constriction sequence, design concepts and stresses are essential to design and execute a durable concrete platform.

In this paper, some of these parameters and their effect of durability of concrete will be discussed.

2 Parameters Affecting Durability of Concrete

2.1 Cementitious Materials and Concrete Mix Design

Cementitious materials and concrete mix design, controls permeability and strength of concrete. Water to cementitious ratio (w/c) is probably the most important parameter [3] followed by selection of type of cementitious materials (Ground Granulated Blast-furnace Slag (GGBS), or Silica fume (SF) replacement). The w/c ratio required for full hydration of cementitious materials is about 0.22–0.25 [4]. The excess water acts as capillary pores providing a path for chloride ion ingress to rebar location.

The main hydration products of Ordinary Portland Cement (OPC) is calcium hydroxide and calcium silicate hydrate. The main contribution of GGBS and SF as partial replacement of OPC is to consume calcium hydroxide from OPC reaction and to produce more calcium silicate hydrate. With this, the pore structure of concrete is more refined and chance of drying shrinkage is reduced.

The secondary reactions of GGBS and SF are important; however, if initial w/c ratio is high, this refinement of microstructure is localize and will have little effect on total permeable pores. Our in-house studies showed that an OPC based mix with $w/c = 0.35$ has much better water absorption coefficient due to capillary action [5] than a concrete mix with 10% SF replacement with $w/c = 0.45$.

The specification of mix design is another important consideration. Most specifications are combination of performance based and prescriptive requirements that do not yield a good outcome. For example, some specifications call for Grade 40 MPa concrete with $w/c < 0.4$ and 7% SF. However, in order to achieve Grade 40 MPa with SF, a w/c as high as 0.55 is needed (for Grade 52.5 cement) which has a poor durability performance yet the mix design will achieve the required strength. Therefore, specifying a good mix design with proper tools for quality control is very important.

With a low w/c ratio, the effect of GGBS/ SF replacement is more pronounced resulting in significantly denser concrete with much lower permeability.

2.2 *Cracks in Concrete*

Concrete is brittle in nature and is prone to cracking. The main types of cracks are plastic shrinkage crack, drying shrinkage crack and constraints cracks. These cracks have a detrimental effect on concrete durability [6] and often designers do not have any specification or precaution to limit or eliminate these cracks. The plastic shrinkage cracks are deep cracks often to first layer of reinforcement or sometime through the concrete section happens in first few hours of casting when concrete is in a plastic state. When the rate of evaporation is more than 1 kg m^2/h, plastic shrinkage crack has a high chance of occurring [7] and often designers neglect to define concreting condition and protection requirements to prevent plastic shrinkage cracks. Drying shrinkage cracks are also common and can be mitigated by proper additives [8] or selection of the appropriate aggregate type. Figure 1 shows drying shrinkage cracks on Grade 40 MPa concrete with granite aggregate compared to Grade 40 MPa lightweight concrete after three months of casting. Two modules are identical and cast in a same time. Constraints cracks happen when part of the concrete is cast against old concrete or a joint, which cannot move/ shrink and therefore it will crack. Construction sequence is very important factor to prevent or minimize the constraint cracking. Figure 2 shows restraint cracking on fresh concrete cast in-between two precast members.

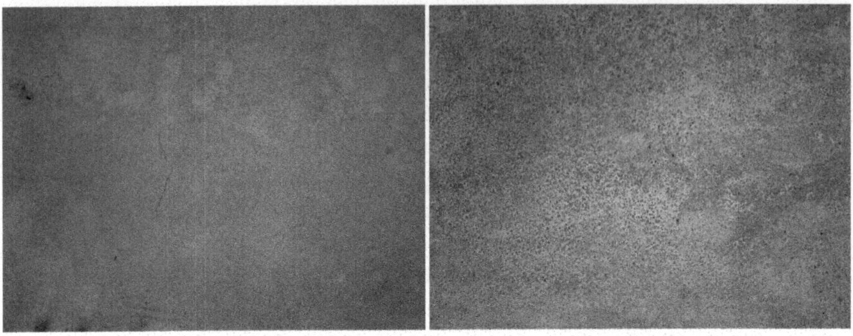

Fig. 1 Drying shrinkage crack on normal Grade 40 MPa concrete with granite aggregates (left), no crack on Grade 40 MPa lightweight concrete (right)

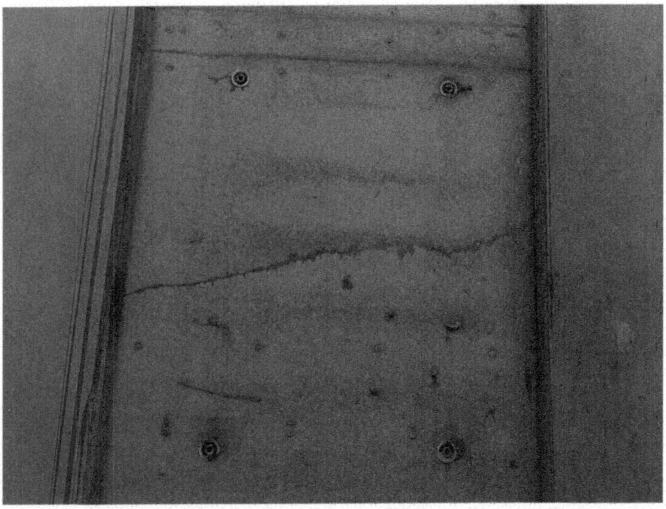

Fig. 2 Restraint cracking on fresh concrete cast in-between two precast members

2.3 Concrete Cover

Concrete cover is one the important factors often not appropriately considered by designers. If the concrete cover is too little or too much, both have a negative effect on concrete durability. A thick layer of unreinforced concrete is easily cracked with small tension.

According to ACI 318-14 [9], concrete cover in reinforced concrete is the least distance between the surface of embedded reinforcement and the outer surface of the concrete. Most designer assume this is cover to main reinforcement surface and often stirrups, ties reinforcement is ignored. Also most designers ignore that reinforcement is threaded and not smooth. Threads for 20 mm nominal reinforcement is 3 mm and for 12 mm reinforcement is 2 mm. Also, there is always gap of 1–2 mm between stirrups/ties and main reinforcement. Let us assume that the main reinforcement is 20 mm and stirrups/ties reinforcement is 12 mm, the designer is proposing 30 mm cover to main reinforcement. The actual executed cover to surface of stirrups/ties reinforcement is 30–1.5 (half of T20 thread)—2 mm (gap between main reinforcement and stirrups/ties reinforcement)—14 (stirrups/ties reinforcement and its thread) = 12.5 mm. For marine concrete structures, the cover to stirrups/ties reinforcement for durability is very important.

In addition, the accuracy of reinforcement placing follows a normal distribution. A typical standard deviation (SD) of about 5 mm is observed. It means 99% of reinforcements have at least design cover minus 2 × SD. For the case above with 30 mm cover, 1% of link reinforcements will have less than 12.5 mm − 10 mm = 2.5 mm of cover. This is unacceptable in marine environment. The ACI code clearly mentioned that concrete cover is measured to the outer edge of stirrups, ties, or spirals if transverse reinforcement encloses main bars. Optimum cover of 50 mm to the outer surface of embedded reinforcement and not the main bar should be considered.

3 Design Concept

Concrete structures are often designed to the ultimate limit state (ULS) which involves excessive cracking and deformation in concrete section. This is not acceptable for marine environment. Marine structures should be designed to limit state of cracking. This limit can be increased by post tensioning and addition of fibers to achieve strength hardening with limited crack propagation.

Design of Floating Concrete Dry-Dock The author was part of a team to build a large floating concrete dry-dock of 138 m × 46 m and a lifting capacity of 9500 tons in Batam Indonesia for Marisco Ltd. GL E&C was the contractor. For this platform, a Grade 85 MPa concrete with superior performance was proposed. According to specification, durability of concrete shall be measured according to ASTM C1202 [10] for its ability to resist chloride ion penetration using the accelerated test. In Rapid Chloride Permeability (RCP) test, the total charge (Coulombs) passes through concrete in 6 h of testing shall be less that the limit specified in Table 1. Alternatively, concrete shall be tested according to NT Built 492 [11] for rapid chloride migration or for its electrical resistivity, by using four points Wenner Probe as described in ACI 222 [12]. Either of limits presented in Table 1 shall be considered.

Figure 3 shows Wenner probe reading of 186.6 kΩ cm beyond the limit specified in Table 1. The in situ strength was also measure using Schmidt rebound hammer as shown in Fig. 4. The values recorded were in range of 93–110 MPa.

For success of this project, many considerations and small details had to be considered. This includes, large slab casting at night (to minimize plastic shrinkage crack), two step quality control which includes retempering of concrete to ensure all ready

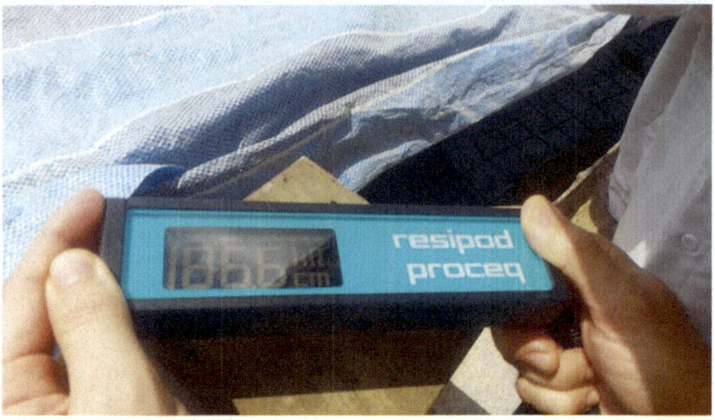

Fig. 3 Wenner Probe reading to measure concrete mix durability

Fig. 4 Testing of in situ concrete strength using Schmidt rebound hammer

Table 1 Limits on rapid chloride migration and electrical resistivity

RCP according to ASTM C1202	Rapid chloride migration test (NT Built 492)	Electrical resistivity using 4 points Wenner Probe as specified by ACI 222
Bellow 300 coulombs	$D < 0.4 \times 10^{-12}$	>160 kΩ cm

Fig. 5 Completed concrete dry-dock 1st May 2017

mix trucks having slump of 650 ± 50 mm, and special considerations for curing and to ensure no cold-joint when there are long delays in concrete arrivals. Many lessons were learned and correction were made during this construction.

The structural framing of this platform is closely spaced honeycomb system, which slabs and main walls are post-tensioned. Post tensioning and close spacing of cells will make this platform work under compression for most loadings, which will reduce cracking. Use of flat slab and avoiding sharp corners will minimize anode-cathode formation, which is important consideration in structural design and well as durability design. Figure 5 shows the completed dry-dock before delivery to owner.

4 Concluding Remarks

In summary, there are many factors from concrete mix design, structural design, construction sequence and construction quality affecting achieved quality of concrete platform and most of these parameters are often ignored or unplanned when designers or builders are not familiar with construction in marine environment. It is important to differentiate between design of concrete in marine environment and design of normal concrete buildings.

References

1. Mehta, P. K. (1986). *Concrete. Structure, properties and materials*. Englewood Cliffs, NJ, USA: Prentice-Hall.
2. Markeset G., & Myrdal, R. (2008). Modelling of reinforcement corrosion in concrete—State of the art, COIN P4 operational service life design SP 4.1 F Service life modelling and prediction COIN Project Report 7.
3. Zivica, V. (2003). Influence of w/c ratio on rate of chloride induced corrosion of steel reinforcement and its dependence on ambient temperature. *Bulletin of Material Science, 26*(5), 471–475.
4. Pang, X. (2015). The effect of water-to-cement ratio on the hydration kinetics of Portland cement at different temperatures. In *Proceedings of the 14th International Congress on Cement Chemistry*. Beijing, China, https://doi.org/10.13140/rg.2.1.4526.2800.
5. BS EN 772-11. (2011). Methods of test for masonry units. Determination of water absorption of aggregate concrete, autoclaved aerated concrete, manufactured stone and natural stone masonry units due to capillary action and the initial rate of water absorption of clay masonry units, BSI. ISBN: 9780580702587.
6. Shaikh, F. U. A. (2018). Effect of cracking on corrosion of steel in concrete. *International Journal of Concrete Structures and Materials, 12*(3). https://doi.org/10.1186/s40069-018-0234-y.
7. Ghourchian, S., et al. (2017). A practical approach for reducing the risk of plastic shrinkage cracking of concrete. *RILEM Technical Letters, 2*, 40–44.
8. Nangy, N., et al. (2013). Crack-free concrete, shrinkage-compensating concrete for the European construction industry. *European Coating Journal*, 11.
9. ACI 318-14. (2014). *Building code requirements for structural concrete* (*ACI 318-14*) *commentary on building code requirements for structural concrete* (*ACI 318R-14*). Reported by ACI Committee 318.
10. ASTM C1202. (2018). *Standard test method for electrical indication of concrete's ability to resist chloride ion penetration*. Developed by Subcommittee: C09.66.
11. NT Built 492. (1999). Concrete, mortar and cement-based repair: chloride migration coefficient from non-steady-state migration experiments. UDC 691.32/691.53/691.54.
12. 222R-01. (2010). *Protection of metals in concrete against corrosion*. ACI Committee 222.

Design and Construction of the Floating Concrete Pier in Golden Harbor, Incheon

Kwanghoe Jung, Sanghyu Lee, Heesung Kim, Yoonho Choi and Sara Kang

Abstract There are two 200 m long floating piers for berthing car ferries in the Golden Harbor, Incheon, South Korea. Each floating pier has almost the same dimensions, but one of these piers is made of steel whereas the other pier is made of concrete. This paper focuses on the design and construction of the floating concrete pier. In the design of the floating steel pier, the 198 m long pier has two inner moorings and comprises three separated steel modules (with each module length L = 66 m). In contrast, the floating concrete pier composed of a single structure (of length L = 200 m) without inner mooring dolphins. This concrete pier design not only maximizes the use of the top-side space but it also minimizes the number of mooring dolphins. The construction method involves fabricating concrete segments on land and assembling them as modules on the seawater. This modular construction method was adopted instead of the conventional method of using a big floating dock because it increases the construction efficiency and enhances the concrete quality of the modules. More detail considerations of this design and construction are explained in this paper.

Keywords Floating pier · Steel · Concrete · Modular construction method · Module

1 Introduction

The Golden harbor is an international ferry terminal in Incheon, South Korea. It consists of new trade complexes and buildings as well as large scale wharfs for berthing cruise and car-ferry ships as shown in Fig. 1. It will be developed as a home port for many cruise ships as well as a trade hub of South-East Asia. This area is famous for a dramatic tidal variation of approximately 10 m because it is located in the western coastline of South Korea. In order to overcome this large tidal variation, several floating structures were applied instead of general gangway systems attached to fixed quay walls. There are two large T-shape floating piers having over

K. Jung (✉) · S. Lee · H. Kim · Y. Choi · S. Kang
Hyundai Engineering and Construction, Seoul, South Korea
e-mail: jkh@hdec.co.kr

C. M. Wang et al. (eds.), *WCFS2019*, Lecture Notes in Civil Engineering 41,
https://doi.org/10.1007/978-981-13-8743-2_16

Fig. 1 Floating piers in Golden harbor, Incheon

200 m length in this port which can move up and down with the tidal levels so that the required freeboard is maintain at any time. These facilities have almost same dimensions and connected to the land by an access bridge (L: 60 m) but they are made from different materials; one pier is made of steel and the other pier is made of concrete. Built later, the concrete floating pier was proposed as an alternative to a floating steel pier so that more innovative and efficient design considerations and construction methods may be brought to this harbor project [1]. This paper presents the innovative design and construction methods of the floating concrete pier in detail.

2 Design Considerations for the Concrete Floating Pier

This section provides the improved design considerations for the concrete floating pier as compared to the floating steel pier.

2.1 Design Concept

Figure 2 shows the design concept for the two T-shape floating piers used for both sides berthing of 30,000 ton class car-ferry ships (L: 196 m × W: 27 m) having a 6.7 m draft. Basically, each floating pier has floating bodies composed of a sub-module (L: 200 m × W: 30 m) and a main module (L: 85 m × W: 30 m), and several mooring dolphins as a station keeping system. However, the number of dolphins and sub-modules are different. The conventional concept for the floating steel pier comprises

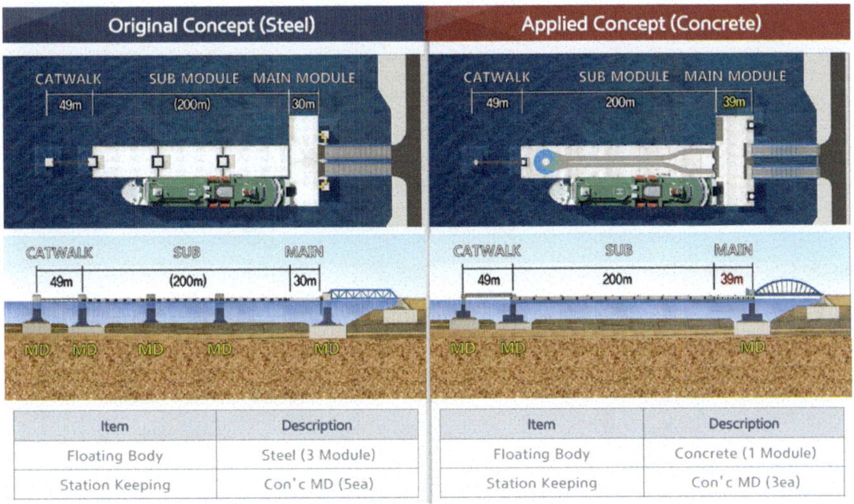

Fig. 2 Design concepts for floating piers

three sub module parts (L: 66 m × 3 ea) and six mooring dolphins including two inner dolphins as shown in Fig. 2. In contrast, the concrete floating pier has only one sub-module (L: 200 m × 1 ea) without the two inner dolphins in order to fully utilize the topside space. So, the floating concrete pier design has a total of four mooring dolphins, and the width of a main module can be extended to 39 m because it has no gaps between sub-module parts in the case of the floating steel pier. To provide a suitable freeboard of about 2 m for berthing a car-ferry ship, the floating steel and concrete piers are required to have 2.9 and 5.0 m of total height, respectively. This is because the floating concrete pier has heavier self-weight and high stiffness as compared to the floating steel pier [1]. Finally, it can be summarized that the main design concept of the floating concrete pier is not only to maximize the use of topside space but it is also to minimize the number of mooring dolphins.

Figure 3 presents operational plans for these floating piers with respect to tidal levels. They are connected to the land by access bridges (L: 60 m) having hinge supports on the land and roller supports on the main module. When the tidal level goes up to the maximum level, the sub and main modules move up together and access bridges reach the flat level with respect to the modules. On the other hand, when the tidal level goes down to the minimum level, the sub and main modules move down together and access bridges are inclined as shown in Fig. 3.

Figure 4 presents the applied systems in this design for free up-down movement. In order to safely go up and down according to the tidal levels, special devices such as rubber rollers, frictionless pads can reduce the friction between dolphins and modules are considered. In case of the floating concrete pier, frictionless pads are attached on the surfaces of the concrete dolphins as shown in Fig. 4a. The roller systems of access bridges go forward and backward with the up-down movement of floating

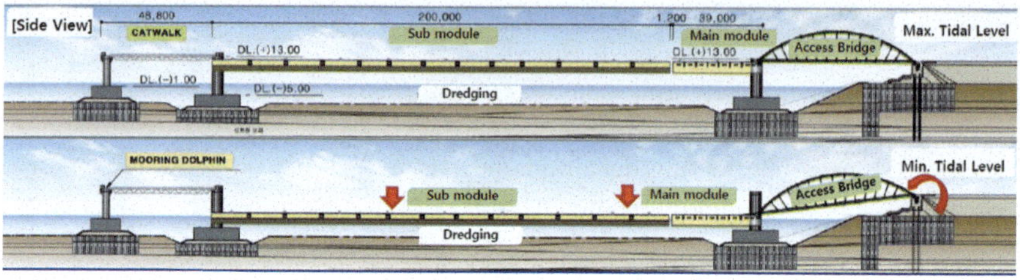

Fig. 3 Operational plans by tidal levels

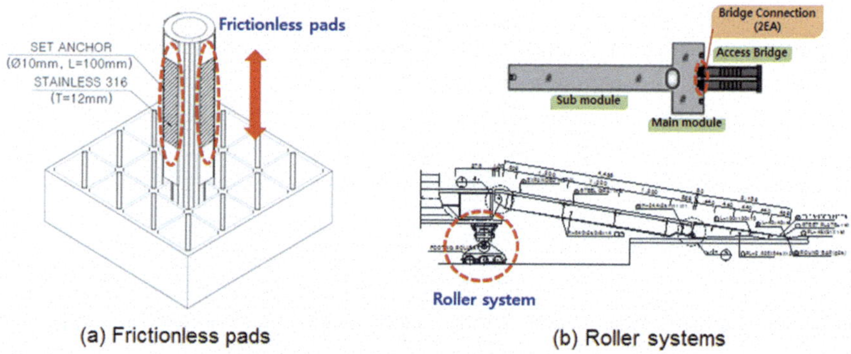

Fig. 4 Systems for free up-down movement

pier as shown in Fig. 4b. At the same time, the hinge systems on the land are rotated so that whole floating pier can be safely inclined.

2.2 *Environmental Conditions and Design Loads*

The environmental conditions such as wave, current and wind can be directly related to the stability and safety of a floating structure. Table 1 shows the environmental conditions applied in this project. The wave condition (height: 1.5 m, period: 4.5 s, 65° with sub module direction) based on 100 year return period and the current velocity (0.9 m/s) are presented as a calm sea condition because these floating piers are located in the inner harbor surrounded by breakwaters. However, there is a big tidal variation of 9.27 m in this area which is the water level gap between the highest high water level of (+) 9.27 m and the lowest low water level of 0.0 m. The water depth is 18.3 m based the mean sea level of (+) 4.64 m. It is found that the wind velocities at mooring and non-mooring states are about 32.8 and 55 m/s, respectively.

Figure 5 presents all design loads applied in this project, which can be classified into environmental loads, hydraulic loads, payloads and mooring loads [2]. First, the environmental loads caused by above wave, current, and tidal conditions (Table 1)

Table 1 Environmental conditions

Item	Quantity
Wave height, *H*	1.5 m
Wave period, *T*	4.5 s
Wave direction	65°
Current velocity	0.9 m/s
Wind velocity	32.8–55.0 m/s
Tidal variation	9.27 m
M. S. L.	(+) 4.64 m
H. H. W.	(+) 9.27 m
L. L. W	0.00 m
Water depth	18.3 m

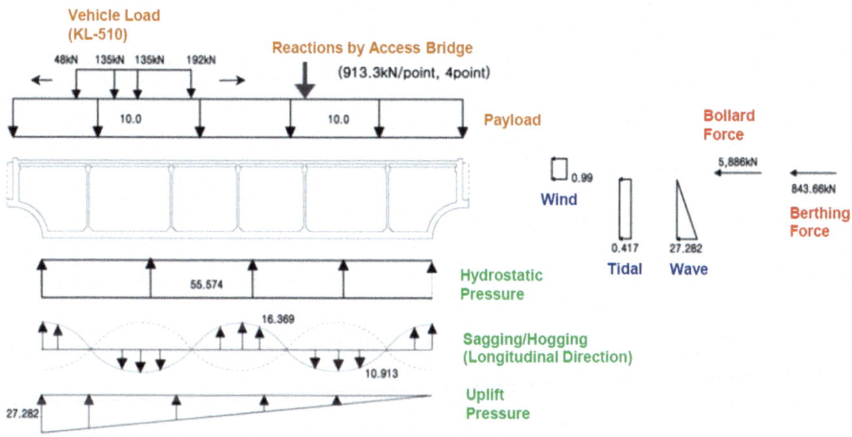

Fig. 5 Design loads for concrete floating pier

are applied to the side face of this floating concrete structure. Second, the hydraulic loads such as hydrostatic pressure, uplift pressure occurred on the bottom surface and sagging/hogging forces in the longitudinal direction are considered. Third, on the topside of the floating structure, vehicle loads according to the Korean standard truck (KL-510), reaction forces by access bridges as well as uniform payloads are applied. Finally, bollard forces and berthing forces by 30,000 ton class car-ferry ships are regarded as essential design loads. Based on these design loads, all possible load combinations in the ultimate and service limit states were considered [2].

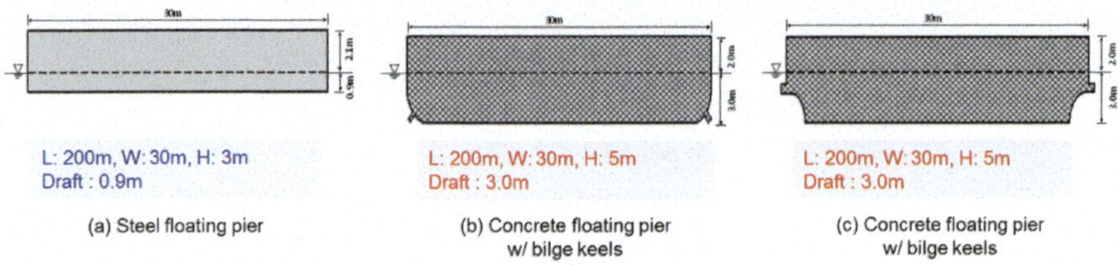

Fig. 6 Considered Cross Sectional Shapes

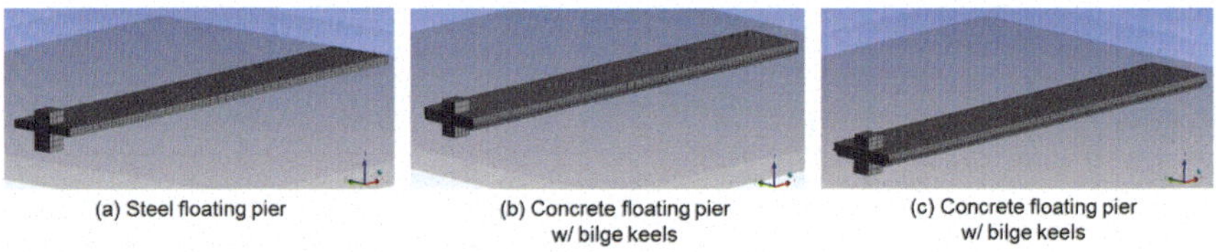

Fig. 7 Hydrodynamic analysis model of each cross section

2.3 *Optimization of the Cross Section*

Basically, the steel and concrete floating piers have a rectangular cross section like a general pontoon, but the heights are different. The heights of steel and concrete cross sections are 3 and 5 m, respectively, so as to maintain the required free board of about 2 m.

Among the six motions of the floating body, it was expected that the heave and roll motions can directly affect the entire stability of this long floating structure [3, 4]. In order to acquire more stable conditions at the service stage, the cross section of the floating concrete pier was optimized. The typical shapes such as bilge keels and semi-circular section below the sea are examined by hydrodynamic analyses to reduce the roll motion of this floating structure.

Figure 6 shows the cross sections considered herein. Figure 6(a) presents the cross section of the floating steel pier. Figure 6b and c present the cross sections of concrete floating pier. Figure 6b shows the pier having bilge keels while Fig. 6c shows the pier having a semi-circular cross section with bilge keels.

In order to compare the roll motions of these cross sections, hydrodynamic analyses were performed by using the commercial software ANSYS AQWA. Figure 7 shows the hydrodynamic model of each cross section. By applying design wave and wind conditions as shown in Table 1, all motions of floating body were examined. The roll and heave are directly related to whole stability of this floating structure.

Figure 8 presents the rolling ranges of three cases. It is clear that the rolling angle of Fig. 8a (−2.86°–2.60°) is relatively larger than that of Fig. 8b (−2.27°–2.44°) and Fig. 8c (−1.96°–2.13°), which implies that the floating concrete pier is more stable than the floating steel pier. It is clear that the rolling angle of Fig. 8b (−2.27°–2.44°) is relatively larger than that of Fig. 8c (−1.96°–2.13°), which means that the semi-circular cross section with bilge keels below the sea is helpful to reduce the roll motion

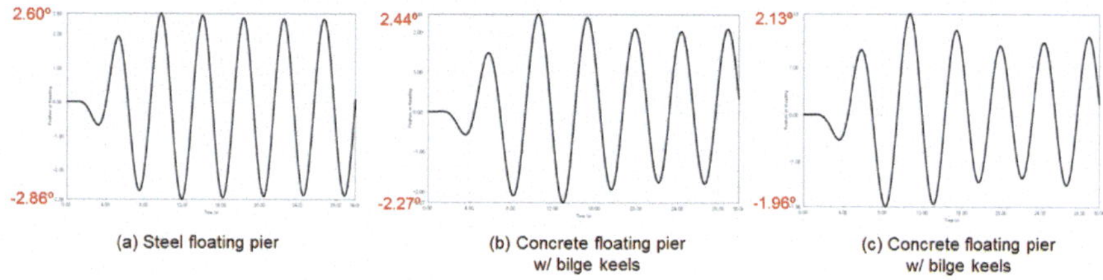

Fig. 8 Roll motion of each cross section

Table 2 Material properties

Material	Specification	Note
Concrete	50 MPa	Cylinder strength
Reinforcement	SD400	fy = 400 MPa
Prestressing tendon	SWPC 7B, Ø15.2 mm	fpu = 1900 MPa, fpy = 1600 MPa
Prestressing bars	SBPR 785/1030, Ø40 mm	fpu = 1900 MPa

of the floating concrete pier [3, 4]. Finally, Fig. 6c was selected as the optimal cross section for the design.

2.4 *Material Properties and Structural Design*

Table 2 shows the material properties used in this structure. In order to guarantee the 50 year design life in the marine environment condition, all concrete cover depth exposed to seawater is basically 100 mm and high strength concrete of 50 MPa is used. In order to meet the design assumption that non-cracking section should be maintained for service limit state [5], longitudinal prestressing tendons (SWPC 7B, Ø15.2 mm) are applied in this design. The prestressing bars (SBPR 785/1030, Ø40 mm) are used for segment connections.

Structurally, the prestressing tendons must be enough to resist to external bending moment in the ultimate limit state. Figure 9 shows the analysis model and results for this floating structure using commercial software MIDAS. The whole structure is modelled as frame elements with cross section (see Fig. 6c) and subjected to all external design loads as shown in Fig. 5. In the structural analysis results, the maximum and minimum bending moments in the ultimate limit state are 113,536 kN-m and −155,690 kN, respectively. To resist these ultimate moment, a set of 12EA tendons (Ø15.2 mm) are arranged at 1.9 m spacing on the upper and lower slabs.

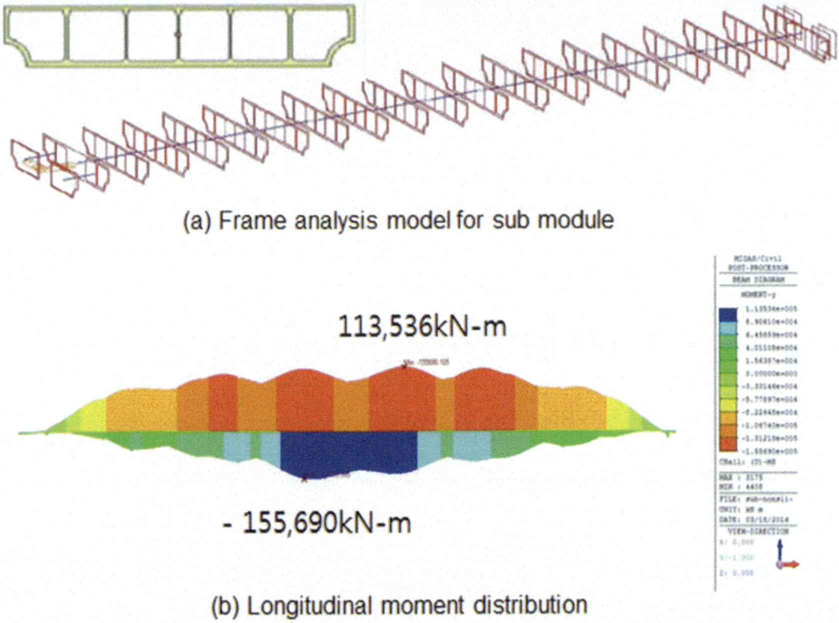

Fig. 9 Structural model and analysis results

3 Construction Method for Floating Concrete Pier

3.1 Conventional Construction Method

It was a big issue on how to efficiently fabricate and safely install this huge floating concrete structure. It was estimated that self-weights of main and sub modules are approximately 18,000 and 9000 ton, respectively. Conventionally, the construction method is to use a large floating dock having 50,000 ton class capacity (W: 51.5 m × L: 298 m). Figure 10 shows the construction steps of this method. The 200 m long sub module is composed of 5 segments (L: 40 m) because concrete works cannot be done in one time. In the first step, Segment 1 is initially fabricated on this floating dock, and then two segments (Segments 2 and 3) at both sides of Segment 1 are also fabricated together. For the second step, these three segments are longitudinally prestressed (1st prestressing). In the third step, last segments (Segments 4 and 5) at both sides of Segments 2 and 3 are fabricated and the five segments are longitudinally prestressed (2nd prestressing). In the fourth step, the concrete sub module is launched using this floating dock on the sea and installed at the destination. If this unique equipment, a large floating dock, was available at the required period, it would be regarded as a reasonable and safe construction method.

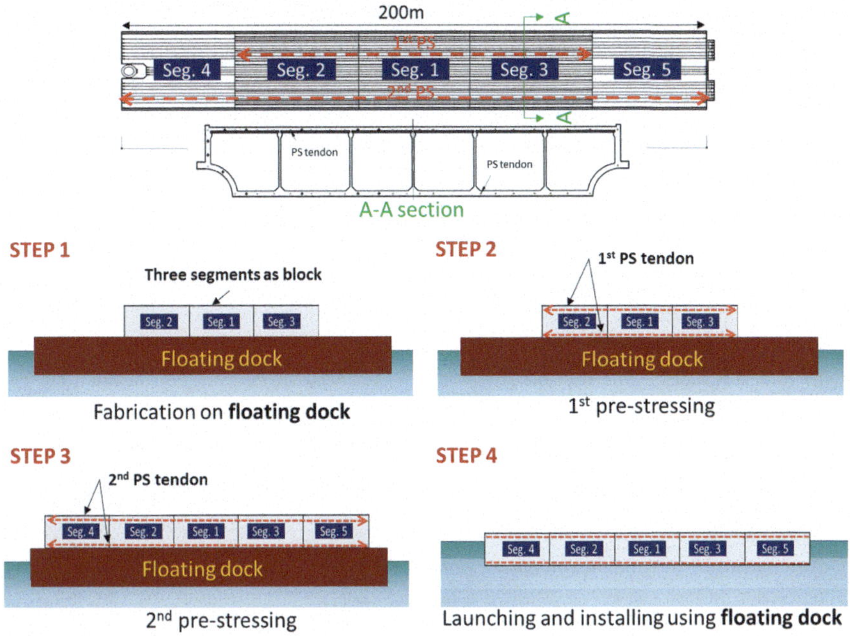

Fig. 10 Conventional construction method using a large floating dock

3.2 *Modular Construction Method*

In the real construction, the conventional construction method was invalid because the 50,000 ton class floating dock was not available during the required period and concrete quality control on that was not easy in the restricted working area. So this modular construction method using a floating crane having 3600 ton capacity was developed and applied in this project. In this modular construction method, all segments were fabricated on the land, lifted by this floating crane, connected on the sea, and installed by tug boats. Figure 11 shows the construction steps of this modular construction method. The 200 m long sub module is composed of 6 segments (L: 33 m) because the dimensions and weight of each segment should be in accordance with the floating crane capacity. In the first step, each segment was fabricated on the land at the same time and respectively prestressed. Each segment has connection holes for PS bars as well as longitudinal tendons to assemble with neighbor segments. In the second step, all segments are lifted and launched by the floating crane one by one. In the third step, all segments were connected to one another by PS bars on the sea to make a 200 m long sub module. In the fourth step, this concrete sub module was towed and installed by tug boats. Although segment connections on the sea are not easy, it was used because it can reduce the construction time and enhance concrete quality.

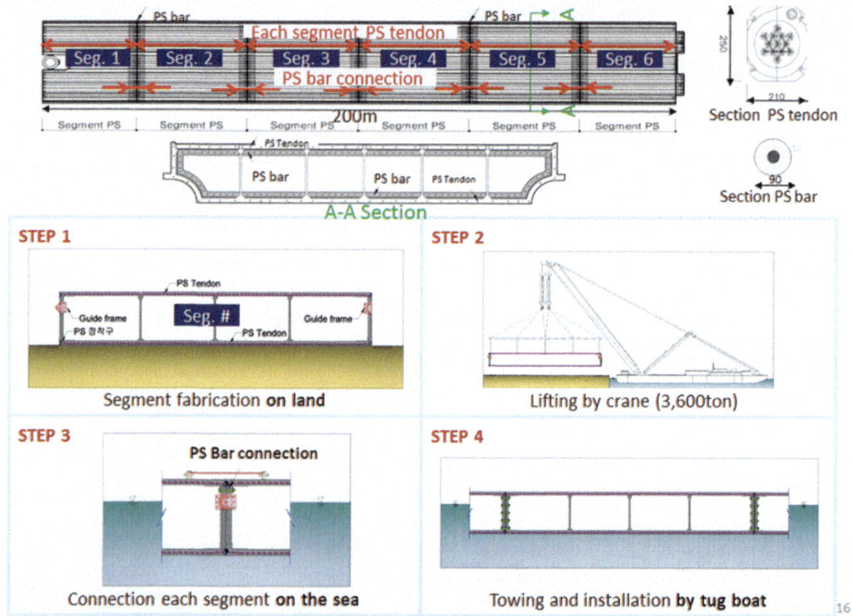

Fig. 11 Modular construction method using a floating crane

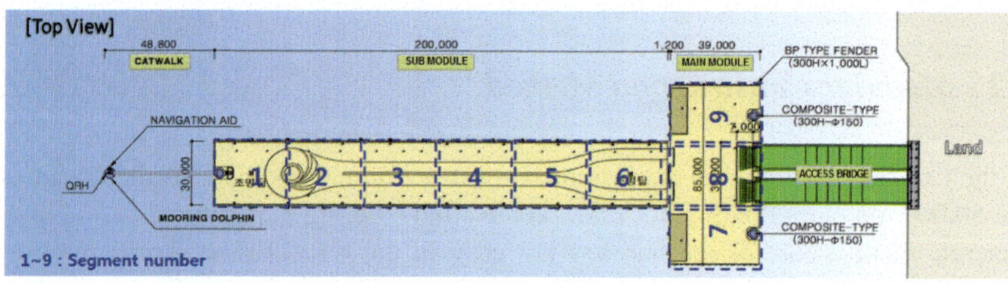

Fig. 12 Segmental plan for modular construction method

3.3 Module Installation of Modular Construction Method

Figure 12 shows the plan view of the floating concrete pier composed of main and sub modules, access bridges, and concrete mooring dolphins. By applying the modular construction method, the main module comprises three segments and the sub module comprises six segments in view of the capacity limit of the floating crane which is 3600 ton, for lifting and launching. So the weight of each segment is approximately 3200 ton to fully utilize this lifting capacity.

Figure 13 presents the module installation procedure of this modular construction. First, nine segments (main module: M1, M2, M3, sub module: S1, S2, S3, S4, S5, S6) are fabricated on the land at the same time. Second, three concrete mooring dolphins

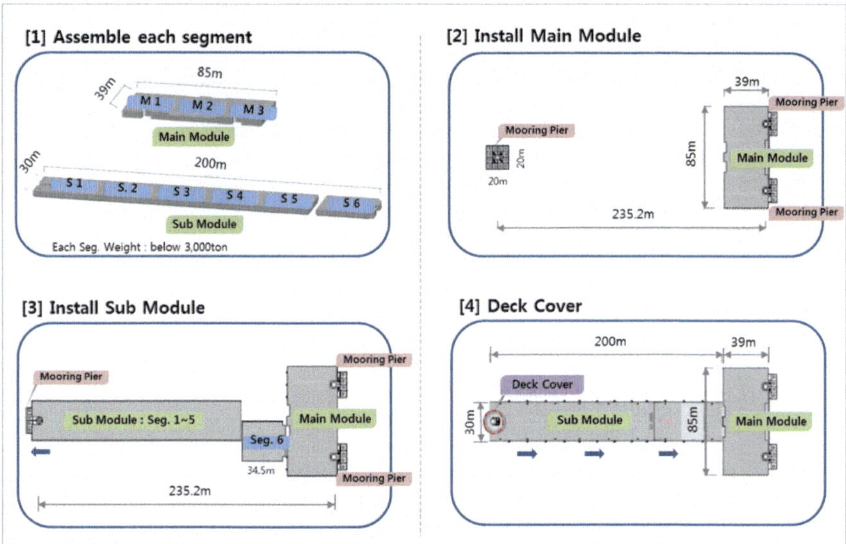

Fig. 13 Module installation procedure of modular construction

(a) Main module installation

(b) Sub module installation

Fig. 14 Main and sub module installation

are firstly installed at the fixed positions. Then the three segments of the main module are installed one by one. Third, five segments (S1–S5) among six segments of sub module are assembled together and then set backward to make space for insertion of the last segment of S6. Finally, sub module sets forward and connected to main module after S6 segment is inserted and connected. The deck cover plate is placed to cover the gap between mooring dolphin and segment S1.

Figure 14 shows photos taken by a drone during main and sub module installations. The photos show the fabrication yard for making segments and a floating crane in the project site. Figure 15 shows the main module connections. When one segment (M1) of main module is installed at the mooring dolphin and hanging by the floating crane through the lifting device, another segment (M2) is towed by tug boats at the proper position to connect each other.

Fig. 15 Main module connections

3.4 Ballasting Plan for Main and Sub Modules

The weight of each segment is approximately 3200 ton. However, the weight distribution and draft of each segment are not equal because the external shape and the inner diaphragm arrangement are different for each segment. Owing to this reason, a careful ballasting plan is used to achieve the desired draft when all segments are connected on the sea.

Figure 16 shows the ballasting plan for main and sub modules. The plan has two stages: the temporary ballasting by water filling before segment connection and the permanent ballasting by concrete filling after segment connection. The former is to maintain flatness of segment itself in order to assemble segments together using ballast water. The latter is to connect main and sub modules as the final state using concrete. Before connecting segments, the drafts for main and sub modules are 2.82 and 3.46 m, respectively. Finally, the final draft is 3.15 m after main module and sub module connection.

3.5 Construction Method Comparison

The construction method for this large floating structure was a critical issue, because it can directly affect the construction period and cost. Table 3 presents the comparison of two construction methods (conventional versus modular). In view of equipment use, the conventional method is required to use a big floating dock (50,000 ton) for six months as all segments are fabricated on it at one time, but the modular method is required to bring a floating crane (3600 ton) for only one month because all segments are made on fabrication yard utilizing project site. In view of workability for segment fabrication and concrete quality control, it can be said that the modular method is better than the conventional method because all segments are made on land at the same time and it can provide enough space for workers and fabrication devices. However, in view of the works on the sea, it can be explained that the conventional

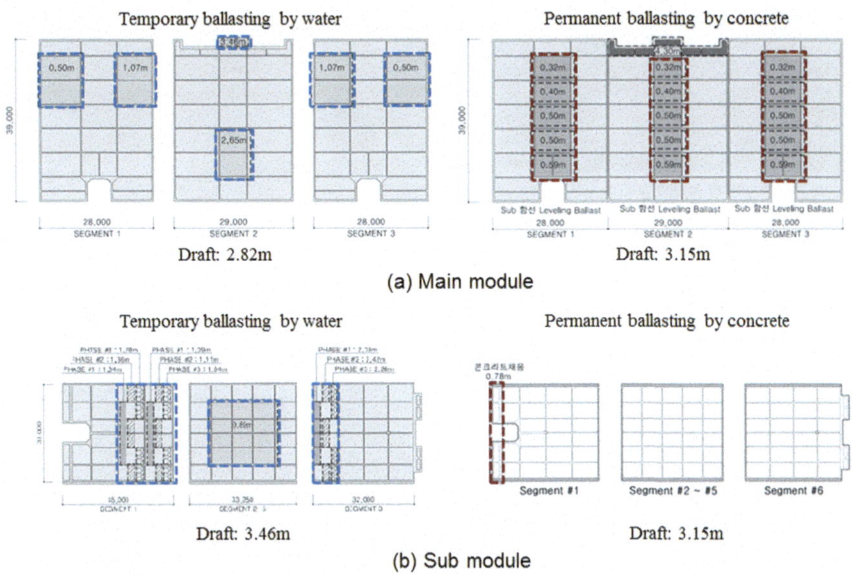

Fig. 16 Ballasting plan

Table 3 Comparison of construction methods

Index	Conventional method	Modular method
Equipment	Floating dock (50,000 ton)	Floating crane (3600 ton)
Period of use	6 months	1 month
Fabrication	On floating dock	On land
Launching	Floating dock	Floating crane
Concrete quality	Hard to control	Easy to control
Workability	Less stable	More stable
Temporary ballasting	X	O
Permanent ballasting	O	O
Segment connection	X	PS bars

method is better than the modular method because it can reduce sea works such as ballasting and segment connections by prestressing bars.

Finally, it is believed that the modular construction method is better than the conventional construction method judging from the fabrication efficiency and concrete quality control as well as construction cost and period [6]. However, it requires very careful construction control to safely connect segments on the sea. To enhance the degree of construction precision for sea woks, 3D simulations using photo geometry data taken by drone were performed before lifting and launching segments. All dimensions and hole locations of each segment were checked by laser scanning and verified by BIM model before connecting segments to enhance efficiency and con-

(a) Installing simulation (b) Laser scanning (c) BIM model

Fig. 17 3D digital construction control

Fig. 18 Floating Piers at Incheon harbor (completion)

struction control as shown in Fig. 17. By using these digital technologies, the modular construction method was successfully applied in this project. Figure 18 shows the 200 m long floating concrete piers at Incheon harbor completed at the end of year 2018.

4 Concluding Remarks

This paper presents the design considerations and construction methods for the 200 m long floating concrete pier in Incheon. The concrete pier design not only maximizes the use of topside space but it also minimizes the number of mooring dolphins because only one concrete structure (L: 200 m) without inner mooring dolphins was proposed as an alternative to the steel floating pier design that has two inner moorings and three separated parts (L: 66 m × 3 ea). In order to acquire more stable conditions at the service stage, the cross section of this floating concrete pier was optimized as a kind of semi-circular section with bilge keels below the sea based on hydrodynamic analysis results.

Compared to the conventional construction method that uses a large floating dock, the adopted modular construction method involves fabricating concrete segments on land and assembling them as modules on the seawater. The construction method not only increases the construction efficiency but it also enhances the concrete quality of modules. To enhance the degree of construction precision, before lifting and launching segments 3D simulations using photo geometry data taken by drone were performed. Before connecting segments, all dimensions and hole locations of each segment were checked by laser scanning and verified by BIM model. Owing to these digital technologies, the modular construction method was successfully and safely applied in this project.

References

1. Hyundai Engineering and Construction. (2016). *Design report on Incheon port passenger terminal*. South Korea: Incheon Port Authority Institute.
2. Jung, M. H. (2014). *Harbor design standard*. South Korea: Korea Port and Harbor Association.
3. Kang, S., Wang, X., & Jung, K. H. (2016). Motion reduced floating response of very large structure under wave loads. In *Proceedings of Korean Concrete Institute Conference*. Seoul, Korea.
4. Kang, S., Wang, X., & Jung, K. H. (2017). Study of pontoon-type very large floating structure (VLFS) supported with dolphin mooring system. In *Proceedings of Singapore Maritime Technology Conference* (*SMTC2017*). Singapore, December 2017.
5. Lee, J. H. (2012). *Concrete structure standard*. South Korea: Korea Concrete Institute.
6. Kang, S., Jung, K. H., Lee, S. H., Kim, H. S., Choi, Y. H. (2018). Modular construction method for concrete floating quay at Incheon harbor. In *Proceedings of the 25th Australian Conference on Mechanics of Structures and Materials* (*ACMSM25*). Brisbane, Australia, December 2018.

Potential of Floating Urban Development for Coastal Cities: Analysis of Flood Risk and Population Growth

B. Dal Bo Zanon, B. Roeffen, K. M. Czapiewska and R. E. de Graaf-van Dinther

Abstract Population growth and urbanization mainly take place in vulnerable coastal areas. This article presents a global overview of these areas with both rapid population growth and high flood risk, in order to identify coastal areas that could benefit most from floating urban development. The analysis focuses on port cities, since they are coastal cities that have both availability of locations and the required expertise (e.g. maritime industry and services) to enable floating developments. After identifying the most promising locations, an implementation strategy is discussed, which favours areas where floating projects are already present to start testing medium and large-scale concepts. Next, a large scale floating maritime spatial project is presented, which integrates urban and ecosystem development with food and energy production in the North Sea. This plan provides a spatial concept for floating urban expansion in front of the coast of the Netherlands.

Keywords Floating urban development · Vulnerability · Port cities · Flood risk · Population growth · Land scarcity

1 Introduction

Rapid growth of coastal cities in vulnerable delta areas introduces many challenges such as increasing flood risk and reduced land and resources availability. In the last decade there has been a rising attention for water management approaches which are not only focused on optimizing the current urban water system, but instead seek to deal with multiple, integrated challenges by establishing an entirely new model of urban development. Examples are Cities of the Future [16], Water Sensitive Urban Design [25] but also Floating Urban Development [7, 14, 24] and Floating Productive Developments [6]. These are developments that are based on floating foundations and

B. Dal Bo Zanon · B. Roeffen · K. M. Czapiewska (✉) · R. E. de Graaf-van Dinther
Blue21, Delft, The Netherlands
e-mail: karina@blue21.nl

R. E. de Graaf-van Dinther
Rotterdam University of Applied Sciences, Rotterdam, The Netherlands

C. M. Wang et al. (eds.), *WCFS2019*, Lecture Notes in Civil Engineering 41,
https://doi.org/10.1007/978-981-13-8743-2_17

Fig. 1 Three examples of floating development concepts, left to right: BlueRevolution [4]; Green Float [18] ; Seasteading Implementation Plan [8]

can adapt to changes in the water level autonomously. Since most cities are located in coastal delta areas that are threatened by sea level rise, floating developments (e.g. Fig. 1) are gaining more interest. In particular for port cities, floating urban development could be an interesting alternative for land-based urban expansion and urban renewal on land. The main reasons are their proximity to water, the availability of sheltered water surfaces, economic activity and maritime expertise. Additionally, the abundance of waste heat, CO_2, and nutrients could provide a source for floating production of food and biofuels. Floating algae production could contribute to achieve a circular urban metabolism at city level, addressing both global and local land scarcity [6]. Earlier research [17] on global land scarcity in 2050 estimated a shortage between 13 million and 36 million km^2 and indicated floating production of food and energy as well as floating urban developments as a potential strategy to address the land scarcity issue.

This paper aims to investigate which port cities could benefit most from floating developments. Two criteria are analysed for this purpose. These are (1) the expected flood risk and (2) the expected population growth. The hypothesis explored herein is that cities that are facing rapid urban expansion, are characterised by high flood risk and require large amounts of space for accommodating their population growth, would be areas with high potential for floating developments. Finally, an implementation strategy and an innovative large scale floating urban project is presented, based on the innovation agendas and policy documents of the Topsectors of the Dutch Economy. These are sectors selected by the Dutch government. Top sectors are areas in which the Dutch business community and research centres excel globally. The business community, universities, research centres and the government work together on knowledge and innovation, internationalization, human capital and on reducing regulatory pressure to make this position even stronger [21].

2 Methods

Two criteria were used to identify coastal cities with high potential for floating urban development. The first was flood vulnerability. Cities that show a high flood vulnerability could benefit from floating urban developments as these developments can adapt to water level changes. The second was urban growth. Rapidly growing cities need more space which is increasingly lacking on land but possibly available on the water through developing floating systems.

The evaluation of the most interesting port cities is done performing a Multi-Criteria Analysis (MCA). Since data used in the MCA often have different unit of dimensions (or no dimension), a rational method is required to compare data across multiple criteria. A way to standardize (or normalize) data in Multi-Criteria Analysis is provided by the z-score method. In statistics, the z-score (also referred to as the 'standard score') is obtained by subtracting the data population mean from an individual raw score, and then dividing the difference by the standard deviation. The advantage of using this method is that it manages to preserve extremes, while the overall dataset is normalized. The spread of values is also captured by the z-score method [13].

For each set of data that is used to quantify certain criteria, the z-score is calculated. Each z-score is then summed in a combined value. The combined values are finally ranked to determine a list of most interesting cities. The z-score method allows for a comparison of different datasets in the MCA. If some criteria are considered more important than others, weighting could be added. Next to a list of cities that could benefit from floating development, a case study is introduced. The BlueRevolution case study shows how floating development may be applied in the Dutch context, integrating the policies and innovation agendas of the Dutch Topsectors.

2.1 Data Sources

The analysis of Hallegatte et al. [11] was used as a starting point. This study investigates the 136 coastal port cities with more than a million inhabitants in their agglomeration in 2005 [12]. The study of Hallegate et al. gives estimations on the mean annual loss (M$) that is expected in these cities in 2050. The expected loss is based on socio-economic changes, climate change, land subsidence and adaptation measures. For this study, the scenario with 20 cm sea level rise, land subsidence and adaptation measures with constant probability was chosen. For the global analysis of rapidly population growth, data from the United Nations [23] on cities larger than 300,000 inhabitants were used. From this data, the 136 coastal port cities were extracted. Data included historical growth figures from 1950 onwards as well as projections of population growth up to 2030. The ranking of the cities was based on the expected population growth in the period 2015–2030.

2.2 BlueRevolution Case Study

The concept of the BlueRevolution was originally coined by Takahashi [19] to develop a pro-active plan for ocean resource development. DeltaSync/Blue21 expanded on this philosophy by proposing floating cities with a positive impact on the planet, including local production of food and energy and ecosystem development [7]. The concept was proposed as a perspective to deal with global challenges such as sea level rise, urban growth and global land scarcity. The Netherlands seems to be a very interesting application area for the BlueRevolution concept. It is a low-lying densely populated country, it is located partly under the sea level and it has several floating projects already implemented. Therefore, the Netherlands, and more broadly the North Sea, could be an ideal demonstration and testing location.

For the spatial concept of the BlueRevolution case study, a literature survey was done of policy documents and innovation agendas of the nine Dutch economic Topsectors [22]. Moreover, key stakeholders of the Topsector Water were involved in the analysis. From these sources the most relevant initiatives, plans and projects were extracted and incorporated in the spatial context using the research by design method [2].

3 Results and Discussion

The first screening results mapped in Fig. 2 provide an indication where floating urban developments could contribute to reduce the increase in flood risk and accommodate population growth. For rapidly growing cities in South East Asia in particular, floating development seems to have potential.

The results in Table 1 show the top 25 cities with most potential for floating development, according to the two criteria. Most cities (64%) are located in Asia. China has 5 cities in the top 25. India follows with 4 cities and the USA with 3 cities. Five cities in the list are from Africa and only 1 is from South America. There are no cities from Europe or Oceania in the top 25.

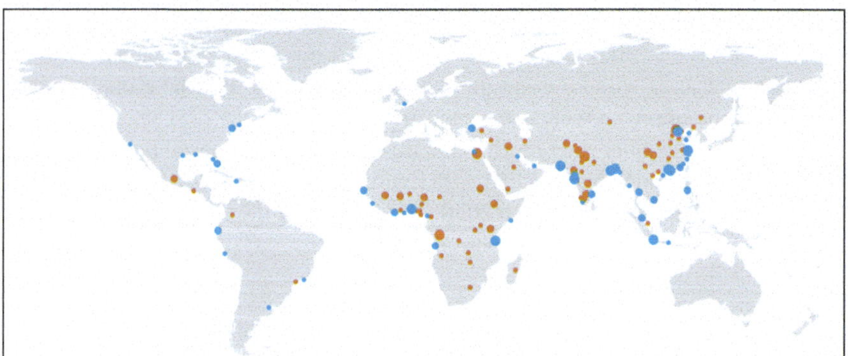

Fig. 2 First screening: rapidly growing cities with high flood risk are indicated in blue, rapidly growing cities without high flood risk are indicated in orange. Own graph based on population and flood risk data [11, 23, 1]

Table 1 Ranking of port cities with most potential for floating developments

Rank	Urban agglomeration	Country	Mean annual flood risk, z-score	Relative population increase, z-score	Combined score
1	Guangzhou	China	9.26	2.09	11.36
2	Mumbai	India	4.34	2.98	7.31
3	Lagos	Nigeria	−0.08	5.34	5.25
4	Dhaka	Bangladesh	0.05	4.61	4.66
5	Kolkata	India	2.11	1.61	3.72
6	Karachi	Pakistan	−0.22	3.77	3.55
7	Shanghai	China	−0.25	3.12	2.86
8	Tianjin	China	1.33	1.19	2.52
9	Shenzhen	China	1.96	0.37	2.32
10	Jakarta	Indonesia	0.95	1.21	2.16
11	Dar es Salaam	Tanzania	−0.28	2.38	2.10
12	Ho-Chí-Minh City	Vietnam	1.10	0.90	1.99
13	Chennai	India	0.36	1.51	1.87
14	Guayaquil	Ecuador	1.99	−0.25	1.75
15	Luanda	Angola	−0.32	1.99	1.67
16	Abidjan	Ivory Coast	0.54	0.90	1.44
17	Manila	Philippines	−0.08	1.39	1.30
18	Surat	India	0.35	0.93	1.28
19	Miami	USA	1.53	−0.27	1.26
20	New York	USA	1.17	0.03	1.20
21	Xiamen	China	0.21	0.67	0.88
22	Bangkok	Thailand	0.21	0.55	0.76
23	Kuala Lumpur	Malaysia	−0.28	0.73	0.45
24	New Orleans	USA	1.03	−0.60	0.43
25	Dakar	Senegal	−0.29	0.69	0.40

The results are not surprising, since most of the rapidly growing cities in vulnerable lowland areas are located in Asia. Including different criteria and perhaps also different weights of these criteria in the analysis, would probably lead to a different list of cities, but the overall picture of geographical distribution of cities with most potential would probably be similar. The presented results should be considered an exploratory analysis. Additional research is needed to further substantiate these findings and critically re-assess them.

4 Implementation

The list of cities included in Table 1 provides an overview of the port cities with most potential for floating development. Although such cities experience considerable flood risk and population growth, in many of these areas floating projects are not yet implemented. In order to start implementing medium and large scale floating concepts, it is probably best to look at the areas where most floating projects are already realized. The abundance of floating projects in a city is an indication of expertise in water and maritime sectors and receptivity to such technology and type of development. When looking at where most floating projects are located, the Netherlands is one of the global frontrunners, with the cities of Rotterdam and Amsterdam. While no Dutch city was listed in the top 25 cities with most potential for floating developments, the Dutch have a long history of adapting to the water. The Netherlands is located in one of the most densely populated delta areas in the world. Almost half of the county is below sea-level. Many of the problems that are present in the Dutch delta are found in delta areas all over the world. A possible implementation strategy could consists of two stages. As a first step: medium and large scale floating concepts could be implemented and further developed in the Netherlands. After that, they could be applied in the locations with most potential that were identified in Table 1.

5 Case Study: Blue Revolution North Sea

In the Netherlands, floating urban development has gained much interest in the past two decades as alternative option for climate adaptation. Already a considerable number of floating projects were realized. Examples are floating neighbourhoods such as Steigereiland in IJburg, Amsterdam (2011) and iconic projects such as the Rotterdam Floating Pavilion (2010). In 2018, more floating neighbourhoods in Woerden, Amsterdam and Rotterdam are completed, have started or are under development.

More recently, floating urban development became part of the 'Blue Route' in the National Science Agenda [15]. A large research project Space@Sea was started as part of the Horizon 2020 innovation and research agenda of the European Commission. The Dutch Topsector Water embraced the concept of floating urban development [20]. The water sector has been designated Topsector by the government because of the strong position of the Dutch in water management and the opportunities that water presents for economic growth. The Topsector aims to facilitate collaboration between the government, companies and researchers to stimulate innovation (Fig. 3).

An iconic large-scale floating project concept was developed by Blue21, in cooperation with Topsector Water. The objective of the Icon project was to develop an integrative concept for the 9 Topsectors of the Dutch Economy by creating a floating urban development with a positive impact and a circular metabolism of nutrients and CO_2. The spatial concept was based on the BlueRevolution concept [3]. Moreover,

Fig. 3 The Floating Pavilion in Rotterdam, the first step for the city to create floating neighbourhoods in the Port of Rotterdam [9]

the project should function as "Icon project" overarching and uniting the different Topsectors of the Dutch economy. Existing initiatives and projects were mapped and an exploratory design was made how floating developments could be added to this. For this purpose, 4 integrative themes were formulated:

1. Netherlands 100% CO_2 Neutral
2. Netherlands 100% Bio Based and Circular
3. Resilient Urban Delta
4. Smart Floating City and Main port Logistics.

For each theme, innovative existing and new projects and initiatives were collected and included in the design as part of an integrative plan. The plan included both floating functions and areas developed by using traditional land reclamation. Table 2 shows an inventory of innovative projects and initiatives formulated for each theme. These projects were included in the plan. The water-based development on the North Sea uses the waste products of land-based cities such as wastewater, industrial CO_2 and waste heat in a productive way by growing floating aquatic biomass such as algae and seaweed. As a result, a circular metabolism and symbiosis between cities on land and water is created. Figure 4 shows how the different initiatives and projects were integrated in a spatial plan for the Dutch coast. The plan answers to many urgent local and global challenges such as sea level rise, land scarcity, food security and CO_2 emission reduction. Moreover, the project 'BlueRevolution North Sea' could serve as a platform to integrate many different fields of technology such as civil

Table 2 Inventory of initiatives, projects and plans sorted per integrative theme. These initiatives were included in the spatial concept for BlueRevolution North Sea

100% CO_2 neutral	100% bio based and circular	Resilient urban delta	Smart floating city and main port logistics
Energy island	Strengthening food and agriculture sector	Protecting existing coast	Floating city of 1 million inhabitants
Smart grids, decentral energy production	Using coastal cities and ports as source for CO_2, heat and nutrients	Building with nature, *Zandmotor* (sand engine)	Living lab for sensors, internet of things, monitoring, drones, autonomous vehicles and ships
Wave and tidal energy	Aquatic biomass	Showcase for light corrosion proof (nano) materials	New economy: testing ground for high tech start-up companies
Blue energy and saltwater batteries	Aquaculture	Artificial reefs	Expand main ports: floating airport and floating sea hub
Offshore wind	Create blue-green jobs; cluster of biotechnology, bio based resources for circular economy	Floating breakwaters	High speed vacuum tube transport connection to Amsterdam and Rotterdam
Gas at sea	Self-supporting protein supply	Closed water cycle	Smart solutions for internal logistics
Floating solar	Use industrial CO_2 in horticulture greenhouses (Kas energiebron/OCAP)	Biggest wetland of Western Europe, fish shelter	

engineering, water management, biotechnology, information technology, energy and food technology. The BlueRevolution concept could be realized first as a showcase in the Netherlands. Since many delta areas all over the world are facing similar problems, the concept could be adapted and applied in many areas, including the port cities that were analysed in this article.

6 Conclusion

With a combined analysis of the two criteria flood risk and expected population growth, an overview was made of port cities that could benefit most from floating developments. A first screening was done as indication where the most promising areas are located. With the combined ranking, an outline was made for the port cities with most potential. Most of the cities in the ranking are located in Asia, in

Fig. 4 Spatial concept for BlueRevolution North Sea, a multifunctional self-sufficient expansion near the Dutch coast with a positive ecological impact. *Source* Blue21 and Topsector Water [5]; design: Bart Roeffen

particular in India and China. None of the cities in Europe is experiencing a growth as rapid as Asian cities. However, countries such as the Netherlands are frontrunner in floating development. Therefore, a possible implementation strategy could consists of two stages. As a first step: medium and large scale floating concepts could be implemented and further developed in the Netherlands. After that, they could be applied in the locations with most potential that were identified in this study. An innovative case study was presented for the Dutch North Sea coast. This spatial concept of BlueRevolution North Sea addresses many global societal challenges such as population growth, CO_2 mitigation, the energy transition and ecological enhancement. The project could serve as a model how to deal with these challenges in delta areas all over the world. The floating developments use the waste products of land-based cities such as wastewater, CO_2 and waste heat in a productive way by growing floating aquatic biomass. Moreover, the project BlueRevolution North Sea could serve as a platform to integrate many different fields of technology such as civil engineering, water management, biotechnology, information technology, energy and food.

References

1. Angel, S., Parent, J., Civco, D. L & Blei, A. M. (2012). *Atlas of urban expansion. Lincoln Institute of Land Policy*. Cambridge, Massachusetts
2. Biggs, M., & Buchler, D. (2008). Eight criteria for practice-based research in the creative and cultural industries. *Art, Design & Communication in Higher Education, 7*(1), 5–18.
3. Blue21. (2016). *7 Reasons why our future is on the water*. Available at: http://www.blue21.nl/7-reasons-why-our-future-is-on-the-water/ visited: March 2019.
4. Blue21 (2019). BlueRevolution. Available at: https://www.blue21.nl/portfolio/blue-revolution-4/. Visited: June 2019
5. Blue21 and Topsector Water (2016). *Icoonproject Atlantis 2050*. Available at: https://www.youtube.com/watch?v=HaALXa3Bp7o *(in Dutch)* visited June 2018.
6. Dal Bo Zanon, B., Roeffen, B., Czapiewska, K. M., De Graaf-Van Dinther, R. E., & Mooij, P. R. (2017). Potential of floating production for delta and coastal cities. *Journal of Cleaner Production, 151,* 10–20.
7. De Graaf, R. E. (2012). *Adaptive urban development: A symbiosis between cities on land and water in the 21st century*. The Netherlands: Rotterdam University Press.
8. DeltaSync/Blue21 (2013) Seasteading Implementation Plan. Final Concept Report. Available at: http://2oxut21weba5oivlniw6igeb.wpengine.netdna-cdn.com/wpcontent/uploads/2015/12/DeltaSync-Final-Concept-Report.pdf.Vistited: June 2019
9. DeltaSync/Blue21 (2014). *Information Brochure Floating Pavilion Rotterdam*. Available at: https://www.slideshare.net/rutgerdegraaf/information-brochure-floating-pavilion-rotterdam. Visited July 2018.
10. De Graaf, R. E., Van de Ven, F. H. M., & Van de Giesen, N. C. (2007). The Closed City as a strategy to reduce vulnerability of urban areas for climate change. *Water Science and Technology, 56*(4), 165–173.
11. Hallegatte, S., Green, C., Nicholls, R. J., & Corfee-Morlot, J. (2013). Future flood losses in major coastal cities. *Nature Climate Change, 3*(9), 802–806.
12. Hanson, S., Nicholls, R., Ranger, N., Hallegatte, S., Corfee-Morlot, J., Herweijer, C., et al. (2011). A global ranking of port cities with high exposure to climate extremes. *Climatic Change, 104,* 89–111.
13. Maskell, D., Thomson, A., & Walker, P. (2018). Multi-criteria selection of building materials. *Proceedings of the Institution of Civil Engineers: Construction Materials, 171*(2), 49–58.
14. Moon, C. H. (2012). A study on the sustainable features of realized and planned floating buildings. *Journal of Navigation and Port Research, International Edition, 36*(2), 113–121.
15. National Science Agenda. (2017). Available at: https://wetenschapsagenda.nl/publicatie/nationale-wetenschapsagenda-nederlands/ Visited: July 2018.
16. Novotny, V., & Brown, P. (2014). *Cities of the future—towards integrated sustainable water and landscape management*. Water Intelligence Online 13.
17. Roeffen, B., Dal Bo Zanon, B., Czapiewska, K. M., & De Graaf, R. E. (2013). Reducing global land scarcity with floating urban development and food production. In *Conference Proceedings: International Water Week*. Amsterdam, 8 pp.
18. Shimizu Corporation (2008). Green Float. Available at: https://www.shimz.co.jp/en/topics/dream/content03/. Visited: June 2019
19. Takahashi, P. (1996). Project blue revolution. *Journal of Energy Engineering, 122*(3), 114–124.
20. Topsector Water. (2016). *Blauwe Revolutie*. Available at http://www.topsectorwater.nl/blauwe-revolutie/. Visited: July 2018 (in Dutch).
21. Topsectoren. (2018). Available at: https://www.topsectoren.nl/.
22. Topsectoren (2016). Available at: https://www.topsectoren.nl/. Visited: July 2018 (in Dutch).
23. United Nations, Department of Economic and Social Affairs. (2014). *World urbanization prospects: The 2014 revision, CD-ROM edition*. Population Division 2014.
24. Wang, C. M., & Wang, B. T. (2015). Large floating structures: technological advances. *Ocean Engineering & Oceanography, 3,* 261–280.
25. Wong, T. H. F., & Brown, R. R. (2009). The water sensitive city: Principles for practice. *Water Science and Technology, 60*(3), 673–682.

An Integrated Floating Community Based upon a Hybrid Water System: Toward a Super-Sustainable Water City

Toshio Nakajima and Motohiko Umeyama

Abstract Over the next century or two, rising sea levels and the increased frequency of extensive natural disasters caused by global climate change will bring about serious problems. Such problems are especially foreseen in lower-lying coastal and riverside areas, located below sea level. The steel barge-type floating foundations were chosen for our research because of their outstanding performance in the Mega-float Project which had been conducted by the Mega-float Technological Research Association of Japan during the 1990s. A 1000 m-long floating structure was constructed under this project for siting an airport complex thereon, with many parts of the smaller floating modular units being welded together at sea using a newly-developed method. Now, our concept is of realizing a new waterfront—developed by excavating the soil in a low-lying ground to create an overall area few meters below the surface level, then flooded to produce an artificial reservoir. The foundations for an urban community are subsequently set afloat upon the artificial reservoir, with buildings and other facilities being constructed on these floating foundations, in a manner similar to the Mega-float Project idea. On the other hand, the construction of substantial sustainable cities not only capable of withstanding global environmental disasters but also simultaneously producing minimum burden on the environment is needed. And to realize a truly sustainable city, we should be mindful of the fact that a multi-water supply system will become more efficient, economical and safe for cities compared with the present status dependent upon a single water supply system. Our paper "a new concept for the safety of low-lying land areas from natural disasters" was therefore selected as one of the 100 top papers from among all papers published by springer-nature in 2015 as a ground-breaking paper that could help humanity while protecting and preserving our planet, under the theme of change the world, one article at a time.

Keywords Sea level rise · Multi-water supply · Natural hazard · Flooding · Lower-lying area · Floating structure

T. Nakajima (✉)
Waterfront Real Estate Co. Ltd., Hyogo, Japan
e-mail: tmunakajima@gmail.com

M. Umeyama
Tokyo Metropolitan University, Tokyo, Japan

C. M. Wang et al. (eds.), *WCFS2019*, Lecture Notes in Civil Engineering 41,
https://doi.org/10.1007/978-981-13-8743-2_18

1 Introduction

1.1 Natural Hazards Due to Global Warming

Numerous investigations confirm that the indirect effects may be more significant than the direct effects from global warming during the 21st Century: rising sea level inundating low-lying land areas and eroding beaches while increasing the salinity of rivers/estuaries [1, 2]. In order to establish a sustainable society, many countries have initiated strategies and regulations to reduce greenhouse gas emission in recent years. However, it seems almost impossible to halt the rising sea level brought about by global warming and the subsequent speed of land disappearing under the sea. Even if the global environment improves in the future as a result of regulations against greenhouse gas and other activities, it does not necessarily follow that there will be a reduction in the frequency of worldwide disasters. In fact, it appears that this phenomenon will accelerate further. Moreover, there are many who believe storm surges, abnormally high waves and localized heavy rain caused by powerful storms that continue to intensify in strength under the influence of global warming, will strike many cities throughout the world and bring about devastating flood damages.

However, many big cities such as Tokyo, New Orleans and so on are built on large areas of low ground below sea level known as "zero-meter sites at or below sea level"; these sites are potentially hazardous in that they carry a high risk of natural disasters due to their proximity to seas and/or rivers. Yet recently, because of global warming, lower-lying land areas in cities in particular have seen flooded streets many times as well. Precipitation amounts surpassing expected levels, as exemplified by sudden torrential showers, are pushing records beyond levels that can be handled by today's sewerage system. Therefore, adaptation measures against these serious natural hazards are urgently required hereafter to preserve existing cities.

1.2 Water Shortage Problems

The water shortage issue has now become serious due to increasing population growth all over the world. This issue is especially crucial in big cities since people gather in such locations and an extraordinary amount of water is required for life in these heavily populated areas. Water conservation in cities is necessary not only for daily life but also during disaster situations. Currently, recognition is being renewed as to the fact that increased demand for water is leading to shortages in drinking water supply as part of emergency water supply, this being vital for firefighting; such situations are underscored by the lessons of the great 1995 earthquake in Kobe. If a modern water supply system were to collapse, people in cities will suffer from water shortage as a result. Under emergency conditions, experiences from the 1995 Kobe earthquake also shows that not only will drinking water becomes a necessity

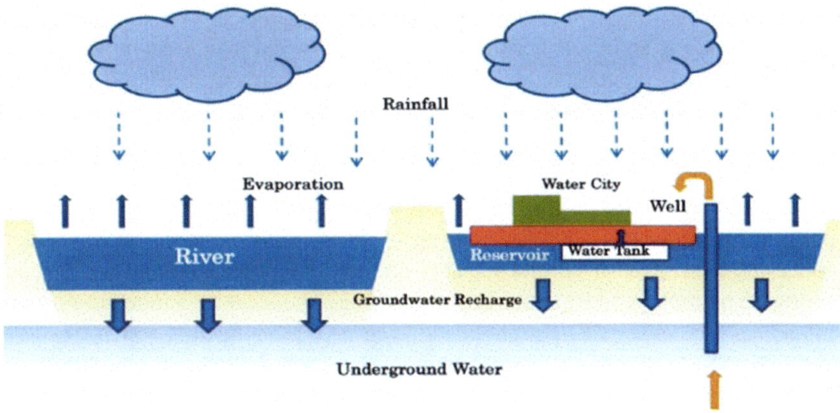

Fig. 1 Conceptual water circulation system in water city

for suffering people but water for other various uses will also become important for life such as for bathing, washing dishes and so forth.

Other serious problems recently are that the modern water supply system has aged and may require replacement, their construction on the average having taken place over half a century ago. Tremendous maintenance costs including repair costs, old piping refurbishments and water leakage detection in old pipes are required today with additional taxes becoming necessary due to increases thereof. This means that the centralized water supply system of modernity has become inefficient over the years to render it unsustainable, so we cannot rely on this water supply system any more. Conventionally, contemporary cities are supported by only this unitary water supply system comprising dams and/or reservoirs at long distance from cities and tremendous numbers of long water pipes carrying freshwater to cities. From the standpoint of cost performance, keeping water near and/or inside cities is much more reasonable; it is a low cost system for maintenance of water supply compared with the existing modern water supply system which already has many difficulties and problems upon sustaining it today, as transport of water is entailed. So we suggest holding large amounts of rainwater in the middle of cities as one of the water supply systems for use by cities while creating numerous reservoirs near urban areas as well (see Fig. 1).

1.3 Problems with Existing Old Civil Structures

Many big cities are built on large areas of low ground below sea level, areas that carry a high risk of natural disasters due to their hazard-prone locations close to seas and/or rivers. So far costly huge levees have been built to avoid water-related disasters such as tsunami and storm surge in cities. However, the levees are not sufficient in saving people from water-related natural hazards. Actually, powerful breakwaters constructed near the shores of northeastern Japan, believed then to be most reliable in the world, had been completely destroyed by huge tsunami in March 11, 2011 and resulted in a huge number of victims. When torrential rainfalls occur inside a city, the highest levees cannot save people from this kind of flooding. Furthermore, the tremendous loss not only in economic matters for cities but also in numbers of casualties once these huge civil structures collapse will be sizable.

To free people who live on lower-lying lands, considered prone to floodwater dangers, from the fear of natural hazards and to turn these areas into safer and comfortable places, we must apply lateral thinking. The weakness inherent in such locations can be converted to strength, offering protection against water disasters by rethinking the way people live in such areas, i.e. by constructing buildings on floating foundations in calm man-made reservoirs rather than on land near rivers and/or seas. Our idea emanates from the International Symposium on Sustainable Urban Environment 2007 (ISSUE2007) [3] which was held, with support from the Tokyo Metropolitan government and the Ministry of the Environment of Japan, at Tokyo Metropolitan University, under the theme of producing a sustainable urban environment. In order to offer a concrete image of our suggestions, we were then provided by our sponsoring organization, the Tokyo Metropolitan University Center for Academia/Industry/Public Collaboration, with funding for FY 2008 and continued to work on more detailed studies on this subject during FY 2009 and 2010.

2 Novel Floater-Based Solution Against Natural Hazards

2.1 Considerations and Objectives

Floating platform technology has become established in the ocean engineering field, thereby is now safe, trustworthy and reliable. In the 1980s, many offshore structures including floating platforms had been built for drilling offshore oil and oil production. During this era, the technology for floating structures was established and advanced drastically in such fields as ocean engineering or naval architecture. We should seriously consider an application of very large floating structure (VLFS) such as the Mega-float on land as a contribution for civil purposes to realize a new proposed countermeasure for extraordinary natural hazards. Between 1995 and 2000, the Mega-float Project was conducted by the Mega-float Technological Research Association established around a core of Japan's shipbuilding industry. In this project, a gigantic floating structure (VLFS) 1,000 m in length was constructed and floated in Tokyo Bay (see Fig. 2). We have since then proposed a novel concept for establishing an urban community exhibiting self-reliance during times of disasters, based on buildings atop the floating foundation situated in lower-lying lands.

We thus decided that, for water-related hazards as exemplified by floods, the answer lies not in dealing head-on with these problems using massive structures like

Fig. 2 Photograph of "Mega-float"

dams and/or dikes and seawalls but taking the indirect approach of simply taming the water, to keep to a minimum the effect of water-related hazards in a more economical manner. As an adaptation measure, the floating structure is therefore considered to be an excellent means of dealing with water-related hazards.

2.2 *Conceptual Methodology for Water City*

The concept is of a waterfront that is developed by excavating the soil in a low-lying land area down to a few meters below the surface and then flooding the area to produce man-made reservoir. The foundations for an urban community float on this man-made reservoir; buildings and other facilities are constructed atop these floating foundations.

Step 1: An area equivalent to approximately half of the total site's area is excavated up to a depth of several meters. Topsoil removed from the site is relocated in order to raise the urban site above sea level (see Fig. 3a).

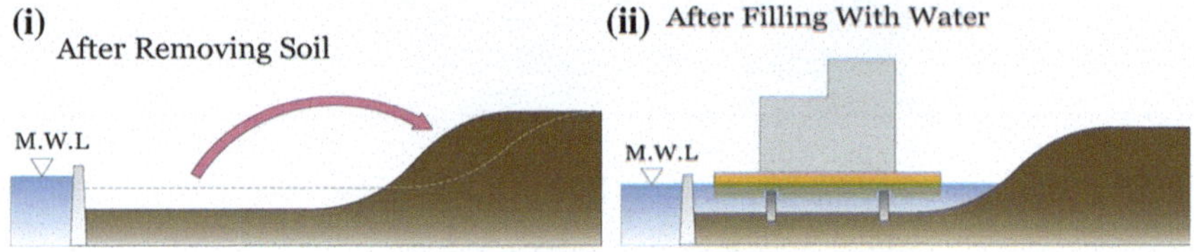

Fig. 3 Construction scheme for water city

Step 2: Concrete levees prevent erosion or destruction of the bank. The area inside the levees is then flooded until the water level inside the levees reaches that of surrounding seas and/or rivers, to produce a man-made inlet and/or reservoir.
Step 3: Floating modular units constructed at shipbuilding facilities are towed to their sites and are combined and welded to one another at their sites. Then the larger floating foundations for the water city are constructed.
Step 4: Finally, buildings are constructed on the floating foundations inside the man-made waterfront. Roads and bridges are constructed to allow passage between the floating foundations and the land, in addition to a water gate (see Fig. 3b).

In principle, these floating foundations stay afloat by themselves but the proposal calls for what is known as the "Soft-Landed System" for the purpose of positioning instead of using a mooring system. With this system, the floating foundations are partially propped by support piles found underneath. The buoyancy provided by the floating foundations serves to lighten the load on the supporting piles. Laterally-mounted piles located on sides of these floating foundations feature rubber fenders to help control their horizontal excursion. Since approximately a tenth of the total load is supported by several concrete piles, the remaining 90 percent for a floating foundation is supported by the buoyancy force of said floating foundation. When there is an unexpected increase in water level due to floodwater from heavy rains and so on, the floating foundation can easily change its position up from the Soft-Landed condition.

3 Water City Project for Koto Ward in Tokyo

3.1 Water City Proposal for Koto Ward Against Water-Related Natural Hazards

The Koto Ward area, which is selected as the project site used to be one of the most prosperous suburban industrial sites in Tokyo, and is one of the higher density urban area. The population in 1957 was 300,000, but by end of 2015 this had grown to more than 500,000. The total land area of Koto Ward is 39.9 km^2 and its population density is 125.3 persons/ha.

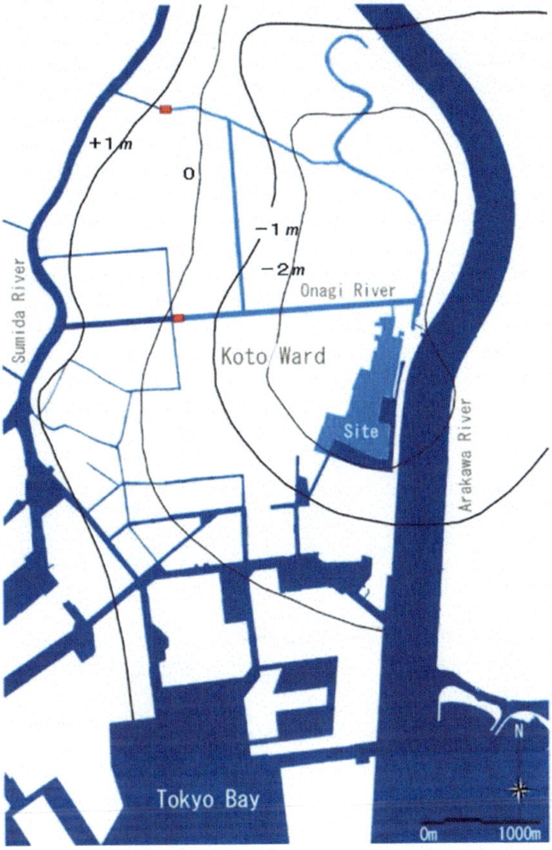

Fig. 4 Site for water city, Koto Ward

Today, as is show in Fig. 4, two-thirds of the total area of Koto Ward is lower than the average sea level of Tokyo Bay due to the fact that lots of underground water around this area had been pumped out during long time before. If a huge tsunami and/or fire due to a large earthquake were to occur, the status here would become highly critical as well. Since the Koto Ward in Tokyo is near the mouth of Arakawa River, there is the serious risk of a tidal bore from Tokyo Bay due to storm surges under typhoons and also tsunami. This levee could be destroyed by the flow, which may cause serious flooding around low-lying land areas of Koto Ward. The planned site in Koto Ward is protected by an old levee along Arakawa River, about 6 m high. This height seems sufficient to deal with storm surge level due to typhoon in Japan. The highest storm surge observed in Japan were recorded as 5.02 m including the astronomical tide under Isewan typhoon conditions in Sept. 26, 1958. Furthermore, even larger storm surges must be taken into account as long as the sea level continues to rise further and/or typhoons become larger due to future global warming.

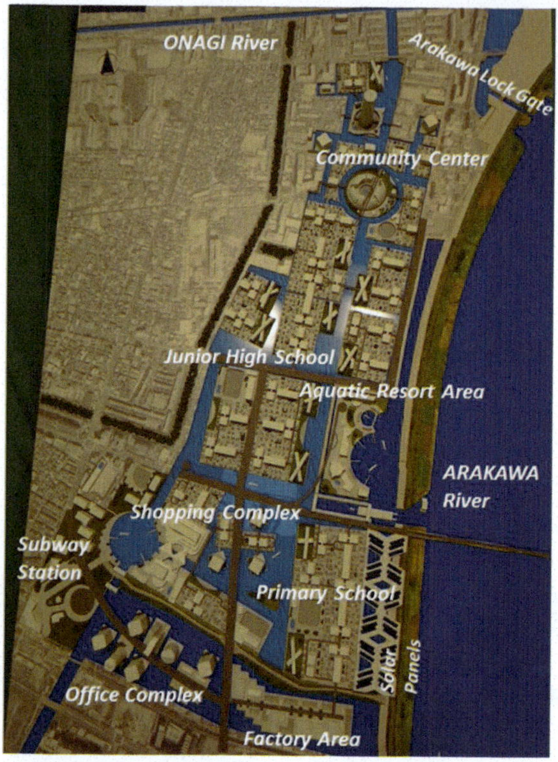

Fig. 5 Various urban facilities in water city

The final stage of the planned entire water city project for Koto Ward, Tokyo covers an area of approximately 2 km from north to south and of between 0.5 and 1.0 km east to west, with a total land area of around 1.25 km^2 [4, 5]. The planned population is close to 27,500 people consisting of nearly 12,600 households, resulting in a population density of 223.6 persons/ha. This planned population, which is based on its current size between 27,000 and 30,000 people, is three times the neighboring communities' size (8000–10,000 people). In accordance with these demographic proportions, the project site constitutes a mid-sized city that will require three elementary schools and one junior high school. The plan includes the development of a mid- to high-rise housing area (about 560 persons/ha) and a waterfront as an open space that occupies some half of the total area being developed (625,000 m^2 or 22.7 m^2/person). In addition to its residential functions, under consideration is a variety of urban facilities to enliven the city. These include: (1) a shopping complex necessary for comfortable living, (2) an aggregated industrial site for small workshops, (3) an aquatic resort space along Arakawa River, and (4) an office complex (see Fig. 5).

Generally, since an entire variegated mega-city cannot be developed all at once, the area under development will be divided into several construction stages as is shown in Fig. 6 and handled on a step-by-step manner. For this adaptation planning for Koto Ward, the first development site involves an area approximately 400 m from east to west and about 600 m from north to south near Higashisuna 3-chome, in Tokyo's Koto Ward. Right to the east of this site, close to Tokyo Bay, is the bank of Arakawa River. The water surface level is designed to match those of Onagi River and the canals located near this site to enable boat movement between the man-made reservoir and the river system. Urban redevelopment at this site based upon our design concept is to be carried out as the first stage and the process outline is described as follows: To begin with, this area is excavated to a depth of 5 m from the existing crowded site shown in Fig. 7a and the topsoil removed. The total area for this region, which will be developed by excavating soil from lower-lying land areas, is approximately 185,000 m^2.

The topsoil removed from the site is to be relocated in order to raise the other sites above sea level and this excavated area is filled with water until a man-made inlet and/or reservoir is made. Secondly, as is shown in Fig. 7b, floating foundations consisting of many standard modular units will be built inside the man-made reservoir. The total number of floating modular units used for this water area is 44 modular units (total area: 11 ha) while those at the final construction stage is about 130 modular units. The steel barge-type floating foundations were chosen in this case because of their outstanding performance during the Mega-Float Project. In general, the floating foundation is to be constructed by welding many pieces of the standard modular units; 100 m in length, 25 m in width and 4.5 m in height (3 m in draft).

Next, residential housings are built on the floating foundation inside the man-made reservoir (see Fig. 7c). The population on the floating foundation will near 4300 people comprising some 1850 households, with a population density of 390 persons/ha (three times larger than the existing city) excluding the open water area at the first stage of the water city. As the next construction stage, the additional reservoir is constructed next to the original reservoir and then the residential floating platforms located in the first phase will relocate to different sites (see Fig. 8). In this way, the future enlargement of water city will proceed with the next stage of construction of additional reservoirs in a step-by-step manner. In addition to the construction of new floating platforms, almost all floating platforms placed at the original reservoir can be reused upon future construction in case of water city that is different from the conventional one on lands as is shown in same figure.

3.2 *On Adaptation Measures of Water City by Arakawa River Against Flooding Hazards*

Having a large river like Arakawa next to it, the levee could be destroyed should the river overflow. When the water level of Arakawa River reaches the levee's higher

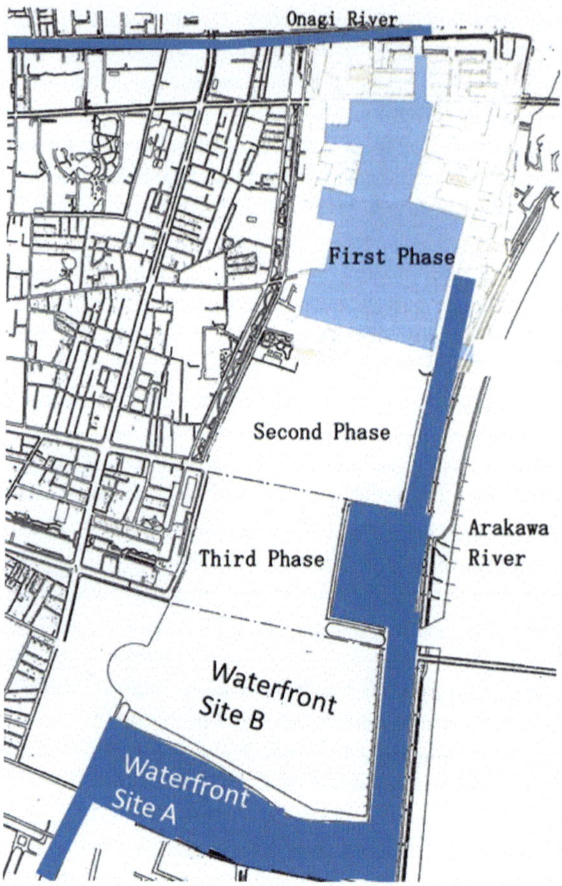

Fig. 6 Construction stages

portion, then the larger water pressure acts on only one side of levee and gradually water invades inside the levee and finally it could collapse at some location due to the water pressure. These levees will collapse cause serious harm to the surrounding low-lying areas in Koto Ward. To avoid such a dangerous situation, we had planned to construct another 3 m high sub-levee along the Arakawa levee paralleling the river (see Fig. 9). The advantage of this measure is that the main levee is seldom breached and safe conditions are maintained even when a high water level of Arakawa River has been reached since both sides of the main levee surface faces the water directly and the horizontal force on the main levee due to water pressure on obverse sides cancel out. Because waterfront Site B connects directly to Arakawa River and Tokyo Bay, the water level in waterfront Site B is almost at sea level and is therefore hardly affected by the discharged water.

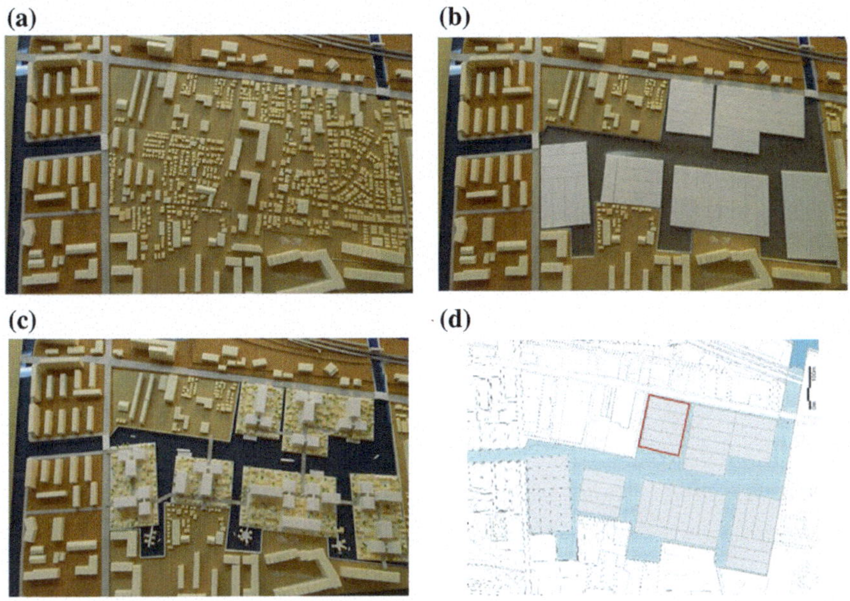

Fig. 7 Construction scheme and floating foundations for water city of Koto Ward

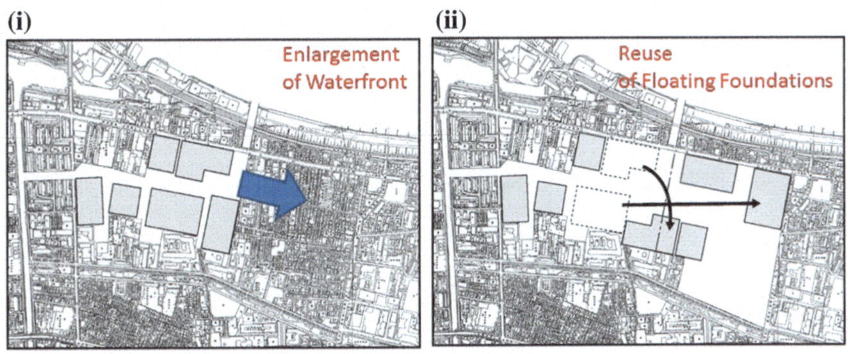

Fig. 8 Construction scheme for enlargement of waterfront area

These sub-levees are located between the water area of Site A and Site B as is shown in Fig. 10. The water level inside Site A is set as 2.344 m below the mean water level of Tokyo Bay, now found for rivers inside Koto Ward. Then the water-borne transport must use the Arakawa Lock Gate to ply between Onagi River and Arakawa River. On the other hand, the water level of Site B changes due to tidal conditions since the water is connected directly to Arakawa River, as long as the watergate is kept closed. In this way, the overall waterfront area to be developed in this Koto Ward Project is divided into two sections. Waterfront Site A (0.85 km^2) on the inland side is completely isolated from Arakawa River and Tokyo Bay. At

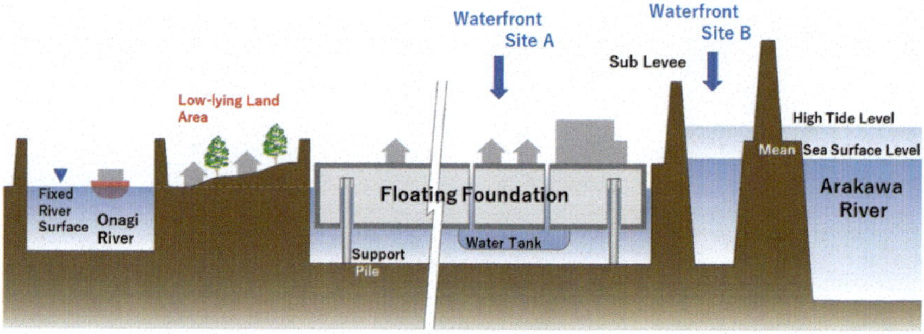

Fig. 9 Sectional view as to water city of Koto Ward, Tokyo

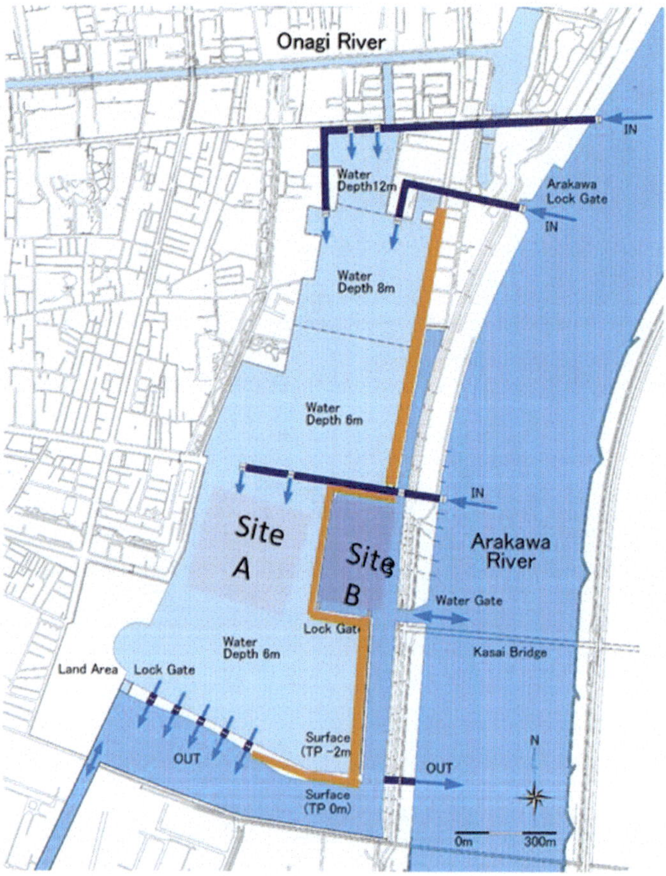

Fig. 10 Water circulation system for water city

present, in the eastern area of Koto Ward where the ground level is very low, the surface level of the water channels is adjusted to some 2 m below the average surface level of Arakawa River. Waterfront Site A is designed to link those channels, including Onagi River. The connection with Onagi River will enable free, extensive transportation of people and goods throughout the community regardless of tide levels. Waterfront Site B (0.34 km^2) is a narrow inverted L-shaped area. Boats travel back and forth between the site and Arakawa River by way of water gates that remain open except during emergencies, such as tsunami tidal bores and/or storm surges. Although the water level in waterfront Site B is altered directly by the tide, boats can move between Arakawa River and Tokyo Bay freely without passing through a lock gate. Consequently, the double levee system gives water city more safety against high water and becomes an excellent adaptation measure against flooding due to water-related hazards such as storm surge due to typhoon, tsunami and so on.

3.3 Multi-water System and Water Supply for a Sustainable Water City

The average precipitation in Tokyo amounts to about 1340 mm per year. We should consider using rainwater as much as possible for the water city of Koto Ward, Tokyo. In addition, by means of the groundwater recharge of rainwater that has once been stored, realization of a highly advanced society that circulates water by utilizing wells, etc. becomes possible. Moreover, with our proposal, we adopted the Grey Water Supply System which adds on an intermediate stage handling grey water to the existing drinking and sewage systems. As for the multi-water system of the water city, four water supply systems are prepared in this study; the freshwater one based on the existing water supply system, the rainwater-based one directly fall on the floating foundations, underground water system using wells and the grey water one made from water inside the reservoir. Rainwater gathered on the floating foundations are used as water for flushing the lavatory while the water inside the reservoir which mainly comes from Arakawa River is used as grey water for washing cars, sprinkling the gardens and so on. As is shown in Fig. 11a, there is a sewage facility near the reservoir and grey water is produced by purification of water inside the reservoir. Other water for living use is to be provided by the existing water supply and/or underground water system the latter primarily through wells.

The water surface area of the reservoir siting 44 floating modular units as a living space total is at approximately 110,000 m^2 (see Fig. 7d). The population for a single floating modular unit is about 90 people so the total water necessary becomes around 27 t assuming one person uses 300 l in a day. The total then for the 44 floating modular units becomes 433,620 t per year. It is said that in Japan general water usage upon living are categorized into five: flushing the lavatory (28%), washing hands (9%), bathing (24%), washing things (16%) and kitchen (23%). The population on five modular units (100 m × 125 m) which is used as a typical prototype of the residential

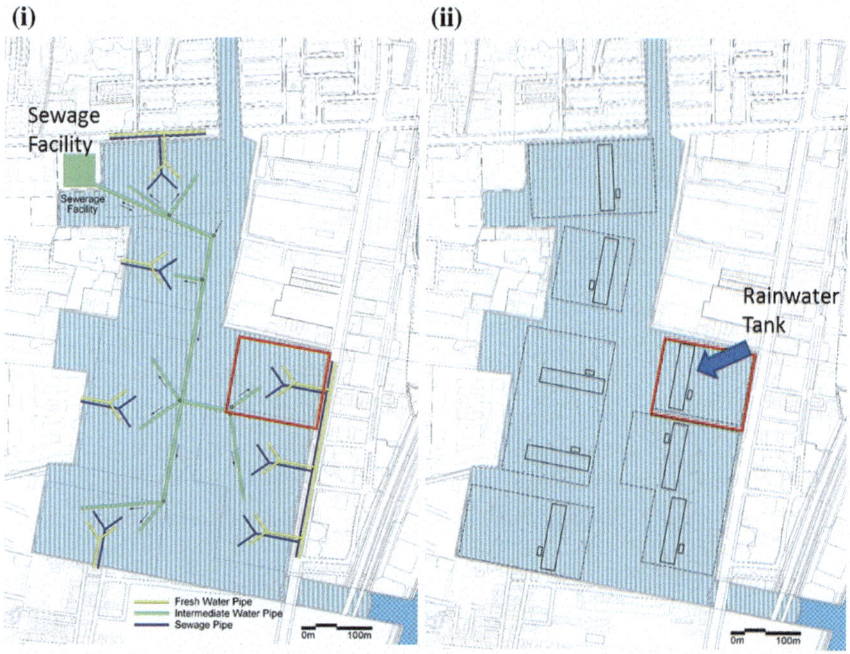

Fig. 11 Multi-water supply system for water city

floating unit showing as an enclosed rectangular line in Figs. 7d and 11 is some 441 people, the amount of water used by one unit's population per day is set at 132.3 t (or 48,290 t per year).

Rainwater falling on the five floating modular units are to be stored in rainwater tanks located underneath the modular units (see Fig. 11b). The rainwater is used to flush the lavatories within the water city. For a floating foundation consisting of five modular units with 12,500 m^2 surface area, a rainwater tank of 2800 m^2 is enough for maximum 250 mm/day precipitation, the heaviest precipitation recorded for the last 35 years, the tank being affixed underneath the floating foundation. Then, the volume of the tank below the floating foundation on the inner reservoir area is designed to cover the maximum precipitation (250 mm per day) of an entire year *in toto* because lots of space is available inside the man-made lagoon and/or reservoir. From this estimation, as is shown in Fig. 12, the tank size was designed to be 90 m in length, 20 m in width and 1.8 m in height while the water tank inside of the floating unit is set to be about 110 m^3 assuming use for flushing the lavatory as being 37 m^3 estimated in a day for 441 people over a 3-day period. In general, the rainwater tank made of quite flexible material such as rubber located underneath the floating foundation need not have a costly steel structure like ones on land. In addition, the largest space for rainwater are made available inside reservoirs.

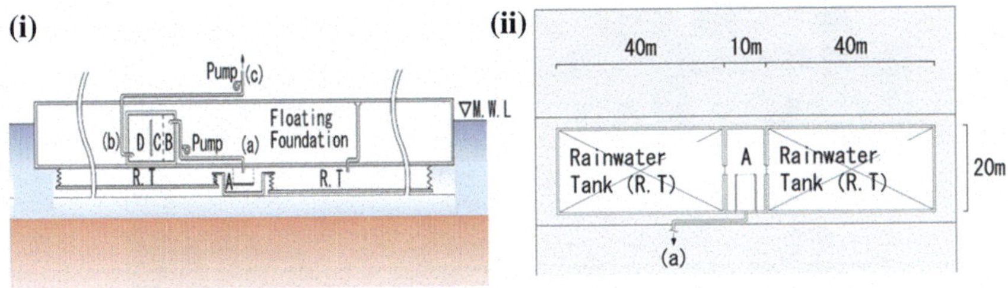

Fig. 12 Sectional plans for rainwater supply system

4 Application of Proposed Approach to Coastal Areas and/or Isolated Islands

In recent years, the "global warming trend" has, while contributing to the sea level rise, not only enabled the growth to inordinate size of hurricanes/typhoons and tornadoes but also led to expansive flooding. Especially, the adaptation measures for coastal areas becomes necessary to avoid the loss of coastal areas in these areas with beaches face the ocean directly and tend to be affected by the sea level rise. Although conventionally flood countermeasures implemented by countries like the Netherlands have been the construction of large-scale civil engineering structures such as dikes, it is impossible to avoid such extraordinary powers of nature using massive civil engineering structures such as seawalls and breakwaters.

We now suggest applying the present methodology to coastal areas by using the so-called "Coastal Aqua-Village System" (see Figs. 13 and 14) [6, 7]. The plan is to change the potential danger in this site to produce an area safeguarded against storm surge due to huge typhoons and massive tsunami. Under this concept, the dangerous coastal zone areas are excavated to a depth of some several meters and the topsoil removed as is shown in Fig. 15a, b. The topsoil removed from the site is to be relocated in order to construct banks in other areas of the seaside site in protecting against a run-up of seawater due to storm surge and/or high waves. Next, these are filled with water including saltwater until the water level reaches sea level, to construct a man-made lagoon and/or inlet. Finally, the floating platforms made of lower cost materials such as lightweight concrete are constructed with housings and even rice paddies/agricultural fields located on floating platforms inside the man-made lagoon and/or inlet (see Fig. 15c).

In principle, the Floaters float by itself also but the proposal calls for what is known as a "Soft-Landed System" (see Fig. 15d). When there is an unexpected increase in water level due to a high tide and/or floodwater from storm surge, tsunami and so on, the Floaters can easily change its position up from the Soft-Landed condition. In addition, tall concrete piles with guiderails are also erected and connected with the Floating Roads to deal with the unexpected extraordinary vertical excursion of floating structures due to a huge tsunami and so forth. For the purpose of keeping costs down and realizing functional flexibility as to change for the floating structure system,

Fig. 13 Bird's-eye view of "Coastal Aqua-Village System"

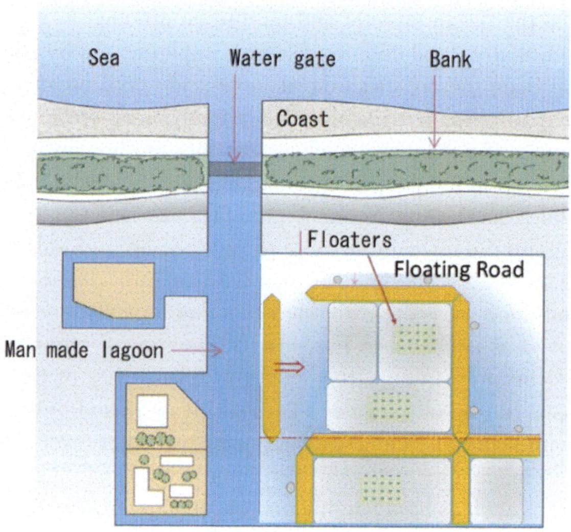

Fig. 14 Construction scheme

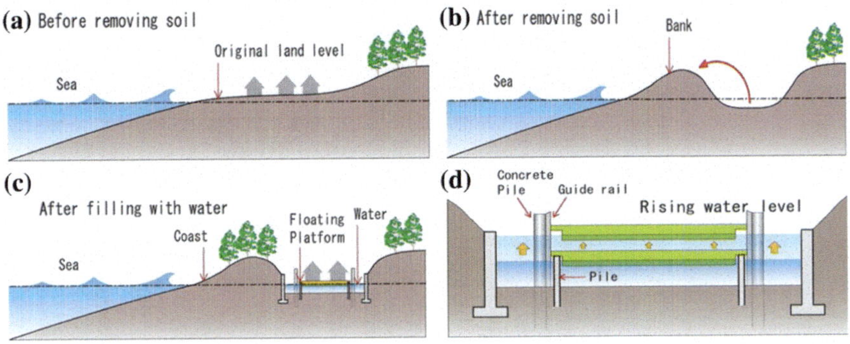

Fig. 15 Floating structure system for "Coastal Aqua-Village System"

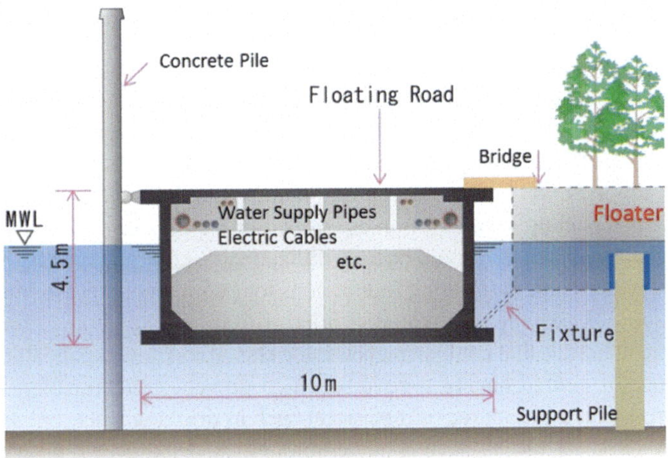

Fig. 16 Detailed sectional view as to floating infrastructural system

we recommend that the floating structure system inside the water area be divided into two components upon planning as is shown. One is the box-type Floating Roads which hold the infrastructures such as water supply pipes, electric cables and so on and the other is the functional Floaters where low-rise housings and/or agricultural fields are built on. Latter structures are bound with the Floating Roads using steel fixtures. Figure 16 shows the detailed sectional view of the Floating Road.

The proposed Coastal Aqua-Village System should stock up on rainwater as much as possible inside the man-made lagoon and/or inlet too, for use in agriculture and such applications since rain falling on the rural landscape is also an important natural resource (see Fig. 17). The precipitation is very precious such for Maldives, Singapore and so on and the proposed Multi-Water Supply System will become extremely useful for supporting water supplies at many places. As an example of isolated islands, there are many being adversely impacted by not only rising sea levels that threaten in

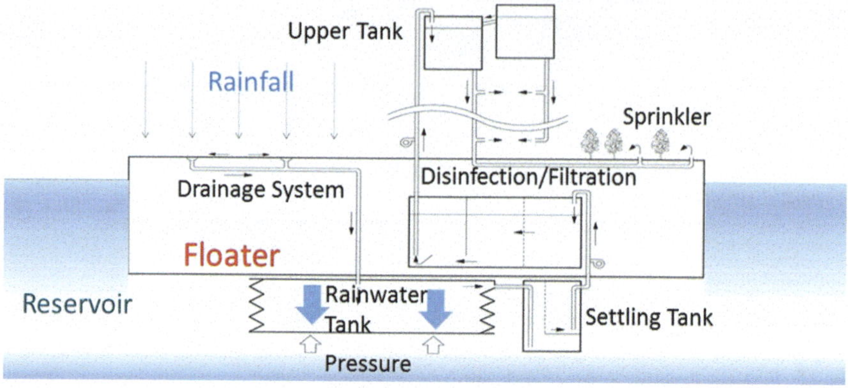

Fig. 17 Schematic sectional plan of rainwater supply system

addition to their shorelines the freshwater supply but other aspects of climatic changes as well like longer dry periods and extreme weather events increasing in frequency. This situation further increases the need to collect rainwater for use in supplementing their water supply. Even if a systematic layout and a link between supplies has been established as a backup system for areas in need of freshwater during dry seasons, there is a lot of rainwater that could be fully utilized through application of "in-water" tanks found underneath the floating foundations to realize an urban-use Multi-Water Supply System. By having water "tanks" filled with freshwater from rain and made from durable materials that could in fact be made elastic and thus expandable, floating in saltwater—of course having a cover on it to prevent contamination and even evaporation—a versatile source of freshwater such as drinking water or base thereof is made available for use by such island communities. As is shown in same figure, the water pressure difference between pressures inside the rainwater tank and outside of the man-made lagoon is almost in equilibrium, consequently, problematical air inside rainwater tank will be pushed out completely due to the water pressure in reservoir.

5 Discussion and Concluding Remarks

Our concept entails excavation of lower-lying land areas in the middle of cities, to produce large reservoirs for placing a semi-floating structure therein and to ready a man-made foundation for placing facilities atop it as is shown in Fig. 18 [8]. This allows for effective use of space above a body of water for realizing living space on the floating foundation. By promoting underground replenishment of water upon storage, a highly advanced society which circulates water by utilizing wells, etc. becomes possible, to turn disaster conditions like flooding into something advantageous. Excavating lower-lying land areas secures space for a large reservoir in the

Fig. 18 Bird's-eye view of sustainable water city "Waterpolis"

middle of cities where precipitation can then be used secondarily for human activities such as flushing the lavatory, sprinkling the garden with water and so forth, as an additional source of water to supplement the existing drinking water and sewage systems in a city. With abundant water resources inside cities, there are many merits. For example, it would be possible to improve the water cycle, to utilize this by directing water into the ground to produce an environment-friendly, sustainable urban community.

The magnitude of natural hazards is likely to keep growing annually along with future climatic changes and cannot be ignored. When considering the rise in sea levels to be unavoidable from the long-term perspective, the height of seawalls must keep increasing. What to do in developing the coastal and riverside areas to safeguard against water-related disasters over the long term? It is urgent that a reliable alternative solution which will ensure a higher degree of safety through mitigation of disasters be found. We believe that the water city concept with the Coastal Aqua-Village System presented here offers a new paradigm as to novel adaptation measures which matches next-generation standards and is applicable to isolated islands (e.g. Maldives and Tuvalu) suffering from sea level rises caused by global warming.

Acknowledgements The authors are grateful to Dr. Hironori Sugimoto, general manager of Marine Technology & Development Dept. at the Shipbuilding Research Centre of Japan, for disclosure of various practical data concerning with the Mega-float Project. Also special thanks to both Mr. C. Ames Pomeroy, Journalist and Mr. Kenzo Nakajima, architectural designer for improving the English language of this paper and for providing illustrations used herein, respectively.

References

1. Intergovernmental Panel on Climate Change. (2014). IPCC Fifth Assessment Report—Climate Change 2014. Intergovernmental Panel on Climate Change.
2. Bender, M. A., Knutson, T. R., Tuleya, R. E., Sirutis, J. J., Vecchi, G. A., Garner, S. T., et al. (2010). Modeled impact of anthropogenic warming on the frequency of intense Atlantic hurricanes. *Science, 327*(5964), 454–458.
3. Nakajima, T., & Umeyama, M. (2007). A proposal for a floating urban communities in the man-made inlets. In *Proceedings of International Symposium on Sustainable Urban Environment (ISSUE2007)* (pp. 27–33). Graduate School of Urban Environmental Sciences, Tokyo Metropolitan University.
4. Nakajima, T., & Umeyama, M. (2013). Water city as solution to escalating sea level rise in lower-lying areas. In *Oceans 2013 MTS/IEEE, International Symposium*, San Diego, California, CD-ROM.
5. Nakajima, T., & Umeyama, M. (2015). A new concept for the safety of low-lying land areas from natural disasters. *Journal of Ocean Engineering and Marine Energy, Springer, 1*(1), 19–29.
6. Nakajima, T., Shintani, T., & Umeyama, M. (2011) A new concept for lower-lying land areas and coastal villages safe from natural disasters. In *Oceans'11, 2011 MTS/IEEE International Symposium*, Kona, Hawaii, 19–22 Sept.
7. Nakajima, T., & Umeyama, M. (2015). Novel solution for low-lying land areas safe from natural hazards—toward reconstruction of lost coastal areas in northeast Japan. *Journal of Marine Science and Engineering, 3*(3), 520–538.
8. Sustainable Water City. Available online: http://www.waterpolitan.com.

Floating Clean Multi-energy Systems Towards Driving Blue Economic Growth

Srikanth Narasimalu

Abstract The blue economy is the recent initiative among Indian Ocean realm association (IORA) countries, ASEAN and Caribbean Islands focusing towards sustainable use of ocean resources for economic growth, improvised livelihoods and jobs and preserving ocean ecosystem health. High-density population resides in coastal and islandic regions of these countries and with climate change, rising sea level is seen an imminent danger to the coastal community in the low-lying area. To counter the rising sea level, countries are looking for technologies to support the displaced coastal population by embarking on floating homes and floating cities in their Blue Growth strategy to support living space and amenities needs. Since these floating homes and cities are powered through fossil fuel there is a need to preserve the Coastal marine ecology from the emissions. This paper discusses about powering the floating homes through ocean based floating clean energy systems as a cost-effective energy system that can be manufactured, assembled and maintained in the onshore and easily towed and deployed at a specific ocean site and assure energy security and resilience even during natural disasters. In addition, the paper discusses the idea of hybrid renewable powered energy systems to exploit the multiple available energy sources at the specific site viz., tidal, ocean thermal, wave, solar and wind that may vary in different proportion of availability to increase overall energy generation density from the ocean site. To illustrate, a case study of floating tidal system is discussed that was designed and deployed in Singapore towards tropical shallow water conditions through systematic resource assessment, device design through simulation and field based assessment.

Keywords Tidal energy · Wave energy · Smart grid · Renewable energy · Floating structure

S. Narasimalu (✉)
Energy Research Institute @ Nanyang Technological University,
Singapore 637141, Singapore
e-mail: nsrikanth@ntu.edu.sg

C. M. Wang et al. (eds.), *WCFS2019*, Lecture Notes in Civil Engineering 41,
https://doi.org/10.1007/978-981-13-8743-2_19

1 Introduction

High-density population resides in coastal regions of different countries. As noted by the International Panel on Climate Change [1], climate change will have many negative effects, increased intensity of storms, floods and droughts; rising sea levels and loss of biodiversity. Sea level rise (SLR) poses a particular threat to countries with heavy concentrations of population and economic activity in coastal regions [2]. Geographic Information System (GIS) based study shows hundreds of millions of population may be affected in more than 84 developing countries in the inundation zones demarcated as 1–5 m of SLR. In order to counter the implications of SLR for population location and infrastructure planning, governments can resort to floating homes and floating cities design concepts to meet the short term and long term coastal regional needs.

In parallel, coastal countries in the Indian Ocean Realm Association, Caribbean Islands, ASEAN region, etc. are focusing to exploit their marine resources through a Blue Economy Growth strategy to utilize their ocean resources towards job growth, energy resilience and economic growth. This paper proposes that the government can resort to Floating homes and Floating cities to meet the short term and long term coastal regional needs in their Blue Economy Growth Strategy to include the implications of SLR for coastal population location and infrastructure planning.

To make this floating homes a reality, the energy needs towards electricity and water even in a remote coastal location should be supported by a credible energy production and supply system. Presently floating homes are supported with fossil based energy systems such as diesel generators, which affects the carbon foot print and marine ecology of the ocean site. Accordingly, clean energy based marine energy systems is seen as a promising energy source to power the various marine industries and operations towards economic growth and help create job creation and increase the energy availability in the coastal region.

Ocean has various energy sources such as tidal energy, ocean currents, salinity gradient, ocean thermal energy gradient, wave energy. Today the corresponding energy generation technologies for deriving electrical power is developing into commercial solution at different pace within various industrial firms and designs are being tested for their technology maturity, performance and reliability in test sites such as European marine energy center (EMEC) in Orkney Islands and Force test site in Halifax (Canada).

Today, land based solar energy systems through integration of smart grids have become a credible clean energy source in terms of ensuring robustness against intermittencies in the regular energy mix of remote inland region and urban cities. In similar lines tidal instream flow currents are highly predictable in ocean conditions and can work well with other energy systems such as floating solar photovoltaic systems, floating wind and floating wave energy systems through smart grid systems that are empowered with machine learning based energy forecasting and load scheduling technology to operate as a credible clean energy supply system. Today, with proper hydrodynamic modeling, these energy systems can be integrated into floating structures and can be designed towards specific design life to structurally support the various energy systems and present as a floating renewable energy platform.

However, in coastal settings due to limited land area for deployment, the present paper proposes the concept of floating renewable energy systems that can wisely utilize the available tidal energy resources, photovoltaic energy, wave and wind energy at a specific site to increase the overall energy foot print at the specific site. For example, in Singapore coastal waters the solar energy can provide ~168 GWh/yr/km^2, Tidal can provide ~166 GWh/yr/km^2 while Wind will provide ~9 GWh/yr/km^2. The concept of the floating energy platform can be achieved through sizing the specific energy systems such as photovoltaic systems, tidal turbines, wind turbines, wave energy systems in tune to the energy resource assessment at a site and structurally integrate on a large floating structure as well as electrically integrating them through a smart hybrid AC–DC nano-grid technology and enhanced with complimentary machine learning based energy forecasting principles and load scheduling techniques to ensure the overall floating hybrid renewable energy system as a clean energy supply system which is trustable in terms of certainty and reliability in power production towards the essential needs such as electricity, water, air conditioning, etc. Accordingly, this paper systematically explains the various steps that includes the resource assessment principles of a coastal site and tailoring the energy systems in accordance to the energy availability in the specific site to maximize the energy production from the available ocean site footprint with minimal skilled manpower and infrastructure and minimal levelized cost of energy.

2 Ocean Based Renewable Energy Systems

Systematic ocean energy resource assessment [3] shows oceans possess various energy sources that exhibit a total energy capacity of:

- Ocean thermal energy gradient (OTEC) amounts to 10,000 TWh/year.
- Wave energy amounts 80,000 TWh/year.
- Tidal range (barrage) amounts to more than 300 TWh/Year.
- Tidal/marine current amounts to more than 800 TWh/year.
- Salinity gradient amounts to 2000 TWh/Yr.
- Offshore wind amount to more than 192,000 TWh/year.

Energy availability varies with geographic location due to earth curvature, spin, water depth, Coriolis forces, uneven solar radiation, and salinity concentration variation. Figure 1 shows the renewable energy resource availability in Southeast Asia.

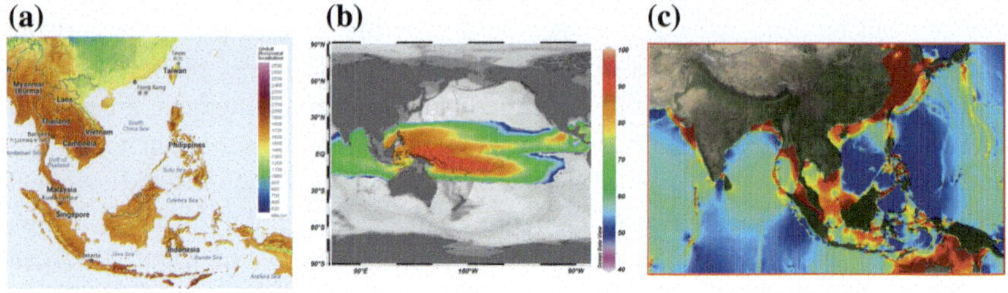

Fig. 1 Renewable energy resource availability, **a** solar energy, **b** Ocean thermal energy gradient, **c** tidal energy

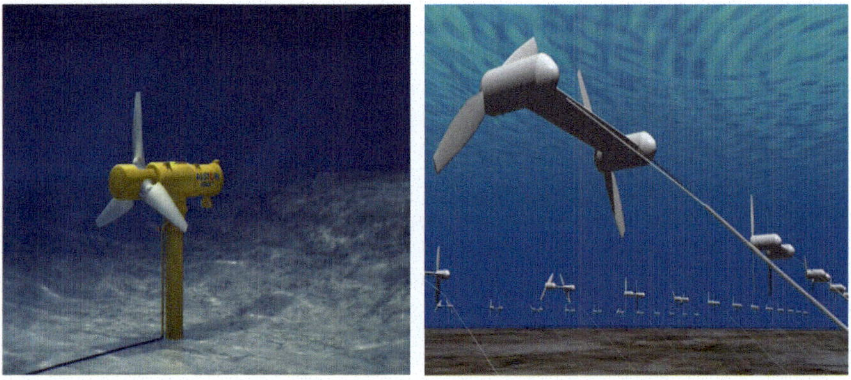

Fig. 2 Seabed mounted tidal turbine and semi-buoyant tidal turbine designs

2.1 Tidal Energy System

Tidal In-Stream Energy (TISE), a type of ocean energy which refers to the potential presence in tidal currents or flow, is a reliable clean energy resource option coastal region's energy needs [3]. TISE is more reliable (because it is periodic, thus highly predictable, since the resource follows the astronomical/earth-sun-moon rhythm), weather-resilient (unlike solar energy which may be disrupted by cloud cover) and is unaffected by climate-change (unlike wind energy). Tidal In-Stream Energy extraction technologies (devices, installations) do not require land area but make use of otherwise dormant underwater sea space (existing installations are seabed-mounted and does not conflict with shipping and navigation) as shown in Fig. 2.

2.2 Wind Energy System

Ocean surface experience higher wind speeds than over land more than five times as much energy as wind turbines over land. This presents an enticing opportunity

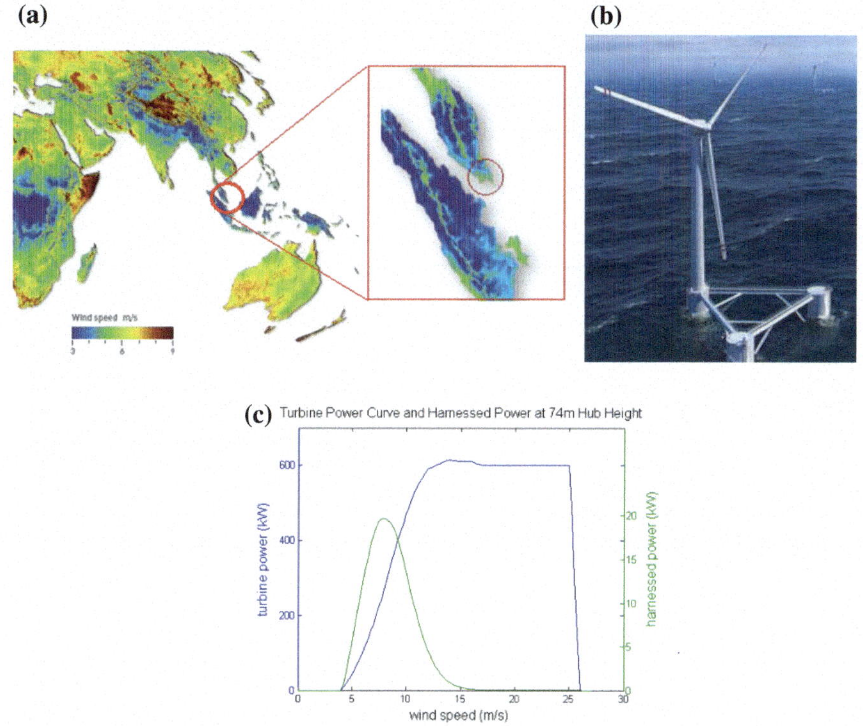

Fig. 3 **a** Typical wind energy flow in Southeast Asia, **b** floating wind turbine, **c** wind velocity Weibull distribution (green) and the power curve of typical wind turbine (blue)

for generating renewable energy through wind turbines through offshore support structures and floating structures. In Southeast Asia, wind speeds are relatively low compared to Europe. In Singapore, the wind speeds range around 6 m/sec at a hub height of 50 m as shown in Fig. 3 [4].

A floating wind turbine, an offshore wind turbine mounted on a floating structure, allows the turbine to generate electricity in deep water depths where fixed-foundation turbines are not feasible. Floating wind farms have the potential to significantly increase the sea area available for offshore wind farms. Sea surface experiences least turbulence and reach stronger and more consistent winds. Hence, commercial installations are towards large wind turbine structures to exploit increasing returns to geometric scaling. Floating wind farm in Hywind Scotland, developed by Statoil and commissioned in October 2017 is a good case study to show the robustness of commercial floating wind turbine which has 5 floating turbines with a total capacity of 30 MW [5].

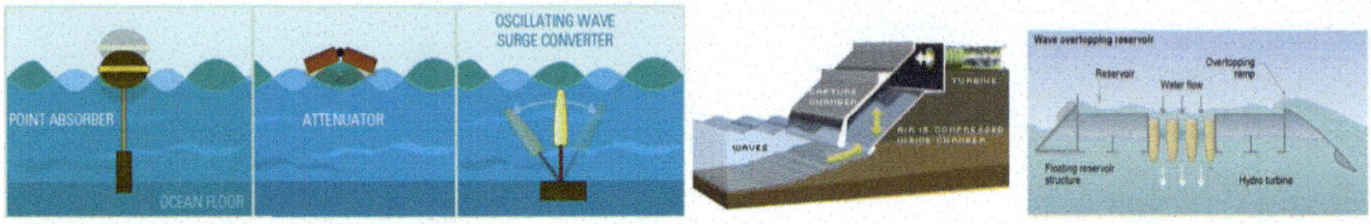

Fig. 4 Typical wave energy converters towards larger wave heights

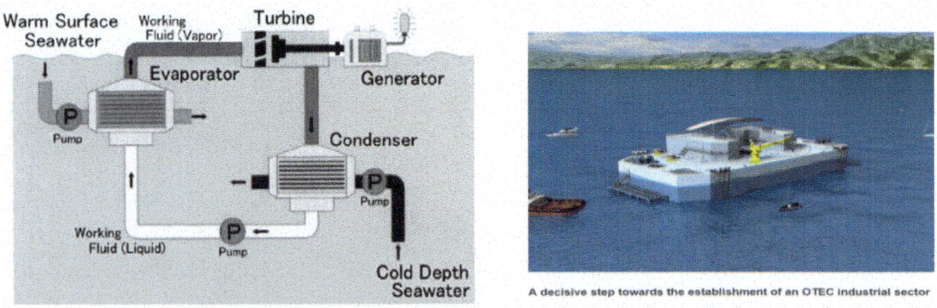

Fig. 5 OTEC working principle and floating DCNS system towards desalination support

2.3 Wave Energy System

A variety of wave energy converter designs have evolved (Fig. 4) and have been tested in European waters. These have been specifically designed towards larger wave heights (>1 m) [3].

However in the tropical belt experience low wave heights (<1 m) and hence in a recent study, the author had demonstrated that a roller based wave energy converter that can be integrated to guide rollers in pontoons can operate and produce electrical energy at a wave height of below 1 m wave height.

2.4 Ocean Thermal Energy Conversion

Southeast Asia has significant deep water regions that possess a thermal gradient of up to 20 °C within a depth of 1.5 km ocean depth from the ocean top surface, in places such as Philippines, Malaysia and Indonesia. This thermal gradient can help in boiling low boiling point liquids, as shown in Fig. 5, such as Ammonia to run a turbine and generate useful power [3].

Figure 6a shows with cumulative production the cost of energy of the different energy harvesting technologies have been reducing and the solutions are maturing at various rates (see Fig. 6b) for these different energy sources. To minimize the capital cost and operating cost, the design can integrate to exploit the common parts among the various systems and present as a hybrid energy system to maximize the energy

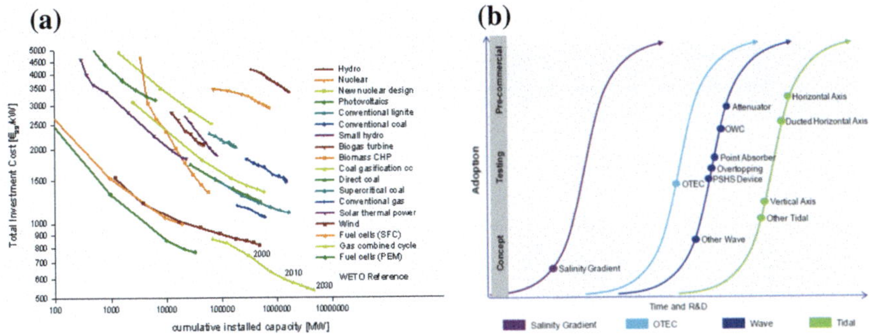

Fig. 6 **a** Total investment cost reduction with cumulative production (also called learning curve), **b** present day technology maturity curves of the renewable energy systems

Table 1 A comparison of different energy sources in an ocean site with respect to LCOE, footprint, predictability, carbon abatement and technology readiness level

Technology	LCOE (USD/kWh)	Footprint (sq m/MW)	Predictability	Carbon abatement	Technology readiness level (TRL)
Fossil fuel	~0.1	–	High	No	High
Wind	~0.15	~6000	Low	Yes	High
Tidal	~0.30	~2500	High	Yes	High
Wave	~0.4	~ 4000	Low	Yes	Medium
OTEC	~0.4	~5000	High	Yes	Medium

foot print and maximize the energy availability at a minimal levelized cost of energy (LCOE) (Table 1).

3 Case Study: Floating Hybrid Tidal Energy System

To elucidate the concept of floating renewable system a case study of floating tidal turbine is discussed in this section. Conventionally, tidal turbines are seabed mounted as shown in Fig. 2 which demands costly erection and deployment and in terms of maintenance procedure due to need of skilled divers' requirements and the necessary permissions as well as need of costly equipment for handling infrastructures such as special cranes and special vessels in the ocean site. In this present study, a systematic approach was made to develop a floating tidal turbine that enables the turbine to be deployed from a floating platform below the water surface and thereby experience high flow speeds as well enable to be lifted as necessary to provide cost effective maintenance. Owing to the towability of the floating platform, the floating renewable system can be dry-docked in the port for long duration repair

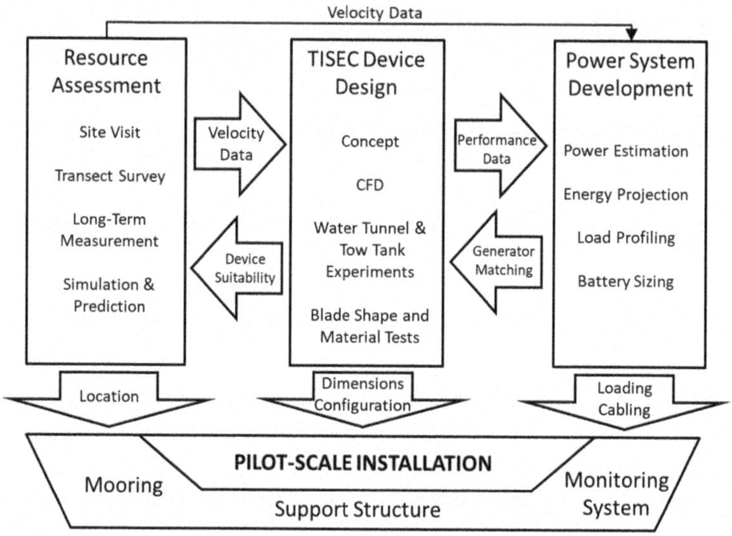

Fig. 7 Various steps in sizing floating tidal turbine towards coastal site

and maintenance. The design and development approach in this study adopted the following steps, as detailed in Fig. 7:

(1) Resource assessment of the ocean site of available tidal and wave energy resource was performed using both simulation and field based sensor deployment.
(2) The right dimensions of the turbine was selected towards optimum energy harvesting to achieve increased annual energy yield, capacity factor, availability factor and lower levelized cost of energy (COE).
(3) Genetic algorithm based hydrofoil selection helped to achieve towards higher lift and minimal drag forces and minimal cavitation factor, flap-wise and edge-wise structural rigidity of blades and hydrofoil robustness towards biofouling surface roughness.
(4) CFD based studies helped to confirm turbine's energy harvesting performance, thrust forces, wake performance and in hydro-acoustics evaluation.
(5) Based on the simulation results the dimensions were finalized and a scaled turbine was developed and was tested in open sea condition.
(6) The floating barge was designed as per naval architecture principles and Hydrodynamics studies were conducted on the floating tidal turbine support system based on the site data.
(7) Fabrication of the Floating tidal turbine was conducted in Singapore Yards.
(8) The turbine was tested under simulated towing test condition to mimic tidal flow conditions and the stability of the floating barge and the dynamic stresses were experimentally measured and were compared against simulation predictions.

3.1 *Resource Assessment in Ocean Site*

Tropical waters have energy resources which includes tidal energy, wave energy and wind energy. Hence in the forth coming paragraph, a macro assessment methodology is shared using simulation based technique. Once upon citing the best energy sites micro energy resource assessment can be made by deploying sensor systems to check and calibrate the simulation model. Using the tidal flow velocities, wave height and wind speed the energy availability can be evaluated and thereby the energy system (such as tidal turbine, wave energy system as well as the floating PV system) can be sized for maximum performance. Further the wind, wave and tidal forces are useful for the structural dynamics of the floating structures and designing the mooring systems.

3.2 *Tidal Modeling and Assessment in Ocean Site Energy*

In the evaluation of tidal power resources cataloguing of appropriate sites and estimation of achievable energy are greatly important [6–9]. Performing tidal energy resource measurements is costly and hence simulation methods are popular to do macro and micro level simulations to provide accurate energy availability. Models are being developed to identify the locations with high flow velocities and later analyzing those areas for the average power density. In this way, sites are being identified for optimizing the tidal turbine design and size to suit the site for optimum energy harvesting before selecting and installing tidal turbine systems. However, the correctness of these models is a function of the accuracy and the resolution of the input data required for these models. Further, the certainty also depends on the hydrodynamic phenomenon being examined by the various models to simulate the ocean flow. Like certain models does 3-dimensional simulation [6] while other does a 2-dimensional depth-averaged simulation [7, 8]. Still, these models serve the purpose of distinguishing the potential sites for tidal energy extraction which can be later studied for specific sites through micro-level siting studies based on field level sensor deployment.

3.2.1 Macro Level Tidal Energy Modeling Methodology

In the present study, a two dimensional depth averaged model was developed which included a full region of Singapore mainland (Fig. 8) and nearby areas basically the computations have been performed in almost whole of Singapore so that the influence of most of the flow characteristics could be apprehended which influences the flow in the Region. The coastlines and the island boundaries have been extracted with 1-minute arc resolution to create the mainland and islands boundaries. The bathymetry of the area under consideration is obtained with 30-second arc resolution dataset and

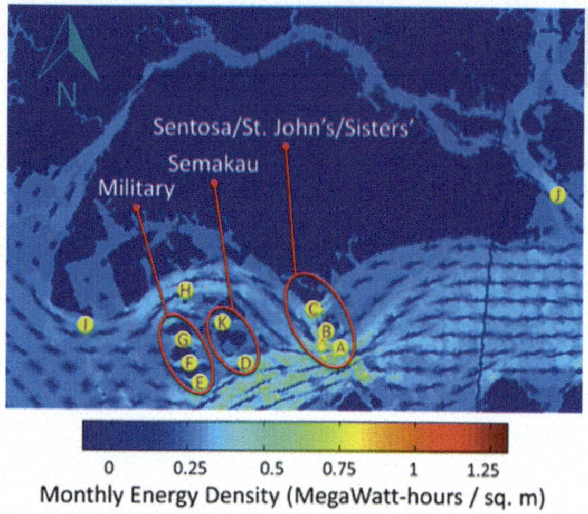

Fig. 8 Tidal instream flow in Singapore waters

the highest values of the depth of the region after interpolation is around 100 m. After generating the domain, the two dimensional depth averaged mesh model is developed using unstructured meshing technique to incorporate the complex coastline features of arbitrary shapes. The model is ensured to capture the seabed characteristics to create the drag force accurately and influence the velocity field. Each node of the model was specified with a certain value of friction coefficient which was predicted in accordance with quadratic shear stress law and the ocean boundaries were forced with tidal constituents of up to eight constants [9].

The tides propagations in the region was solved by ADCIRC which provides temporal and spatially varying velocity scalar and vector field in the region. Further, the vector plots enable to locate the regions which have maximum flow velocities. The simulation was performed for a 30 days time period of the month with a time step of 5 s and a ramp of 5 days. All the results have been shown for full simulation length of 30 days.

Velocity time series shows the transient variation of the scalar velocity at a point in the region of simulation. The velocity time series for a point in Singapore as shown in Fig. 9 has a maximum value of around 1.4 m/s with the friction coefficient of 0.009. The repetitive pattern of the velocity shows the influence of the tidal constant having a time-period of 15 days. Velocity time series was obtained by calculating resulting velocity from u, v components of the velocity field.

Surface elevation basically refers to the tidal heights because of all tidal constituents' interference which was used for modelling the tides. With the tidal constituents as a given boundary condition, for the selected site Fig. 10 shows the surface elevation can range from −1.5 to +1.5 m i.e. a range of 3 m.

The above flow velocity provide estimates of tidal turbine rotor power density by computing the power density per unknit frontal area. Thus the Fig. 11 shows

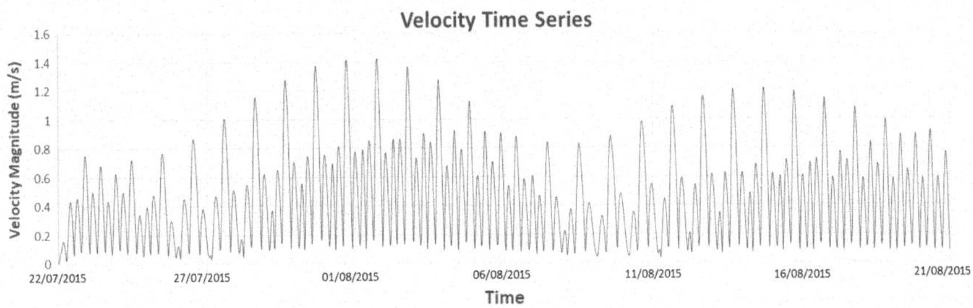

Fig. 9 Velocity time series

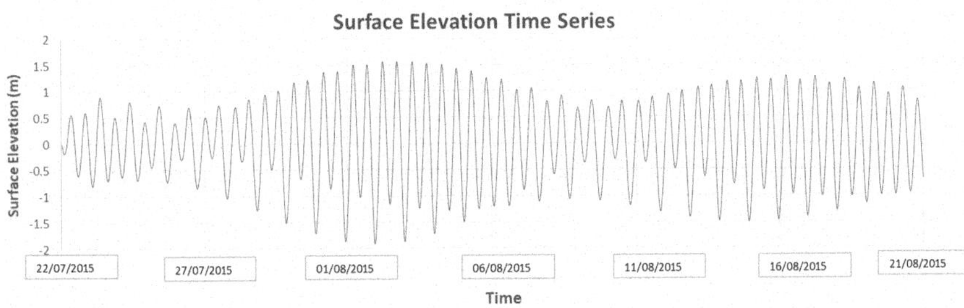

Fig. 10 Surface elevation

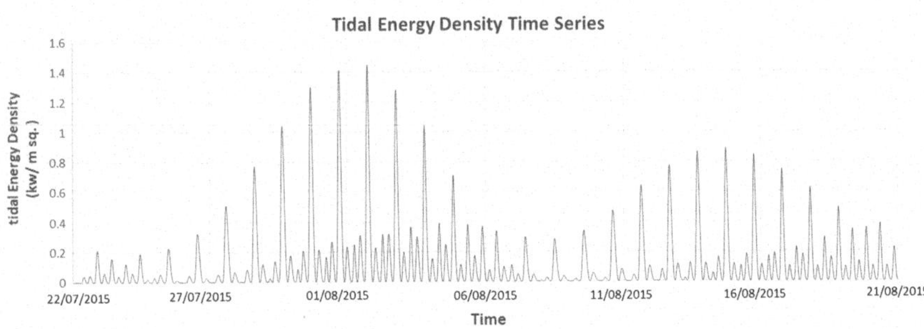

Fig. 11 Average power density

flow condition equates to a maximum annual power density (APD) of approximately 1.5 kW/m^2.

Further analyzing the plots of velocity time series, it can be concluded that the flow velocities at the specific site reach up to 1.4 m/s based on the numerical simulation estimation that the tidal energy will be a viable energy source. The tidal stream amounts to a maximum average power density of approximately 1.5 kW/m^2 of tidal energy density.

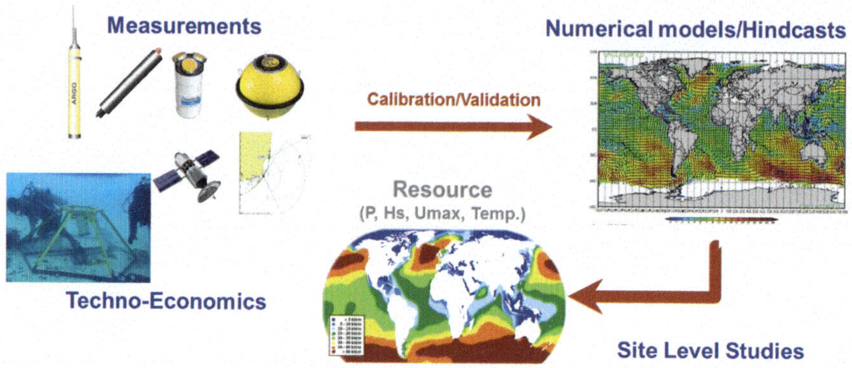

Fig. 12 Energy resource assessment in an Ocean site

3.2.2 Micro-Siting of Tidal Energy Deployment

Based on the macro level citing, high energy sites are identified towards a micro level citing to predict the energy potential distribution which shows the magnitude and the high energy potential locations towards the tidal turbine deployment. The prediction is based on combining the ocean depth information (seabed bathymetry and coastal bathymetry information) and the tidal current measurements as shown in Fig. 12 using simulation methods helps to predict flows and estimate the energy density plots of the site. In order to reduce the cost, the right sizing of the turbines has to be performed in tune to the sites tidal resource potential variation on an annual basis. For example, Fig. 13 shows the energy density maps that are generated and are presented for four sites in the Philippines, namely: Matnog-San Bernardino Strait, Verde Island Passage, Cebu-Santander Strait, and Davao Samal-Talicud Channel. Various estimates are made include annual energy yield, capacity factor, relative levelized cost of energy and availability factor to estimate different turbine operation in the ocean site. Figure 13 shows the results of the 4 sites that clearly identifies the best deployment locations of the optimum tidal turbine and its flow velocity distribution at that location.

4 Floating Tidal Turbine Development

In order to develop a low cost floating tidal turbine system towards remote coastal conditions, the present study focused on the following tasks:

- Tuning the hydrofoil towards low tidal flow conditions for Southeast Asian waters.
- Low inertial blade development with anti-seaweed and biofouling resistive functional coatings.

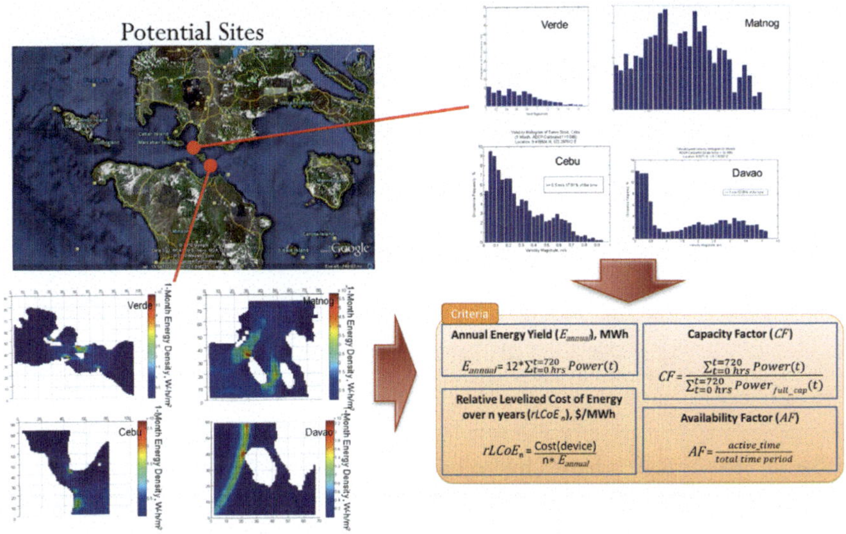

Fig. 13 Tidal energy modeling of specific sites and evaluating the site-turbine matching

- Testing of the turbine system through performance studies and further validation in field conditions.
- To differ from traditional seabed mounted tidal turbines, the present study focused to develop and demonstrate a novel tidal turbine integrated barge with a unique 'A' frame that was developed based on the naval architectural principles that is easily towable and deployable to any coastal location to minimize commissioning and decommissioning time and minimize overhaul and maintenance operations (O&M).

4.1 Hydrofoil Optimization of the Tidal Turbine Blade

Analogous to the wind turbine blade, a hydrofoil shape plays a major role in capturing tidal in-stream energy from the available tide's hydrokinetic flow. Significant efforts have been found in designing new airfoils for wind turbine application especially for megawatt class of wind turbines [10]. In a support for the development of small HAWT's, SG60XX airfoils are generally used [10]. However, it is from the operating conditions that hydrofoils for tidal hydrokinetic turbines the structural loads and the performance requirements differ significantly from the wind turbine airfoils and traditional aviation airfoils. Bigger challenge in designing marine current rotor is to avoid cavitation which causes lift to decrease and drag to increase. Cavitation inception can be predicted using cavitation parameter [11]. As a matter of fact, hydrokinetic turbines are plagued with soiling effects due to erosion and coating

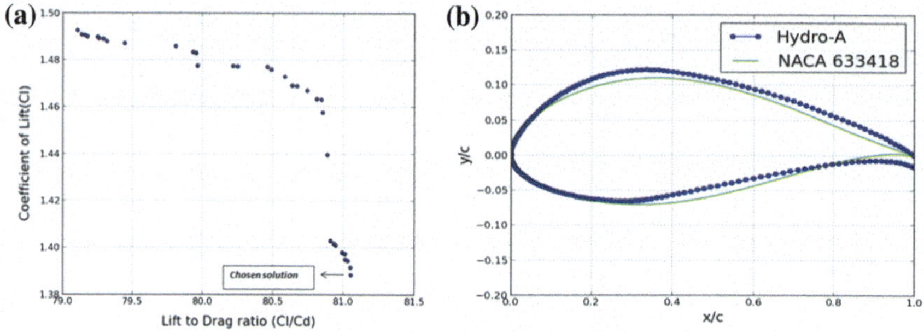

Fig. 14 **a** Pareto solution obtained from NSGA-II, **b** geometric shape comparison between optimized (Hydro-A) and baseline hydrofoil (NACA 633418)

spallation. These effects harms hydrofoil performance which includes reduction in lift and the lift curve slope, progressive aerodynamic stall with higher Reynolds number and surge in pressure drag due to premature transition on the boundary layer. In blade level, this poor performance leads to reduction in energy capture by virtue of drop in aerodynamic efficiency. Furthermore, thrust loads acting on the hydrokinetic blades are higher due to the density of water which is approximately 800 times larger than that of air. Therefore the designer should consider the structural requirements in designing hydrofoils for marine current application.

In the earlier study [11], the authors developed an elitism preserved Non-Dominated Sorting Genetic Algorithm (NSGA-II) can be used to design a hydrofoil by considering specific hydrodynamic and structural requirements. For example, when a NACA 633418 hydrofoil taken as the baseline geometry to start the process can be fit through a four Bezier curves to generate each hydrofoil in the geometry module. Accordingly in the present study, a hydrofoil was optimized for the tip region of a 3 m diameter axial flow horizontal axis hydrokinetic rotor operating at a tip speed ration (TSR) of 4.5. Based on the rotor specifications, optimization were performed at Reynolds number 1.8e6 and at a 6° angle of attack. Newly generated hydrofoil (called Hydro-A) exhibited a maximum thickness of 18.6% of chord and exhibits better performance than baseline airfoil in both free and fixed transition condition with less sensitivity towards biofouling induced surface roughness condition.

In multi-objective genetic algorithm, set of optimal solutions were obtained in a single simulation run which is called Pareto optimal solution as shown in the Pareto plot (see Fig. 14a) which shows optimal solution after 200 aerofoil design generation. Maximization problem is considered in this study and each dot in the Pareto plot represents an optimized hydrofoil obtained for a given design specification. In analogous to wind turbine, tip portion of the hydrokinetic machine contributes more towards the power production. Therefore from the Pareto plot, the hydrofoil that has high lift to drag ratio (termed C_l/C_d ratio which in this study was found to be $C_l/C_d = 81.05$) and with high lift coefficient ($C_l = 1.39$) was chosen as the optimized hydrofoil design solution (named as Hydro-A).

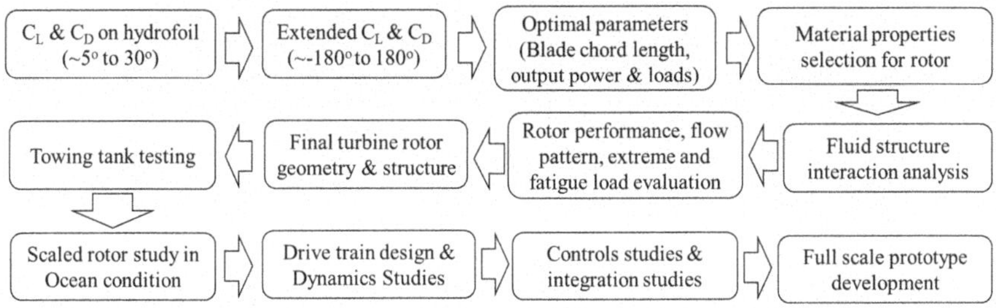

Fig. 15 Tidal turbine blade design methodology

As expected, flat back Hydro-A exhibits higher C_{lmax} both in free and forced transition condition that shows better lift performance at design condition when compared to the NACA 633418 (Fig. 14b). Drag characteristics evaluated in forced transition condition were also investigated. Hydro-A has a C_{lmax} value of 1.79 and corresponding C_d value of 0.051 at around 14 degree angle of attack in forced transition condition. Furthermore, in both free and forced transition condition, Hydro-A shows better post stall characteristics which avoids abrupt stall behavior. Due to the significant increase in C_L, lift to drag ratio value is increased to 81.05 at design condition which is 16% greater than the baseline value. Soiling and roughness effects are the major concern in the development of hydrokinetic turbine and it leads to performance degradation and increases maintenance cost. To mitigate soiling and roughness effects, design of flat back hydrofoil is employed in this study and optimization performed in the forced transition condition. As a result, the optimized hydrofoil Hydra-A exhibits less sensitive towards the roughness condition. Hydro-A exhibited high performance even for the rotors having other Reynolds number at the tip with less prone to cavitation. However as suggested in reference [11] the present study utilized twist modification at tip and usage of fibre reinforced composite which helped to reduce cavitation inception on the suction surface. Figure 15 showed the systematic procedure of the overall turbine blade and rotor development towards the full turbine development.

4.2 Analysis of Turbine Rotor

In order to obtain the optimal turbine power output, the BEM method is used to predict the hydrokinetic turbine performance and loads including rotor speed, torque, power, and thrust at the different current speeds [12, 13]. This analytical method is based on the aerodynamic theory, is easy to understand, and has high efficiency to analyses the turbine behaviour. It is the most widely used theory in practical wind or hydrokinetic turbine design currently. The computing procedure is described in the design methodology outlined in Fig. 15. The occurrence of cavitation is avoided theoretically in this processing by considering the local flow speed and water depth.

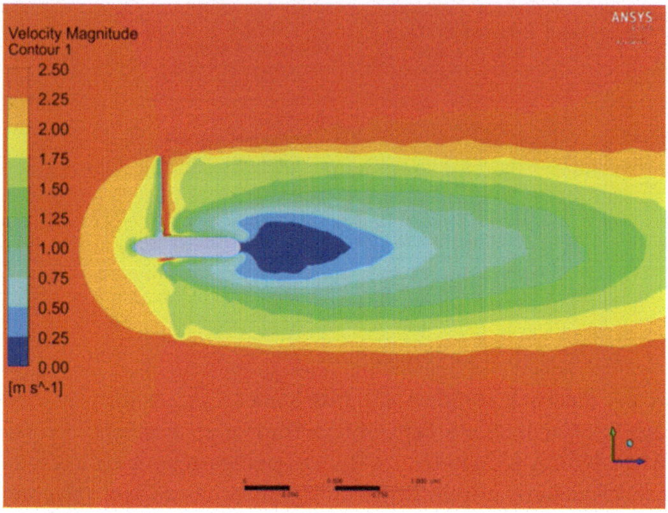

Fig. 16 Velocity-contour around tidal turbine and wave formation behind turbine

Novel blade designs have been analysed with different rake angle, pitch angle and load attenuation features to achieve good energy harvest characteristics and lower blade deflections. Detailed computation fluid dynamics was employed to evaluate the fluid induced torque, thrust force and the wake structure behind the wind turbine (see Fig. 16) and minimal hydro-acoustic noise.

4.3 Tow Tank Studies

Tidal Turbines capture the kinetic energy from hydrokinetic flow and convert it into useful mechanical power. The useful power generated at the turbine is given by

$$P = 0.5\rho A V^3 C_p \tag{1}$$

where P is power produced in Watts, A is the area swept by the rotor in m^2, ρ is the density of sea water in kg/m^3, V is the velocity of incoming flow in m/s, and C_P is the power coefficient which is the ratio of power extracted by the turbine to that contained in the tidal flow.

Towing tank testing is the experimental stage to validate the power coefficient of small scale device study and can mimics real field conditions in inflow direction (Fig. 17a). However it lacks cross-flow behaviour which occurs in field (see Fig. 17b). The initial towing tank studies help to obtain the turbine performance measurement (Fig. 18) which includes the power coefficient characteristics of the rotor and measure the different loads acting on the blades/rotor and to check the vibration from any

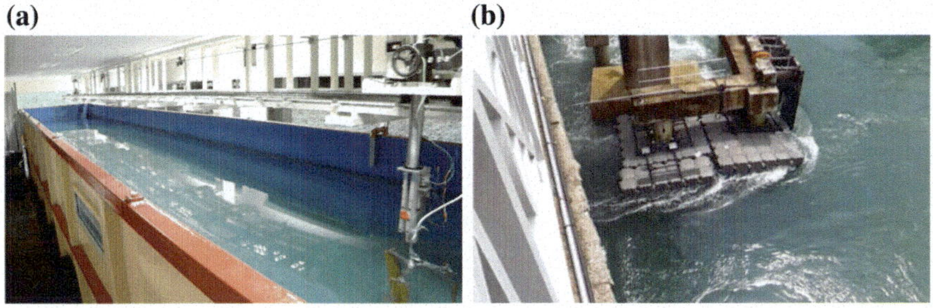

Fig. 17 **a** Typical towing tank to mimic different flow conditions on tidal turbine model in lab condition, **b** actual test bedding site where turbine is mounted on floating barge and tested

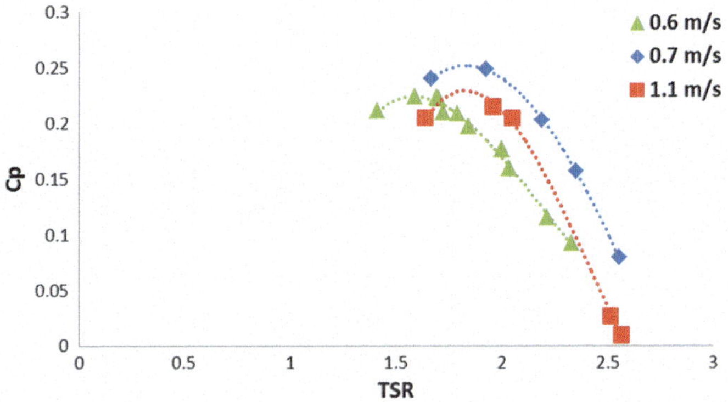

Fig. 18 Typical blade design's performance coefficient (Cp) versus tip speed ratio (TSR)

unbalanced loads. The experimental data is also used to validate the analytical and numerical results on the turbine performance and wake field investigation. This "Lab to Field" approach helps to directly allow the accurate prediction for the behaviour of the scaled up turbine once it is deployed into the sea as shown in Fig. 17b.

4.4 Tidal Turbine Development and Deployment

The initial prototype of the tidal turbine was designed towards a 1 m diameter towards a three bladed horizontal axis tidal turbine configuration. The geometry also includes a hub of about 0.1 m diameter and the nacelle part which is around 0.45 m long. Figure 19 displays the tidal turbine model used for the current study.

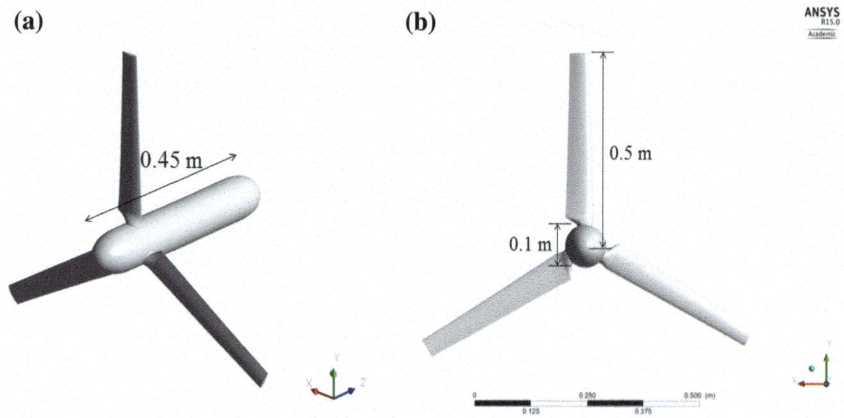

Fig. 19 **a** Isometric view of the turbine, **b** front view of the turbine

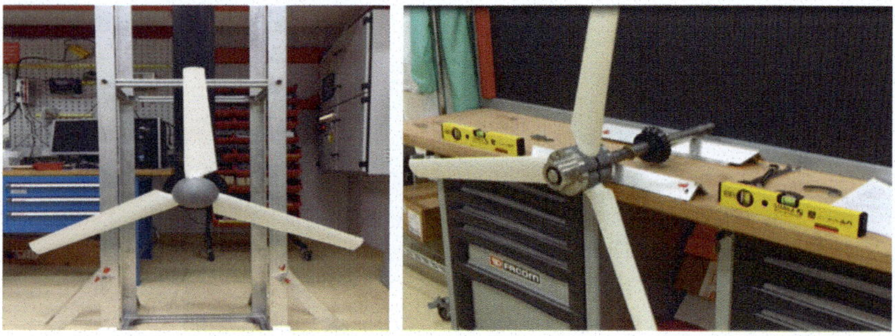

Fig. 20 Testing of balanced blade with rotor assembly

4.5 Blade Manufacturing Method

In this study, a novel manufacturing method for producing blades with moldless method was attempted and achieved to reduce the cost significantly. The blades were achieved using a novel core made by 3D printing process and further enhanced through carbon fiber layup to provide extra load bearing on the skin surface (Fig. 20). The blades were structurally tested for thrust loads (the predominant loading direction for the blade) in the lab using dead weight and further checked for structural damage. Results showed the design prototype can withstand the required loading level with a safety factor of 3.

4.6 Control System Design of Tidal Turbine

The working principle of tidal energy conversion system (TECS) is to use the potential and kinetic energy of the ocean waves caused by the celestial lunar gravitation to drive the turbine that is coupled with a generator to produce electricity. However, due to the physical structure of the generator, the power output is closely related with water speed, direction, load amount, etc., which show highly non-linear characteristics. The water speed and direction of the tidal stream are closely related to the lunar cycle, which can be considered as deterministic factor. However, there are non-deterministic factors such as sea habitat migration, seaweed entanglement, bio-foiling, etc. that can make the power output sub-optimal. In order to extract the maximum potential of tidal power in terms of electricity, the maximum power point (MPP) needs to be monitored and tracked in real time. In the present study, the state-of-the-art MPP tracking (MPPT) of TECS such as optimal tip speed ratio (TSR) method, optimum relation based (ORB) method, and perturb and observe (P&O) method was utilized to control the tidal turbine system [14]. Based on the reviewed methods, the principles, advantages and limitations of the methods can be summarized to show the performance of MPPT based on P&O in TECS with simulated results.

The Turbine was assembled in the Lab for detailed alignment check and friction & noise study was performed by obtaining the frequency response characteristics. Further it was tested to mimic the real field condition through motor actuation (Fig. 21). The electric motor was driven at variable speed to emulate the different speeds of the turbine. A gearbox is necessary to multiply the torque of the electromotor (max torque 7.3 Nm) 4 × times so the PM-generator was put under nominal load. The generated AC-power (variable voltage) was put through the rectifier bridge to get a variable DC-power to the MPPT battery charger. The charger reduced & stabilized the DC-voltage to 24Vdc, while increasing the (battery charge) amperage. Since Power-in is equal to power-out and thereby the losses were found to be within 2% due to power converter. The batteries are charged as per selected pre-programmed set-point schedule.

4.7 Prototype Tidal Turbine Testing at Sentosa Tidal Site

The turbine design was elegant which could be easily assembled on a floating barge and was tested for a significant period at the Sentosa test site in Singapore. The parts shifted were assembled and slide on a guide way design mounted on the floating platform into the water towards testing for performance in the water for a period of one year (see Fig. 22).

Fig. 21 Lab test of drivetrain system

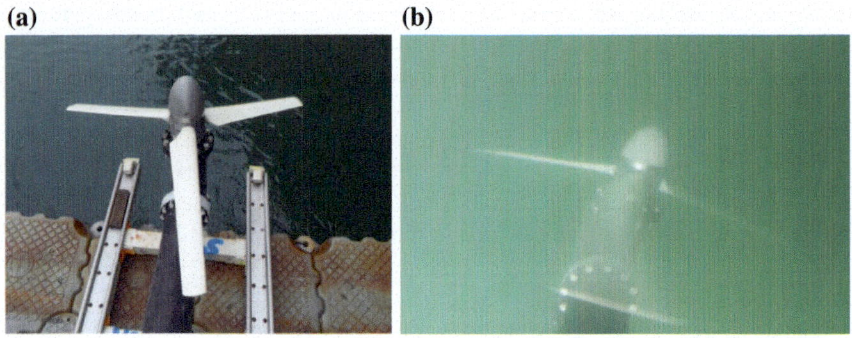

Fig. 22 **a** Tidal turbine ready for deployment, **b** successful deployment in Sentosa waters

4.8 Prediction of Stability of Floating Tidal Turbine Platform Under Towing Conditions

Tow tests were carried out as part of the development process to simulate and validate the behavior of the floating platform under expected operating conditions before deployment. The development of the dynamic model of the floating platform is based on the results of one such tow test that was carried out prior to deployment of the platform.

The dynamics simulation method was used to evaluate the stability of the floating tidal turbine platform. All forces are represented by coefficients, and the final loadings on each object is calculated using the Morison's approach. With the exception of the hull, the model of the floating platform was built within the software using regular geometric objects. The hull was imported as a custom object from a 3D modelling software. The turbine is represented by a disk, to which a drag coefficient equal to

Fig. 23 Illustration of tow setup in simulation model [7] and actual field test

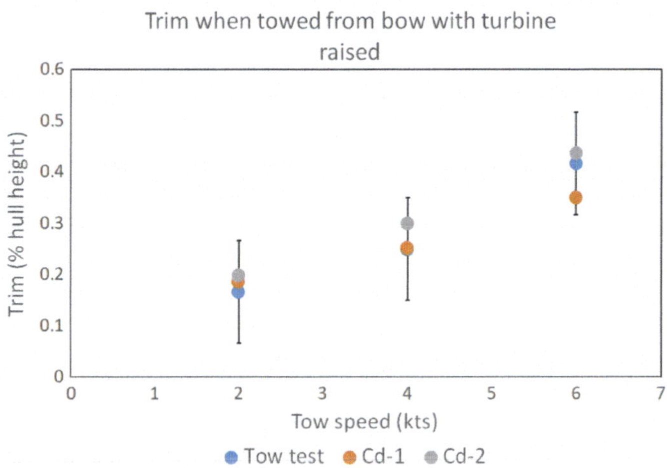

Fig. 24 Trim of model platform compared to tow test results for turbine-raised condition

the thrust coefficient of the actual turbine is applied. The features above the water line are ignored as the wind speeds were found to not be significant in this study. To simulate towing, one end of the tow line was fixed in space, with the other end of line being fixed to the floating platform. The water was then given a velocity to simulate the towing of the floating platform. This is illustrated in Fig. 23.

It is seen that the CG position of approximately 53% of the hull length from the stern produces the best match against the actual platform. Figure 24 shows the results under two drag coefficients tested, Cd-1 and Cd-2. Cd-2 is approximately 60% higher than Cd-1. Both drag coefficients are seen to produce good agreement with the results from the tow test. Cd-1 produced better agreement at tow speeds of 2 and 4 knots, while the Cd-2 produced better agreement at 6 knots.

The model can also be used to determine the maximum allowable wave height and period during towing operations. The model with the turbine raised was tested with 0.25 and 0.5 m wave heights and for wave periods of up to 15 s for flow speeds of 4 and 6 knots. This corresponds to the annual mean significant wave height and the 99% significant wave height found in Singapore. The model was also tested with

Fig. 25 Typical bio-fouling and corrosion in tropical waters

waves in the longitudinal and transverse direction with respect to the model. The 2nd order Stokes wave model was used. The minimum allowable wave period of each wave height was determined with the requirement that significant portions of the hull is not lifted out of the water at any time. The results show that the minimum allowable wave period is generally shorter in the transverse direction than in the longitudinal direction. This is because the platform is shorter in the transverse direction, leading to a shorter minimum allowable wavelength and hence smaller minimum allowable wave period. The minimum wave period is marginally shorter with slower flow speeds. The reason is likely because the shallower trim of the platform at lower flow speeds allows for higher additional pitching movements of the platform due to the wave action. The effect is expected to be more pronounced with higher tow speeds and larger wave heights.

4.9 Environmental Protection

Biofouling and corrosion are the key environmental challenges in tropical waters that can impart the operational efficiency of the tidal turbine (see Fig. 25). Accordingly, a systematic study of marine safe biofouling coating was studied using a dynamic test rig that tested the multi-functional coating in the turbid waters under the turbines and floating barges' operational velocities [15].

4.10 Fully Fabricated Floating Tidal Energy System

Tidal flow models help determine the best flow locations as explained in earlier Sect. 3.2, For example in the southern islands of Singapore, the flow velocity are good as shown in Fig. 8 which is suitable for such floating tidal turbine deployment. Acoustic Doppler Current Profilometers (ADCP) can be deployed to verify the pre-

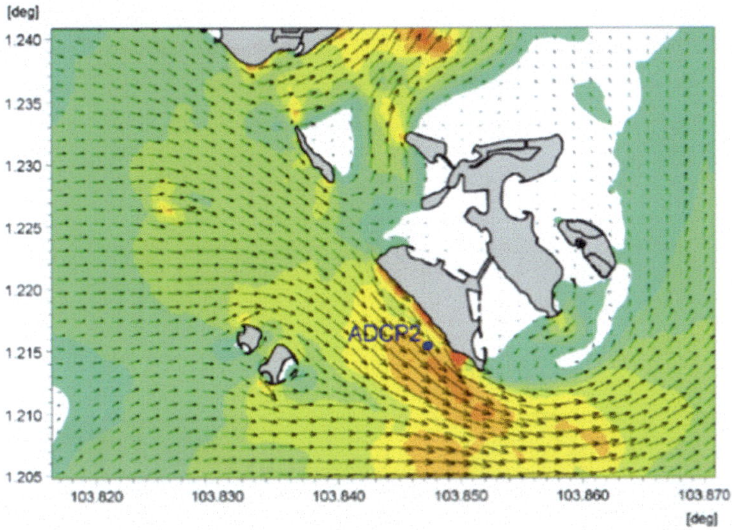

Fig. 26 Tidal flow around Southern Singapore island waters

dicted flow velocities. Using transect analysis the best flow velocity variations can be determined to finalized the optimum tidal rotor diameter. Using the micro-siting principles explained in Sect. 3.2.1, a detailed tidal flow study (see Fig. 26) was performed to identify the best location to deploy the floating tidal system and estimate the rotor size for best energy availability, capacity factor, relative levelized cost of energy and availability factor as explained in Fig. 13.

In the present study, the final floating tidal power platform is of a catamaran design and employed commercial tidal turbine with a rotor diameter of 4 m integrated to a generator that can provide a rated power of 62 kW. The reason being, as seen in Eq. (1) the power production is proportional to the rotor area and hence in low tidal flow conditions the design option is to use larger rotor diameter to compensate for the low velocity. However, in tropical coastal like Singapore waters the water depths are averagely up to 50 m and hence rotor diameters should be carefully sized based on high and low tides and estimates of regions with uniform flow velocities. In the present study the detailed tidal energy assessment was performed for Southern Singapore and the tidal energy density was found to be good in places near St. Johns island, Seringat, etc. as shown in Fig. 8.

Today the capital cost of the commercial small-scale tidal turbine is around 2800 USD/KW and is experiencing a steady cost reduction due to the industry's learning curve of up to 15% per year, as there is knowledge spillover from similar industry such as offshore wind & ship propeller industries.

Conventional ocean energy systems are subject to biofouling and corrosion that demands for regular maintenance. Performing maintenance in the ocean sites is costly than at dry dock conditions. Hence new generation systems are preferred to

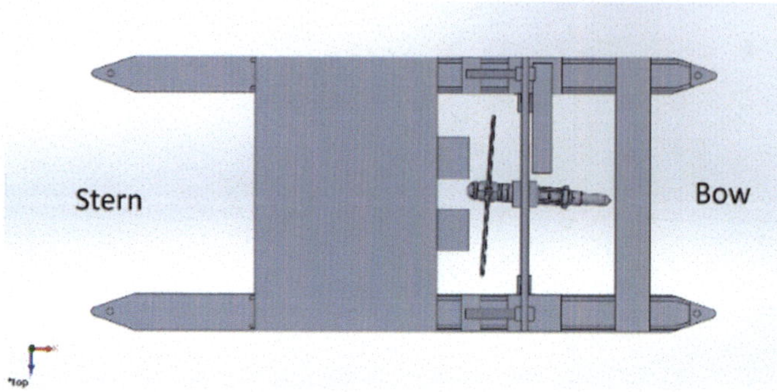

Fig. 27 Planform of floating tidal power platform

Fig. 28 Fully fabricated tidal energy system in Singapore

be floating systems so that it could be towed to any ocean site location and deployed with mooring supports. The mooring systems were designed towards minimal sway and pitching and yawing degrees of freedom [16]. The turbine was designed to be mounted on an A-frame, and the whole structure is capable to be lifted into and out of the water for maintenance using an electrical actuator such as winch and pulleys. Draft markings were painted at the bow and stern of the platform at intervals of 0.1 m, allowing the draft of the platform to be read. The planform of the platform is shown in Fig. 27. Figure 28 shows the fully fabricated floating barge system achieved in Singapore yards and was deployed for 6 months in Sentosa coastal waters to evaluate the stability performance and environment protection.

5 Floating Tidal Energy Concept

The above floating tidal energy system can be enhanced further for high energy harvesting capability per unit foot print by integrating a variety of matured renewable system such as floating solar, vertical axis wind turbine [17–19], wave energy converters [20, 21], ocean thermal gradient energy system [3], salinity gradient based energy system [3], e.t.c., as shown in Fig. 29, based on the ocean site's energy resource availability studies. For example, in sites such as Philippines, Indonesia

Fig. 29 Typical floating hybrid renewable system that integrates various renewable energy systems

and Malaysia there are sites with water depth beyond 1 km that offers ocean thermal energy gradient, tidal currents and wind loads which can be uniquely combined to offer high energy foot print in the floating renewable system. This concept helps to minimize the electrical and structural infrastructure to provide economic reduction in the capital cost and maximize energy availability. Further by deducing the static and dynamic forces in the specific ocean site on the proposed floating structure, the design life, stability and reliability of the floating renewable system and reliability can be evaluated by analyzing against fatigue and extreme loads from wind and wave energy as well as the environmental factors (like corrosion and biofouling) in the chosen site [16].

To evaluate the floating renewable energy system under natural disasters such as typhoons, hurricanes, and tsunamis detailed atmospheric boundary layer interfaced floating studies are being pursued. In reference [4, 22, 23] the tropical wind conditions from radiosonde sensors and LIDAR's are being studied and incorporated through a combination of DHI's MIKE software and WRF models [24].

5.1 *Micro-Grid Design*

The smart micro-grid is capable to combine multiple AC and DC type energy sources and store in multiple energy storage devices which includes, batteries, flywheel and compressed air energy storage. Each energy storage technology differs in its storage capacity and its charging and discharging speed and latency in reacting to external load and charging conditions. Through the smart grid architecture, the various renewable energy can be combined on the floating platform and controlled individually and at system level through artificial intelligence supported control systems.

To demonstrate this concept a smart nano-level grid study was conducted in Tuas Singapore to study the grid architecture of combining multiple AC and DC energy sources towards AC loads of a workers quarters (Fig. 30). The solar PV and wind turbine were directly connected to the 48 V DC bus via DC–DC converters. This demonstration of the nano-grid solution utilized wind and solar hybrid energy generation due to their AC and DC energy characteristics and was combined with Fuel

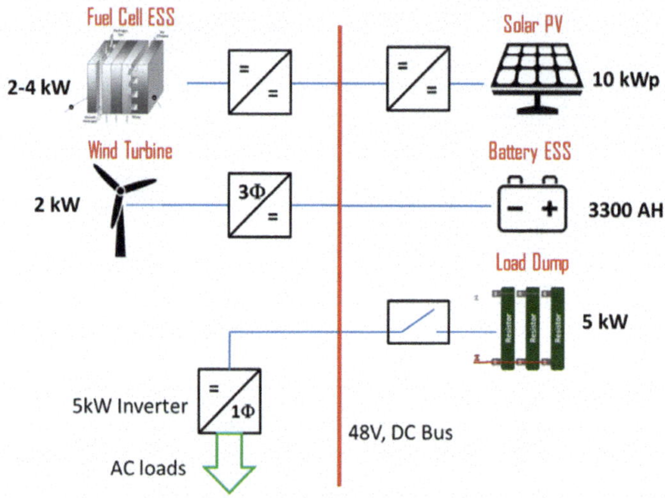

Fig. 30 Smart Nano grid architecture studied in Tuas Singapore towards combining AC and DC energy sources and support AC electrical loads

Cell storage and Battery based energy storage system (BESS). The fuel cell system was of 3–5 kW with one high pressure electrolyser, along with a hydrogen storage. The energy storage systems (Fuel cell and battery) were connected to the DC bus via bi-directional converters as electricity was flowing in and out of the energy storage systems. A load dump was incorporated into the power system to provide to burn out excess energy and as a means to protect battery against overcharging. The electrical loads (single phase) were connected to the DC–AC inverter. Upon commissioning the system was setup to power local loads of the temporary worker quarters such as street lights, fridge and water heater. Further the project focused in demonstrating novel inverters and a nano-grid controller. A supervisory control system was incorporated in the project to the power system along with the energy storage system [14].

To enhance the certainty and reliability of power production machine learning based energy Forecasting methods was utilized [16, 25–27]. The author and other team members have utilized a variety of machine learning techniques to predict wind, wave and tidal energy sources and have found its capability to predict the sharp ramp in energy variations and the intermittencies [16, 25–27]. The capability to forecast helps to implement a feed forward control schema to the individual energy generation system.

Recently a survey was done in Southeast Asia to understand the energy availability in remote coastal and island region and how ocean based renewable energy sources can support the livelihood of the people [28, 29]. The survey focused also to understand the barriers to renewable energy adoption in these remote sites, which included the techno-economic challenges, the lack of understanding in terms of renewable energy resources available in southeast Asian remote coastal locations, lack of skilled personal, environmental interactions such as doubts in terms of underwater noise from ocean energy systems on the nearby fish farms and other sensitive marine industries [28]. In reference [29] it was clearly shown that Ocean based renewable sources are capable to power remote islands and coastal locations. Presently efforts are in progress to study the environmental impacts of the floating renewable system on the marine ecology to ensure there is minimal impact on the marine mammals and fishes.

The present study has shown that the systematic development of the floating hybrid renewable energy system can exploit the presence of various renewable energy sources in a given ocean site. This can be achieved through the advancements in smart grid science to combine various relevant ocean energy systems towards an ocean site and powered through novel controls that are supported with machine learning methods to overcome the inherent intermittencies of renewable energy sources to assure quality power with high certainty.

6 Conclusion

Blue economic growth is seen as a promising strategy to utilize the ocean resources towards economic growth without exploiting the ecology through setting up various marine industries including deep sea fishing, seabed mining, maritime transport, desalination. Development of floating homes and cities will be a key industry in this Blue Economy and Growth strategy as it addresses the imminent need for solutions towards sea level rise effects on inundation zones' coastal dwelling. In any such remote coastal location, multiple energy sources exists viz., solar, tidal and wave energy sources which can displace the use of fossil fuel and can ensure the pristine marine ecology through effective use of renewables. Remote coastal site has challenges of less available skilled people and hence the idea of floating renewables with modular design helps easy means to bring the system to main land for dry-dock conditions and achieve necessary repair and maintenance at least cost and necessary quality.

In this paper the concept of hybrid energy system is proposed that could be housed in a floating platform system and integrated with smart grid and energy forecasting and energy storage methods to become a smart power system in a marine condition. This will ensure credible power supply to marine industrial use and floating homes in any remote coastal sites with high energy availability and support essential human needs such as electrical, water, air conditioning and other energy requirements towards marine operations. The floating power plant will ensure necessary energy resilience and electricity and water security towards a floating home or a floating cities through a uniquely combination of all available energy resources at an ocean site.

This paper also shared how the field based seabed and coastal bathymetry surveys could be uniquely combined with tidal current & wave measurements through simulation to perform energy resource assessment, and assess structural integrity through hydrodynamics, fluid-structure interaction studies of these floating energy systems and load studies under real ocean conditions. In addition, through detailed resource mapping and device performance studies the best site locations can be identified for the ocean device deployment to achieve optimum levelized cost of energy, maximum availability and maximum capacity factor. Thus this paper has elucidated that the hybrid floating energy system can be a viable power plant towards tropical coastal and island regions to support remote floating homes' energy needs through clean energy solutions with greater certainty and power quality.

Acknowledgements The author wishes to thank his organization Energy Research Institute @ NTU and Economic Development Board of Singapore for the generous research fund (grant no. S14-1067-NRFEIPO-EIRP-IHL) during this research. The author greatly appreciates the excellent support and dedication provided by his research team: Mahesh Ramanathan, Kannappan Lakshmanan, Michael Abundo, Ly Duy Khiem, Alex Ng, Martin Koh, Manoj Sekar, Brindha Chellaiah, Paul Hibbard, Pranava Sai during this in-stream tidal energy development and deployment.

References

1. Dasgupta, S., Laplante, B., Meisner, C., Wheeler, D., & Yan, J. P. (2007). *The impact of sea level rise on developing countries: a comparative analysis (English)*. Policy, Research working paper; no. WPS 4136. Washington, D.C.: World Bank.
2. Church, J. A., Gregory, J. M., Huybrechts, P., Kuhn, M., Lambeck, K., Nhuan, M. T., Qin, D., & Woodworth, P. L. (2001). Changes in sea level. In: J. T. Houghton, Y. Ding, D. J. Griggs, M. Noguer, P. J. Van der Linden, X. Dai, K. Maskell, & C. A. Johnson (eds.), *Climate Change 2001: The Scientific Basis: Contribution of Working Group I to the Third Assessment Report of the Intergovernmental Panel.*
3. https://www.ocean-energy-systems.org/. Accessed January 31, 2019.
4. Bhat, K., Negi, P., & Narasimalu, S. (2016). Wind resource assessment for urban renewable energy application in Singapore. *Renewable Energy*.
5. Hywind—The World's Leading Floating Offshore Wind Solution. https://www.equinor.com/en/what-we-do/hywind-where-the-wind-takes-us.html. Accessed January 31, 2019.
6. Satheesh Kumar, J., & Balaji, R. (2017). Estimation of tidal current energy along the Gulf of Khambhat using three-dimensional numerical modelling. *The International Journal of Ocean and Climate Systems*, 1–9.
7. Sutherland, G., Foreman, M., & Garrett, C. (2007). Tidal current energy assessment for Johnstone strait, Vancouver island. *Proceedings of the Institute of Mechanical Engineers, Part A: Journal of Power and Energy, 221*(2), 147–157.
8. Blunden, L. S., & Bahaj, A. S. (2006). Initial evaluation of tidal stream energy resources at Portland Bill, UK. *Renewable Energy, 31*(2), 121–132.
9. Pal, R. K., Narasimalu, S., & Lakshmanan, K. (2018). *Tidal Resource Modeling: Alderney Race, ACEPT IEEE Conference*, Singapore, October 2018.
10. Giguère, P., & Selig, M. S. (1998). New airfoils for small horizontal axis wind turbines. *Journal of Solar Energy Engineering, 120*(2), 108–114.
11. Jesudasan, R., & Narasimalu, S. (2015). Design and optimisation of hydrofoils for hydrokinetic turbine using multi-objective genetic algorithms. In *Eleventh European Wave and Tidal Energy Conference (EWTEC 2015)*, Nantes, France, September 7–10, 2015.
12. Batten, W. M. J., et al. (2006). Hydrodynamics of marine current turbines. *Renewable Energy, 31*(2), 249–256.
13. Dhivya, S., & Srikanth, N. (2016). Hydrodynamics of semi-buoyant tidal turbine support system, AWTEC 2016.

14. Wen, M. X., & Srikanth, N. (2017). Maximum power point tracking for tidal turbine generator. In *IEEE Conference ACEPT 2017*, Singapore.
15. Manoj, S., Mahesh, S., & Srikanth, N. (2018). Review of biofouling paints on the marine vessel. In *ACEPT IEEE Conference on Asian Clean Energy and Power Transmission*, Singapore, Oct 2018.
16. Ren, Y., Suganthan, P. N., Srikanth, N., & Amaratunga, G. (2016). Random vector functional link for short term electricity load demand forecasting. *Information Science Journal*, Jan 2016.
17. Palanisamy, M. K., Mohan Ram Surya, M., Srikanth, N., & Lim, T. C. (2018). Experimental and numerical investigation of novel Savonius wind turbine. *Wind Engineering*.
18. Palanisamy, M. K., Srikanth, N., & Lim, T. C. (2017). On the self-starting of Darrieus turbine: An experimental investigation with secondary rotor. In *Asian Conference on Energy, Power and Transportation Electrification (ACEPT)*, October 2017.
19. Palanisamy, M. K., Srikanth, N., & Lim, T. C. (2017). Performance assessment of Darrieus turbine with modified trailing edge airfoil for low wind speeds. *Smart Grid and Renewable Energy, 8*(12), 425–439.
20. Xiong, M., Ukil, A., Zhao, J., & Narasimalu, S. (2016). Modeling and performance assessment of pontoon roller wave energy converter in Singapore. In *IEEE Tencon Conference*, Singapore, November 2016.
21. Ly, D. K., Aboobacker, V. M., Abundo M. L. S., Srikanth, N., & Tkalich, P. (2014). Wave energy resource assessment for Southeast Asia. In *5th International Conference on Sustainable Energy and Environment (SEE 2014): Science, Technology and Innovation for ASEAN Green Growth*. Bangkok, Thailand, November 19–21, 2014.
22. Srikanth, N., Narasimhamurthy, R., & Kannan, A. (2018). Geospatial model to estimate wind energy resource potential in remote locations. In *ACEPT IEEE Conference on Asian Clean Energy and Power Transmission*, Singapore, October 2018.
23. Tieo, J. J., Koh, T. Y., Skote, M., & Srikanth, N. (2018). Variance characteristics of tropical radiosonde winds using a vector-tensor method. *Energies Journal*.
24. Garg, N., Ng, E. Y. K., & Srikanth, N. (2018). The effects of sea spray and atmosphere-wave coupling on air-sea exchange during a tropical cyclone. *Atmospheric Chemistry and Physics, 18*(8), 6001–6021, April 2018.
25. Ren, Y., Suganthan, P. N., & Srikanth, N. (2015). Ensemble methods for wind and solar power forecasting—A state of the art review. *Renewable and Sustainable Energy Reviews*, Oct 2015.
26. Anh, N., Suresh, S., Pratama, M., & Srikanth, N. (2019). Interval prediction of wave energy characteristics using meta-cognitive interval type-2 fuzzy inference system. *Knowledge-Based Systems*, Jan 2019.
27. Anh, N., Prasad, M., Srikanth, N., & Sundaram, S. (2018). Wind speed intervals prediction using meta-cognitive approach. In *Procedia Computer Science*, Jan 2018.
28. Quirapas, M. A. J., Srikanth, N., Abundo, M., & Lin, H. Electricity from the ocean: A path to secure energy in Southeast Asia. In *ASEAN Green Growth*, 19–21. Bangkok, Thailand.
29. Srikanth, N. (2015). Ocean energy as an energy source to power remote islands. In *Asian Clean Energy Summit (ACES)*, ACES Conference 2015, Singapore.
30. https://www.oceanunite.org/issues/marine-plastic-pollution/. Accessed January 31, 2019.
31. Tangler, J. L., & Somers, D. M. (1995). NREL Airfoil Families for HAWTs. In *Other Information: PBD*, January 1995.
32. Koh, M., & Ng, A. (2018). Prediction of the stability of a floating tidal turbine platform under towing conditions. In *Asian Clean Energy Summit (ACES)*, Singapore, Oct 2018, ACEPT 2018, Singapore, Organizer: Sustainable energy association of Singapore (SEAS).
33. Wala, Abdulqadir Aziz, Ng, E. Y. K., & Srikanth, N. (2018). A Beddoes-Leishman-type model with an optimization-based methodology and airfoil shape parameters. *Wind Energy, 21*(7), 590–603.
34. Low, C. M., Ng, E. Y. K., Srikanth, N., et al. (2018). Numerical modelling of seabed impact effects on chain and small diameter mooring cables. *Applied Ocean Research, 80*, 248–277.
35. Chelliah, B., Manoj, S., & Srikanth, N. (2018). Renewable powered desalination system for remote islands in Singapore. In *ACEPT IEEE Conference on Asian Clean Energy and Power Transmission*, Singapore, Oct 2018.

Floating Offshore Wind Turbines in Goto Islands, Nagasaki, Japan

Tomoaki Utsunomiya, Iku Sato and Takashi Shiraishi

Abstract Offshore wind energy resources in Japanese EEZ are now considered to be huge. In order to utilize the huge amount of energy located in relatively deep water areas, Ministry of the Environment, Japan funded a demonstration project on floating offshore wind turbine (FOWT). In the project, two FOWTs have been installed. The first FOWT mounted a 100 kW wind turbine of downwind type, and the length dimensions are almost half of the second FOWT. The second FOWT mounted a 2 MW wind turbine of downwind type, and was referred to as the full-scale model. The FOWTs consist of PC-steel hybrid spar which is cost-effective and are moored by three mooring chains. The half-scale model was installed at the site (Kabashima, Goto Islands, Nagasaki prefecture, Japan) on 11 June 2012. The half-scale model was attacked by a very severe typhoon Sanba (1216). The behavior of the half-scale model during the typhoon attack was recorded, and compared with the computer simulations, indicating the validity of the design method. After a successful demonstration test of the half-scale model, the full-scale model was designed, constructed and installed at the same site. The demonstration test for the full-scale model was also successful. After completion of the demonstration project, the full-scale model was moved to a different site off Fukue island, where future expansion as a floating wind farm is planned. There, the full-scale model is operating as a commercial floating wind turbine, providing valuable data and experience for operation and maintenance toward commercial-scale floating wind farms.

Keywords Floating offshore wind turbine · FOWT · Hybrid spar · Full-scale demonstration

---

T. Utsunomiya (✉)
Department of Marine Systems Engineering, Kyushu University, Fukuoka, Japan
e-mail: utsunomiya@nams.kyushu-u.ac.jp

I. Sato
Energy Division, Toda Corporation, Tokyo, Japan

T. Shiraishi
Power Production Management Division, Power Business Unit, Hitachi Ltd., Hitach, Japan

C. M. Wang et al. (eds.), *WCFS2019*, Lecture Notes in Civil Engineering 41,
https://doi.org/10.1007/978-981-13-8743-2_20

1 Introduction

In order to mitigate greenhouse gas emission, it is necessary to increase renewable energy production. In Japan, the production of renewable energy is still very limited. However, offshore wind energy resources in the Japanese EEZ (Exclusive Economic Zone) are now considered to be huge. In particular, the offshore wind energy resource in deeper waters (where the water depth is greater than 50 m) is remarkable [1]. For developing deep water wind energy, the use of floating-type foundations is considered to be more economical than the use of bottom-fixed foundations [2]. However, there existed only one multi-megawatt floating wind turbine in 2010; that was the Norway's Hywind, built and installed in September 2009 [3]. The second multi-megawatt floating wind turbine, WindFloat—a semisubmersible-type floating wind turbine, was installed in Portugal in October 2011 [4]. Since then, Japan's Ministry of the Environment kicked-off a demonstration project on floating offshore wind turbine (FOWT). The project took six years; beginning from September 2010 to March 2016. In this project, two floating offshore wind turbines have been installed and tested. Some details are given in [5, 6]. This paper summarizes the demonstration project and states current situations.

2 Outline of the Demonstration Project

The ultimate objective of the demonstration project is to reduce the greenhouse gas emission through commercialization of FOWT in the Japanese EEZ. Towards the commercialization of FOWT, a mandatory and important step is to demonstrate its technical feasibility. In particular, a demonstration of installation and operation of a multi-megawatt floating offshore wind turbine at sea is the primary objective of the project. Although the installation/operation of a multi-megawatt FOWT is the main target of the demonstration project, a step-by-step approach is generally preferred, in order to reduce possible risks. Also, social acceptance may be gained by such a step-by-step approach. Thus, the half-scale model, of which its length scale is almost half of the full-scale model, was planned to be installed before the installation of the full-scale model.

Figure 1 shows the master schedule of the demonstration project. The project continued over six years. In the first year, the offshore site for the demonstration was selected. In the site selection, obtaining permissions/agreements from the local fishery cooperative and the local community was the most critical matter. After establishing the demonstration site, the meteo-ocean measurement and the survey for environmental impact assessment got started. The half-scale model was designed, constructed, installed, operated, and finally removed. Following the half-scale model test, the full-scale model was designed, constructed, installed, operated, and finally relocated to a different site for commercial power generation. The meteo-ocean

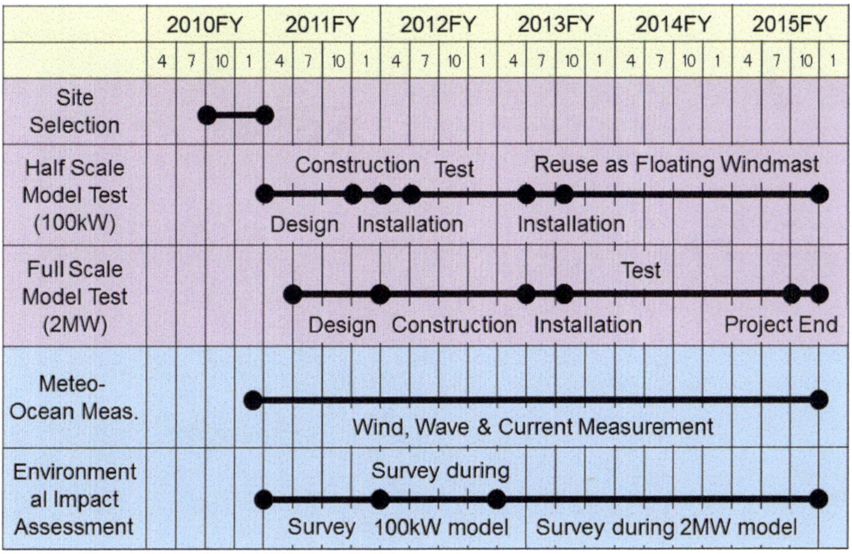

Fig. 1 Master schedule of demonstration project

measurement and the survey for the environmental impact assessment had been made through the demonstration project.

Figure 2 shows the location of the at-sea demonstration site. The site is about 1 km offshore of Kabashima Island, Goto city, Nagasaki prefecture, Japan. The mean water depth is 97.2 m (at mean sea level; MSL). The site is surrounded by the island on the north-western side, but it is open in the south-eastern side. The marine cable has been installed for the grid-connection. The distance to the shore from the FOWT along the marine cable is about 1.8 km.

3 Half-Scale Model

3.1 Description

Figure 3 shows the outline and main dimensions of the half-scale model. In Table 1, the main specifications of the model are presented. The floating foundation has a slender cylindrical shape (spar), with a draft of 37.05 m, an outer diameter of 3.8 m at the bottom and an outer diameter of 2.375 m at the sea level. The bottom half of the floating foundation is made of precast PC (pre-stressed concrete) segments whereas the upper half is made of ring-stiffened steel. In order to mitigate the yaw motion, four straight fins are attached along the PC part. Conventional catenary chain system is used as the mooring system. Three stud-link chains with a nominal diameter of

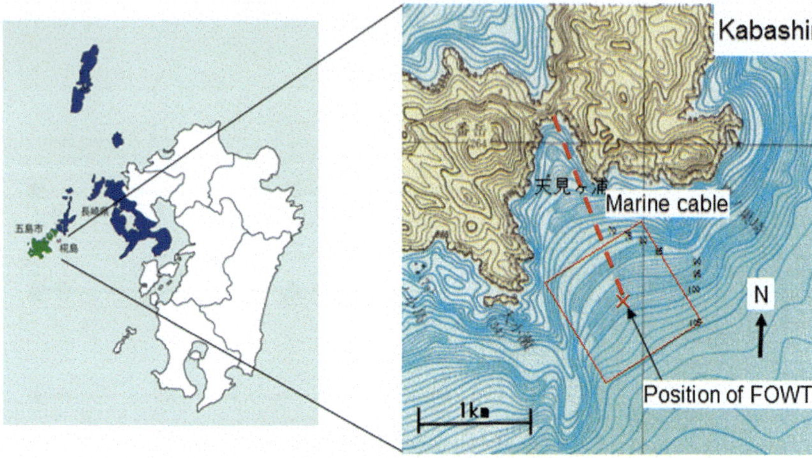

Fig. 2 Demonstration site: Kabashima, Goto city, Nagasaki prefecture

Table 1 Main specifications of half-scale model

Item	Specifications
Wind turbine	SUBARU22/100 (down-wind type)
Rated power	100 kW
Rotor speed	18–72 rpm
Cut-in wind speed	3–4 m/s
Cut-out wind speed	20 m/s
Rated wind speed	7.5 m/s (when power limited to 40 kW)
Dimensions of fins	0.475 m (width) × 20 m (height)
Number of fins	4
Mooring chains	Nominal diameter of 56 mm (Grade 3) × 3
Concrete sinker	Weight in air: 200 tf × 2
Danforth anchor	Weight in air: 10 tf × 1

56 mm are used. For the anchors, two concrete sinkers and one Danforth-type anchor were selected after considering the sea-bed condition.

The wind turbine has a rated output of 100 kW. However, during the demonstration test, the maximum power was limited to 40 kW so as to increase the possibility of occurrence of wind speed above rated wind speed, where pitch control of blades is made. The wind turbine was modified to a down-wind type from the original up-wind type design. The down-wind type turbine is considered to have advantages such as weather-vane effect in yaw direction and the rotor axis being horizontal when the tower is tilted due to wind action.

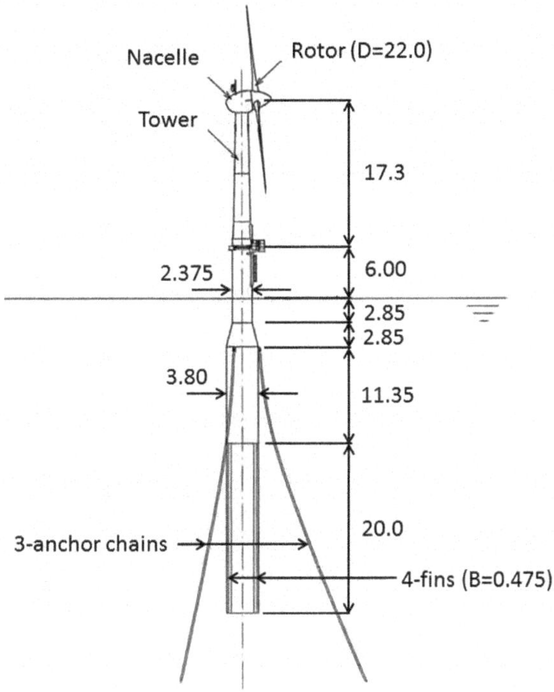

Fig. 3 Dimensions of half-scale model (in m)

The structural design of the floating wind turbine was made by relying on the time-domain numerical simulations. Some details of the numerical simulations and the experiments used for validation are presented in [7, 8].

3.2 Construction and Installation

The PC part and the steel part of the floating body were constructed separately by different manufacturers. PC precast ring segments, with an outer diameter of 3.8 m and length 2.0 m, have been produced by a centrifuge manufacturing process. Upper steel part was fabricated at a shipyard. The ring segments and the steel part were then transported to a quay yard. Figure 4 shows the assembly of the floating body and the tower section at the quay yard. They were assembled in the horizontal position.

Figure 5 shows the installation procedure. The completed floating body with tower and the wind turbine was transported by a barge to the northern side of Kabashima Island, where the weather is rather calm with respect to wind/wave conditions compared to the southern side. The entire floating body was upended by using a large floating crane. After the successful upending, the floating wind turbine was towed to

Fig. 4 Assembly at quay yard

Fig. 5 Installation of the half-scale model

the southern side of Kabashima Island. At the demonstration site, the mooring chains were hooked-up to the floating body. The mooring chains with the sinkers/anchor were pre-laid at the site. The hook-up of the mooring chains were completed on 11 June 2012. Figure 6 shows the general view of the half-scale model with an access boat used for maintenance.

3.3 Response During Typhoon Attack

During the demonstrative experiment of the half-scale model, the FOWT was attacked by two separate severe typhoons, that is, by Bolaven (international designation: 1215) and Sanba (international designation: 1216). Sanba (1216) was a record-making typhoon event. It was closest to the at-sea experiment site at around 5:00am on 17 September 2012. At that time, the central atmospheric pressure of the typhoon was 940 hPa. During the typhoon, several data were obtained, that include the wind speed, wave height, motion of the floating body, strains of the tower and the floating body, tension of a mooring line. The details of the measurement can be found in [9]. In the following, some details during the typhoon attack are presented (for details, see [10]).

Figure 7 (left) shows the 10-min average wind speed measured at the top of the nacelle by the cup-type anemometer. The maximum 10-min average wind speed

Fig. 6 General view of the half-scale model with an access boat

was 36.8 m/s taken during 05:00 a.m.–05:10 a.m. on 17 September 2012. The design wind speed corresponding to the return period of 50 years as the 10-min average wind speed for the FOWT is 48.3 m/s at the hub-height (23.3 m above sea level). Thus, the maximum wind speed during the typhoon event was 24% lower than the design wind speed. Figure 7 (right) shows the significant wave height and the significant wave period measured by the wave measuring buoy. The maximum significant wave height was 9.5 m taken during 05:00 a.m.–06:00 a.m. on 17 September 2012, and the maximum significant wave period was 13.0 s taken also at the same time. The maximum significant wave height of 9.5 m exceeded the design wave height of 8.4 m (for 1-h reference period), but the maximum significant wave period was below the design value of 14.0 s.

In order to validate the numerical simulation method, the dynamic responses have been re-produced by using the measured wind and wave time series data. Table 2 shows the numerical calculation cases. Figure 8 shows the minimum and maximum values of the platform motion in pitch. The reference point of the motion is at 6 m above the sea level (at the tower base). Basically, good agreement can be seen between

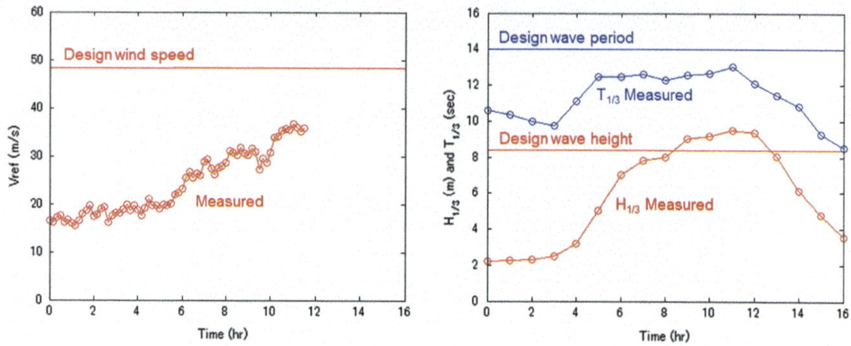

Fig. 7 (left) 10-min average wind speed measured on top of nacelle. Time 0 h corresponds to data for 18:00–18:10 on 16 September, and 11 h to 05:00–05:10 on 17 September. (right) Significant wave height and significant wave period. Time 0 h corresponds to data for 18:00–19:00 on 16 September, and 11 h to 05:00–06:00 on 17 September

Table 2 Numerical calculation cases

Item	C_D of fins	Mooring dynamics	Yaw damping (%)
Case 1	1.5	Quasi-static	2.0
Case 2	10.0	Quasi-static	2.0
Case 3	1.5	Dynamic	0.0

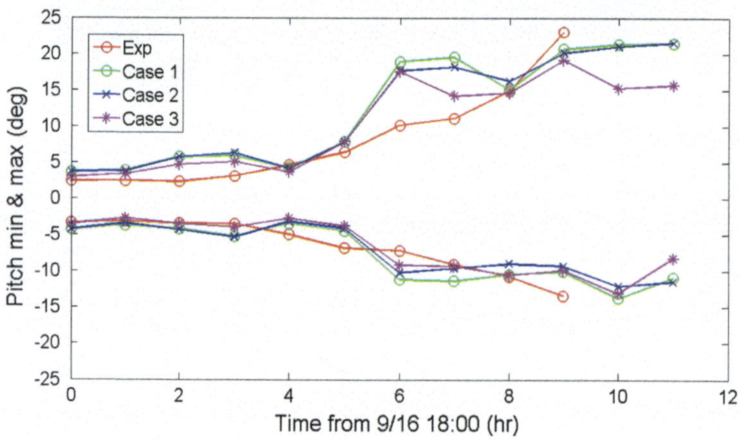

Fig. 8 Minimum and maximum values of platform motion in pitch

the experimental (Exp) and simulation results (Cases 1–3). It should be emphasized that although the significant wave height of 9.5 m exceeded the design wave height of 8.4 m that corresponds to the return periods of 50 years, the FOWT experienced no damage by the typhoon attack.

3.4 Response During Power Production

The dynamic response of the FOWT at power generation is presented below. Nielsen et al. [11] reported that for a spar-type FOWT, a conventional wind turbine control scheme may induce a magnification of the pitch motion. The effect may be referred to as "negative damping effect". The wind turbine control scheme for the FOWT at Kabashima was modified to suit a floating wind turbine so as to avoid the negative damping effect. In order to confirm the effectiveness of the control scheme, the field measurement at power generation has been made, and the results are presented below (for details, see [12]).

Figure 9 shows the turbulence intensities of the wind data used for comparison herein. Figure 10 shows the mean values of the pitch response. A fairly good agreement between the simulation results and the measured values is observed. The effect of turbulence intensities is insignificant for the mean values. Figure 11 shows the standard deviations of the pitch response (pitch SD). As it can be seen in Fig. 11, the numerical simulation results show that the turbulence intensities affect the pitch SD considerably. Thus, a large scatter of the pitch SD in the measured values is due to the large scatter of the turbulence intensities themselves (see Fig. 9). The simulation results of pitch SD predict fairly well the upper bound of the measured data for the corresponding turbulence intensity class (IEC category A).

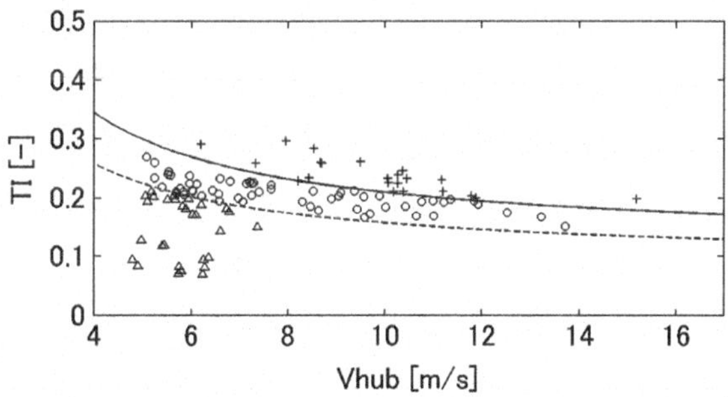

Fig. 9 Turbulence intensities (T.I.) (Solid line: IEC category A, Broken line: IEC category C)

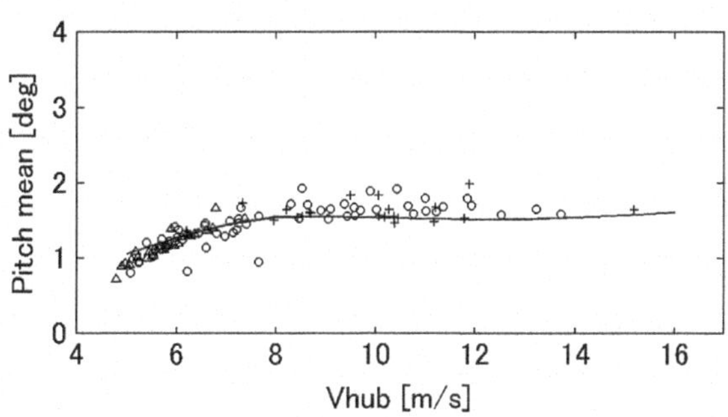

Fig. 10 Mean values of pitch response

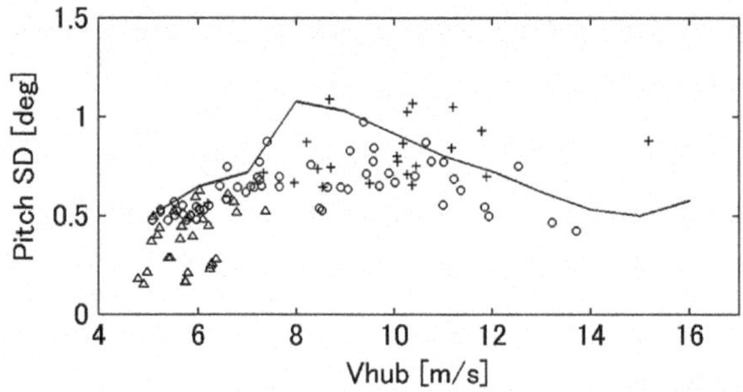

Fig. 11 Standard deviations of pitch response

4 Full-Scale Model

4.1 Design, Construction and Installation

The half-scale model was removed from the demonstration site as scheduled. Subsequently, the full-scale model was installed at the same site. Figure 12 and Table 3 show the main dimensions and specifications of the full-scale model, respectively. Basically, the design concept of the spar platform is the same as the half-scale model except for the scaling up to about twice in the length dimensions. Some details of the design procedure, construction and installation can be found in [13].

The installation procedure can be summarized as follows:

(a) The PC parts and the steel part were connected at a quay yard in the horizontal position.
(b) The completed floater (hybrid spar) was towed to the northern part of the Kabashima Island by using a barge (dry tow).
(c) The floater was then upended by using a large floating crane. The procedure is basically the same as the half-scale model.

Table 3 Main specifications of full-scale model

Item	Specifications
Wind turbine	Hitachi HTW2.0–80 (down-wind type)
Rated power	2 MW
Rotor speed	11.1–19.6 rpm
Cut-in wind speed	4 m/s
Cut-out wind speed	25 m/s
Rated wind speed	12 m/s
Mooring chains	Nominal diameter of 132 mm (R3 studless) × 3
Drag anchor	Weight in air: 12 tf × 3

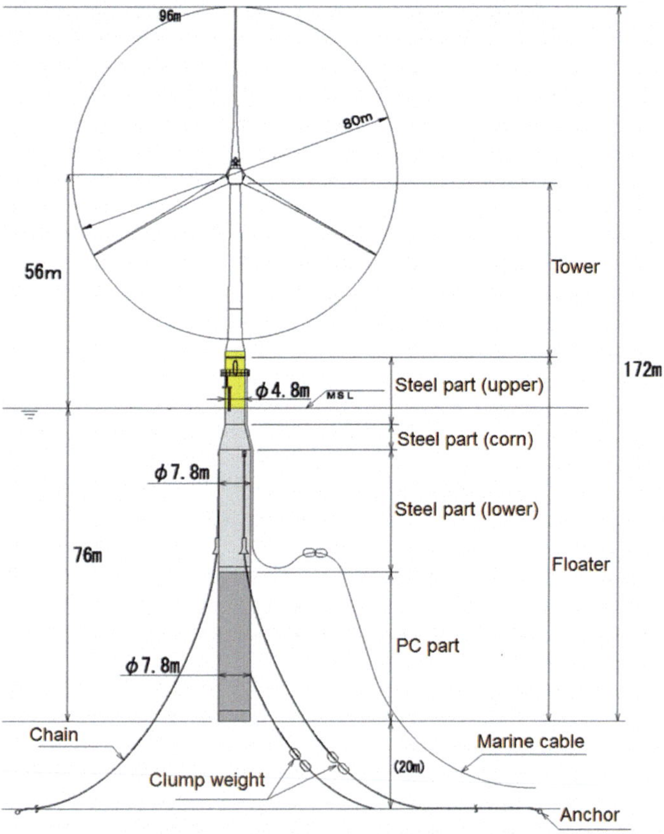

Fig. 12 Main dimensions of the full-scale model

(d) The tower (in two pieces), the nacelle, and the rotor were assembled to the floating body.
(e) After completion of the floating wind turbine, it was towed to the southern part of Kabashima Island.
(f) Finally, the anchor chains were hooked-up to the floating body. Here, the anchor chains had been pre-laid. The final hook-up was completed on 18 October 2013.
(g) After the electric marine cable hook-up and the grid-connection, the opening ceremony, as the first multi-megawatt floating wind turbine in Japan, was held on 28 October 2013.

4.2 Operation and Relocation

The full-scale model was operated and tested at the Kabashima demonstration site for more than 1 year. During the test, several measurement was made, and they were checked against the numerical simulations. Table 4 shows the natural periods of the full-scale model, and their comparison with numerical simulations. As can be seen, close agreement is observed except for yaw (for which accurate data could not be obtained). More details can be found in [14].

The demonstration project completed successfully in March 2016 as scheduled. In order to utilize the full-scale model for further demonstration toward commercialization, it was removed from the site, and relocated to 5-km offshore of Fukue Island. There, a full 2-MW power generation is possible, and future expansion to a floating offshore wind farm is expected. Figure 13 shows the general view of the full-scale model at the relocated site during power production. Due to the thrust force acting on the rotor, steady inclination and elastic deformation of the blades can

Fig. 13 General view of the full-scale model during power production

Table 4 Natural periods of full-scale model

DOF	Simulation	Experiment	Sim./Exp.
Surge	105	120	0.88
Sway	111	120	0.93
Heave	26.5	27.3	0.97
Roll	29.4	28.6	1.03
Pitch	29.1	28.6	1.02
Yaw	19.0	–	–

be observed. Currently, the full-scale model is used as a commercial wind turbine, selling its product with FIT-based revenues.

5 Concluding Remarks

This paper introduced the demonstration project on floating offshore wind turbine held at Kabashima Island, Goto city, Nagasaki prefecture. The half-scale model was attacked by a very severe typhoon, but it survived with no damage. Comparison of the measured data during the typhoon event and the power production with the simulation results confirmed the validity of the simulation tool and the design method.

The full-scale model has also been installed with a great success. It also demonstrates the feasibility of the spar-type FOWT in Japan. The success of the hybrid spar combining lower PC precast segments and upper steel part has opened up the realization of a low-cost floating wind turbine system in near future.

Acknowledgements This work is a part of the floating offshore wind turbine demonstration project funded by the Ministry of the Environment of Japan.

References

1. Ministry of the Environment. Study of basic zoning information concerning renewable energies (FY2011). [Online]. June 2012. Available at: http://www.env.go.jp/earth/report/h24-04/. Accessed December 12, 2018.
2. EWEA. Deep water-The next step for offshore wind energy. [Online]. July 2013. Available at: https://windeurope.org/about-wind/reports/deep-water/. Accessed December 12 2018.
3. Byklum, E., & Atcheson, M. (2016). State-of-the-Art. 2 Hywind. In J. Cruz & M. Atcheson (Eds.), *Floating offshore wind energy—the next generation of wind energy*. Switzerland: Springer.
4. Roddier, D., Cermelli, C., & Weinstein, J. (2016). State-of-the-Art. 1 WindFloat. In J. Cruz & M. Atcheson (Eds.), *Floating offshore wind energy—The next generation of wind energy*. Switzerland: Springer.
5. Utsunomiya, T., Sato, I., Shiraishi, T., Inui, E., & Ishida, S. (2015). Floating offshore wind turbine, Nagasaki, Japan. In C. M. Wang, Wang, B. T. (Eds.), Large floating structures—*Technological advances*. Singapore: Springer.
6. Utsunomiya, T. (2016). State-of-the-Art. 3 Goto Island Project. In J. Cruz & M., Atcheson (Eds.), Floating offshore wind energy—the next generation of wind energy. Switzerland: Springer.
7. Utsunomiya, T., Yoshida, S., Ookubo, H., Sato, I., & Ishida, S. (2014). Dynamic analysis of a floating offshore wind turbine under extreme environmental conditions. *Journal of Offshore Mechanics and Arctic Engineering, 136*(2), 020904-1-11.
8. Kokubun, K., Ishida, S., Nimura, T., Chujo, T., Yoshida, S., & Utsunomiya, T. (2012). Model experiment of a SPAR type offshore wind turbine in storm condition. In *Proceedings of the 31st International Conference on Ocean, Offshore and Arctic Engineering*. Rio de Janeiro, Brazil, 1–6 July 2012. OMAE2012-83993.
9. Ishida, S., Kokubun, K., Nimura, T., Utsunomiya, T., Sato, I., & Yoshida S. (2013). At-sea experiment of a hybrid SPAR type offshore wind turbine. In *Proceedings of the 32nd Inter-*

national Conference on Ocean, Offshore and Arctic Engineering. Nantes, France, 9–14 June 2013. OMAE2013-10655.
10. Utsunomiya, T., Sato, I., Yoshida, S., Ookubo, H., & Ishida, S. (2013). Dynamic response analysis of a floating offshore wind turbine during severe typhoon event. In *Proceedings of the 32nd International Conference on Ocean, Offshore and Arctic Engineering*. Nantes, France, 9–14 June 2013. OMAE2013-10618.
11. Nielsen, F. G., Hanson, T. D., & Skaare B. (2006). Integrated dynamic analysis of floating offshore wind turbines. *Proceedings of the 25th International Conference on Ocean, Offshore and Arctic Engineering*. Hamburg, Germany, 4–9 June 2006. OMAE2006-92291.
12. Utsunomiya, T., Yoshida, S., Kiyoki, S., Sato, I., & Ishida, S. (2014). Dynamic response of a spar-type floating wind turbine at power generation. In *Proceedings of the 33rd International Conference on Ocean, Offshore and Arctic Engineering*. San Francisco, California, USA, 8–13 June 2014. OMAE2014-24693.
13. Utsunomiya, T., Sato, I., Kobayashi, O., Shiraishi, T., & Harada, T. (2015). Dynamic response of a spar-type floating wind turbine at power generation. In *Proceedings of the 34th International Conference on Ocean, Offshore and Arctic Engineering*. St. John's, Newfoundland, Canada, May 31–June 5, 2015. OMAE2015-41544.
14. Utsunomiya, T., Sato, I., Kobayashi, O., Shiraishi, T., & Harada, T. (2019). Numerical modelling and analysis of a Hybrid-spar floating wind turbine. *Journal of Offshore Mechanics and Arctic Engineering, 141*(3), 031903-1-5.

The Dawn of Floating Solar—Technology, Benefits, and Challenges

Haohui Liu, Abhishek Kumar and Thomas Reindl

Abstract Floating solar, which is the installation of photovoltaic (PV) systems on water bodies, is a nascent, yet fast-growing PV deployment option, with a terawatt-scale market potential globally. Singapore has contributed significantly to the research in floating solar application on inland fresh water reservoirs by building the world's largest floating solar testbed. In this paper, we give an overview of floating solar technologies, and highlight some learnings and research findings from the testbed. We quantified the cooling effect on water and its benefits on energy yield of the PV systems. We also discuss some issues and pitfalls to avoid. Going forward, offshore floating solar is the next frontier with significant opportunities.

Keywords Floating solar · Photovoltaic · Reservoirs · Near-shore · Electricity · Energy

1 Introduction

Floating solar or floating PV (FPV) refers to the installation of photovoltaic systems on water bodies, such as lakes, reservoirs, hydroelectric dams, mining ponds, industrial ponds, water treatment ponds, near coast lagoons and other often underutilized water bodies. It is realised by mounting PV panels on pontoon-based floating structures, anchored to the bank or the bottom of water body.

Recently, the development of floating solar has gained traction. This is largely spurred by the limitation or the cost of available land in countries like Japan, South Korea or Singapore, as well as innovations in cost-effective floating solutions for large PV arrays. This new way of deployment avoids potentially competing land use with agriculture and in many cases allows deployment of large PV installations near load centres, hence reducing the cost of transmission infrastructure. The history of floating solar is relatively short. The first floating PV system was built in 2007 in Aichi Japan. Since then, several small pilot projects were installed, mainly in the

H. Liu (✉) · A. Kumar · T. Reindl
Solar Energy Research Institute of Singapore, Singapore, Singapore
e-mail: liu.haohui@nus.edu.sg

C. M. Wang et al. (eds.), *WCFS2019*, Lecture Notes in Civil Engineering 41,
https://doi.org/10.1007/978-981-13-8743-2_21

US, Italy, France, Spain, Japan and Korea [1]. In the past few years, the interest on FPV grew rapidly, with large FPV plants at tens of Megawatt scale being installed or planned, especially in China and Southeast Asia [2–4]. For instance, the relevance of FPV is high in eastern China, because the region is highly populated with limited land availability, but has abundant water bodies. In Southeast Asia, the option of FPV can unlock huge additional capacity in the many existing hydropower plants, for example along the Mekong River. Globally, Floating PV has the potential to unlock a Terawatt scale opportunity and is poised to become the third pillar of the PV industry, after ground-mounted and rooftop installations.

Despite its increased popularity and relevance, so far there are not sufficient studies that rigorously assess in detail the technical implications, the economics, and the environmental aspects when deploying FPV systems in a larger scale. As an effort to address this, Singapore spearheaded the research and launched the world's largest floating PV testbed in October 2016. It is a collaborative initiative by the Singapore Economic Development Board (EDB) and PUB, Singapore's National Water Agency. The Solar Energy Research Institute of Singapore (SERIS) at the National University of Singapore (NUS) is acting as project manager who took complete ownership of design, construction, testing, commissioning and the scientific evaluations of the project.

In this paper, we give an overview of this new type of solar deployment, in terms of technology and the current development status. Subsequently, we highlight some research findings from the Singapore floating solar testbed.

2 Technology Overview and Current Status

The floating platform and its anchoring is an essential part and enabler of floating solar technology. Most large-scale FPV plants have pontoon type floats, upon which PV panels are mounted at a fixed tilt angle. The floating platform usually consists of a matrix of specially designed HDPE floats connected together, or floating pontoons with supporting metal frames. A schematic showing the system composition is illustrated in Fig. 1. Figure 2 shows one design example of the floats that support PV panels and at the same time provide maintenance access. Apart from PV panels, other electrical components, such as combiner boxes and string or central inverters, can also be floating on water. The floating platforms are held in place by the anchoring and mooring system. Depending on the site conditions, the anchoring can be made either to the shore of the water body (bank anchoring), or to the water bed (bottom anchoring). The design of the floats and the anchoring needs to take into account factors such as wind load, float type, water depth, and variations in water level. This often involves careful mechanical design and structural analysis using tools such as Computational Fluid Dynamics (CFD) and Finite Element Methods (FEM). Currently, several companies are supplying specially designed floats for FPV.

Other than the standard fixed tilt floating PV array configuration outlined above, tracking and concentration is also achievable with floating solar. Tracking can be

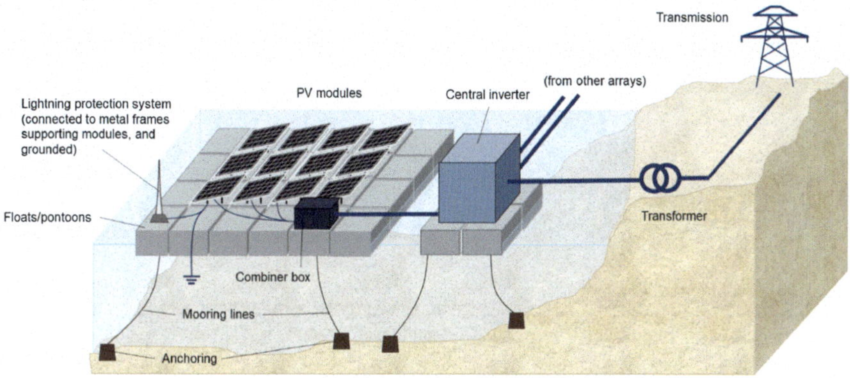

Fig. 1 Schematic of a typical large-scale FPV system, showing key components (*Source* SERIS and World Bank Group [5])

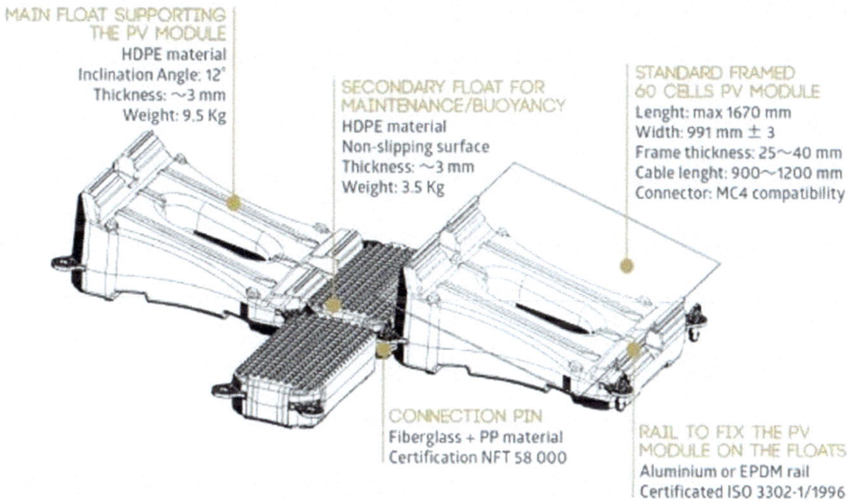

Fig. 2 Floating structure design from Ciel & Terre International (*Source* Ciel & Terre International)

achieved by rotating the entire floating platform to follow the sun from east to west. This type of vertical-axis azimuth tracking is particularly relevant for FPV, since it is relatively simpler to move an array on water (with its lower resistance) than on land. In addition, because alignment with the sun's position does not need to be completely accurate, the disturbances caused by wave movements are of minor impact [6]. Other tracking mechanisms, including dual axis tracking, are possible too [7]. Concentration increases the conversion efficiency of solar panels, and can be achieved using mirrors or Fresnel lenses. For example, light can be concentrated to a horizontal PV panel using V-shaped mirrors. Concentration in FPV systems pairs naturally with tracking, as indicated by the so-called floating tracking cooling

Fig. 3 Concept drawing of a floating tracking cooling concentrator system with vertical 1-axis azimuth tracking (left, adopted from [8]), and a system that adopted this design in a wastewater facility, Jamestown, Australia. (right, *Source* Infratech Industries)

concentrator system (Fig. 3). However, these concepts are still at an early stage of development, and do not have much commercial relevance yet.

There are several proposed benefits for floating solar [5, 9–11]:

- No or reduced land usage: This is important for regions where land resources are scarce, land acquisition costs are high, or land use is undesirable for PV installations (e.g. when PV conflicts with agricultural use).
- Easy installation and deployment: No civil work is needed to prepare the site. Typical floating platforms in the market are usually modular in nature and easy to assemble.
- Reduction in temperature loss: Evaporative cooling effect from the water (thus lower operating module temperatures) is likely to increase the energy yield of the PV modules, particularly in hot climates. Improvement of above 10% as compared to land-based PV systems had been reported in some early FPV projects [1, 12].
- Less shading: FPV is less prone to shading due to more open area and flat environment.
- Less soiling due to dust: water bodies tend to be less dusty than other typical locations for PV deployment (e.g. cities, deserts).
- Synergy with existing electrical infrastructure: Many inland freshwater bodies, especially reservoirs of hydropower plants have nearby grid connections. When utilizing those, the initial capex can be substantially reduced.
- Complementary operation with hydro: There is great potential for the combined operation of hydropower stations with FPV as hydro-PV hybrid systems, not only for the diurnal cycle (i.e. generating solar power during the day and hydropower at night), but there are often also seasonal complementarities, whereby dry seasons with less water flow correspond to period of high solar insolation and vice versa. In addition, instantaneous irradiance variability can be largely compensated by fast-responding hydro turbines.
- Environmental benefit: It is suggested that a reduction in algae growth can be expected because of less sunlight reaching the water body [9].
- Reduction of evaporation loss: This is especially important for drinking water and irrigation reservoirs where the water is very precious. In general, the reduction is

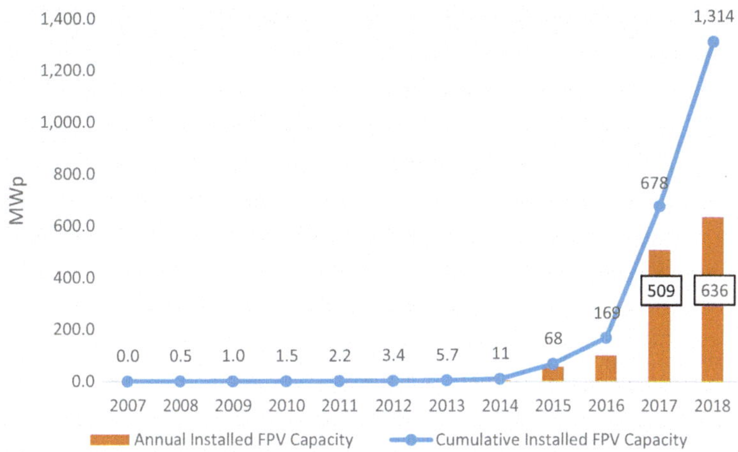

Fig. 4 Global installed floating PV capacity until the end of 2018 (*Source* SERIS and World Bank Group [5, 13])

suggested to be largely proportional to the water surface covered, however, the exact extent of this effect remains to be proven with field data.

- Additional benefits: Potential integration of floating PV with aquaculture and fish farming.
- Huge market potential: there are 400.000 km^2 of man-made reservoirs on the planet. Only covering 10% of those would already unlock a potential in the Terawatt scale.

The market for Floating solar is growing exponentially. The installed floating solar capacity has surpassed 1.3 Gigawatt-peak and continues to grow (Fig. 4), thanks to large installations in Asia, particularly China. Currently, most floating solar deployments are on inland fresh water bodies. This is mainly because (1) there is less challenge from wind, waves and currents for the anchoring and mooring system, and (2) there is less corrosion and degradation stress for floats as well as electrical components. However, there is a strong interest to install FPV systems offshore (or rather near-shore). The more stringent requirements for floats, anchors, and components imposed by the harsher environment may necessitate a different platform design or the use of different technologies. However, the vast experience of the well-established marine and offshore industries should make it possible to meet the challenges. There are currently a few start-ups exploring offshore floating solutions, but still at small scale.

More information about technology overview and status of floating solar can be found in the upcoming report titled "Where Sun Meets Water: Floating Solar Market Report—Market Report" from the World Bank Group, ESMAP, and SERIS [13].

3 Singapore's Floating Solar Testbed and Research Findings

Singapore's floating solar testbed, located at Singapore's Tengeh Reservoir, is composed of ten systems of different floating technologies and system designs, with a total capacity close to 1 MW_p (see Fig. 5). The objectives are to study the economic and technical feasibility, as well as the environmental impacts of deploying large-scale FPV systems on inland water surfaces. To study the performance and reliability of the various floating systems, SERIS measures an extensive set of meteorological and electrical parameters, including solar irradiance, air temperature, air humidity, water surface albedo, wind speed and direction, module temperatures, platform motions, DC output on string level and AC energy output per sub-system. This enables in-depth analyses, for instance, to quantify the amount of module temperature reduction, to determine the advantages of using bi-facial modules on floating systems, and to assess the increase in energy yield compared to normal PV systems.

Through a comparison of readings of operating conditions from the testbed and a nearby rooftop reference system, it is found that the on-water environment has slightly lower ambient air temperature, higher humidity, and higher wind speed. Deeper analyses of the collected data show that the evaporative cooling effect depends on the type of floating struc-ture used. From measured ambient air temperature, wind speed, in-plane solar irradiance level and module temperature, the so-called "heat loss coefficient" of each PV system type is determined, which indicates the effectiveness of module cooling by the environment. The comparison clearly shows a dependence on the way the PV panels are mounted. Higher values of the heat loss coefficient correspond to better cooling, and thus lower module temperatures, leading to a better electrical performance. The floating structures in the testbed are roughly

Fig. 5 The 1MW_p floating solar testbed at Singapore's Tengeh Reservoir (*Source* SERIS)

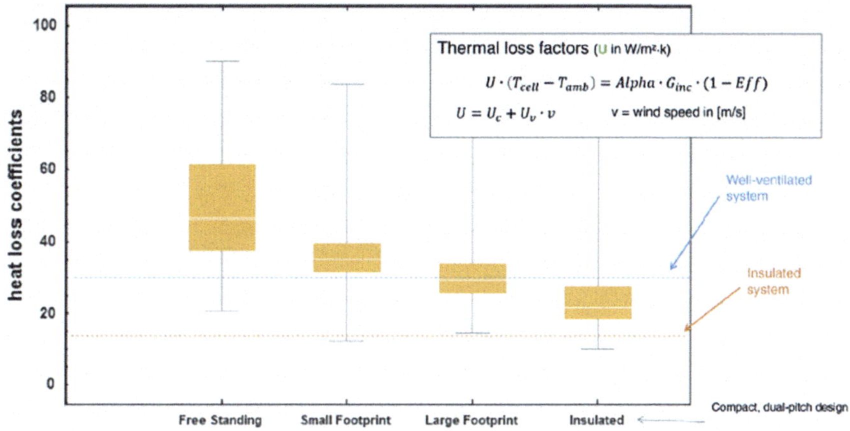

Fig. 6 Extracted heat loss coefficients for different types of floating structures from the floating PV testbed in Singapore. Higher values correspond to better cooling. Also indicated on the graph are U-values normally assumed for a well-ventilated and an insulated ground-based or rooftop systems in PV simulation (Figure adopted from [14])

categorised into "free-standing" types (with PV panels open to the water surface), and the "pontoon" types differentiated by the extent of water surface coverage beneath the modules (from "small footprint" to "large footprint"). The "insulated" configuration corresponds to a large footprint pontoon type structure with modules arranged in a compact, back-to-back east-west facing array.

The module temperatures for the free-standing systems are found to be much lower than those in the rooftop reference system. The heat loss coefficients for those systems are generally in the range of 40–50 W/m^{2}K, which is 30–60% higher than the typical values of 30 W/m^{2}K for a well-ventilated rooftop system. This is primarily due to lower ambient air temperature and higher wind speed on the water. For floating structures that have PV panels mounted close to the water surface, the cooling effect is dependent on the water footprint as well as module arrangement, as can be seen from Fig. 6. Please note that "well-ventilated" refers to a ground mounted installation that has good ventilation, while an "insulated" system would refer to one with little exchange of heat with its environment (e.g. direct mounting on rooftop).

PV system performance can usually be described by its specific energy yield and performance ratio (PR) [15]. The annual specific yield and calculated PR for eight systems that started stable operation from 2017 is shown in Fig. 7. For the period of investigation, the total incoming solar irradiation is 1600 kWh/m^2. The majority of the floating systems, under normal operation without major downtimes, reached PR values of well above 80%, which is not easily achieved in Singapore's hot climate conditions. This is higher than the average of rooftop PV systems in Singapore, which is indicated as a red line in the figure. The system with the highest PR (system H) has a free-standing type as floating structure, as expected. Overall, the floating systems clearly benefit from cooler module temperatures.

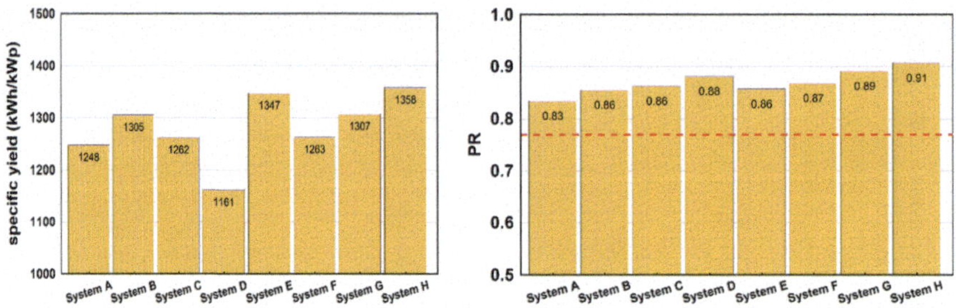

Fig. 7 First year final specific yield of the floating PV systems (left), and the first year PR of those systems without including major downtimes. Typical PR for rooftop systems in Singapore is indicated by a red dashed line

4 Potential Challenges

Many commercial projects as well as the Singapore testbed had demonstrated that floating solar is a viable solution. However, best practices are important to ensure technical quality. In the Singapore testbed, we also study the engineering aspects, and have derived several valuable learnings on the deployment and Operation and Maintenance (O&M) of floating solar. Below, we summarize a few issues encountered in the testbed.

- Bird droppings
  Although less soiling from dust can be expected, soiling due to bird droppings can sometimes be a serious issue for floating PV systems. It leads to partial shading and a reduction of current. In severe cases, it can even lead to hot spots and accelerate module degradation. Bird droppings were occasionally observed in the Singapore testbed, and were mostly washed away by rain. However, some systems experienced heavier soiling than others, and the accumulation of droppings led to significant reduction in energy generation (Fig. 8). In one episode, the current, and thus PR, dropped by over 10% in a course of around 10 days. The fast accumulation of bird droppings can significantly impact revenue if not noticed and cleaned in time. It may be therefore worthwhile to study bird behaviours, install bird deterrent devices, and schedule maintenance accordingly.
- Mechanical wearing due to platform movement
  Constant platform movements can cause challenges for mechanical connections and joints. This is especially the case for platforms where relative movements between modules are frequent. One example is the breaking of equipotential bonding wires/tapes. Equipment grounding is important for the electrical safety of personnel, and equipotential bonding is used to ground modules and frame structures. During the period of operation, we observed several instances of breaking and snapping (Fig. 9), even for cases where there is sufficient slack. Therefore, it might be necessary to implement improved wire management systems on floating platforms.

Fig. 8 Severe soiling due to bird droppings in one floating PV system (Figure adopted from [14])

Fig. 9 Examples of breaking and snapping of equipotential bonding tapes (*Source* SERIS and [14])

- Insulation faults
  Due to the high-humidity environment and the proximity to water, the insulation resistance of the system can sometimes drop significantly. This is especially the case if cables or connectors come into contact with water. Low insulation resistance can lead to electrical leakage to the ground, which poses safety hazards to personnel and equipment. Inverters usually check insulation resistance during start up, and do not turn on when this value does not meet the minimum requirement. In the Singapore testbed, frequent instances of inverters turning on late in the morning are observed for some systems. Figure 10 shows such an instance with the period affected indicated in red. This leads to a non-negligible loss in energy production. Therefore, good cable and connector quality, careful cable management, as well as proper platform design are important to avoid this loss.

The floating solar technology is still in its early development and commercialization phase, and there are several broader challenges to consider in addition to the technical issues aforementioned:

- Higher capital expenses (around 10–15% premium over ground-mounted installations, although this will reduce over time).

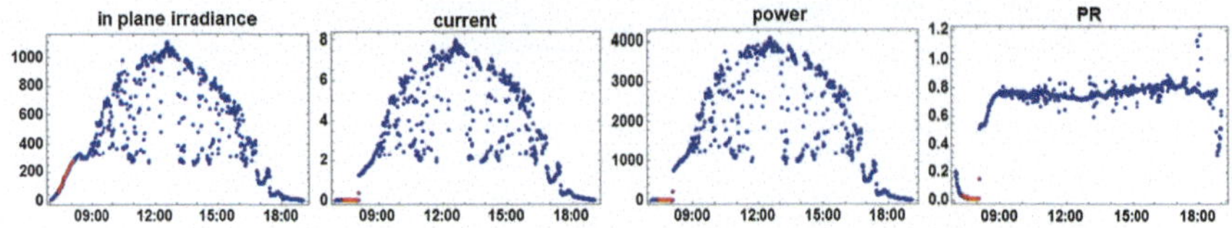

Fig. 10 Illustration of inverter downtime on a particular day. Inverter started late because insulation test failed

- Complex anchoring and mooring may be required in some cases.
- O&M may be more difficult due to accessibility.
- Electrical safety and long-term reliability of system components.
- Logistics and transportation of floats.
- Lack of long-term operational experience and international standards.
- Potential environmental impact on water ecology.

Additional technical challenges need to be solved for floating solar systems to be deployed offshore on a commercial basis. This would involve optimization of the floating platforms and anchoring systems to withstand a harsher environment, while remaining economically feasible. Also, challenges such as biofouling need to be addressed.

5 Conclusion and Outlook

Floating solar is a fast growing PV market segment with a Terawatt-scale potential globally. Numerous large-scale commercial projects as well as Singapore's floating solar testbed have demonstrated that the technology is economically and technically viable. It promises several attractive benefits, such as improved energy yield and reduced land use in highly populated areas, where land is scarce and/or precious. Singapore has contributed significantly to the research in floating solar applications on inland fresh water reservoirs. SERIS manages the world's largest floating solar testbed, and obtained valuable learnings and research findings. We quantified the cooling effect as well as the energy yield gain and differentiated it for different floating platform types. Floating Solar is a nascent, yet fast-growing deployment option and hence there are still technical issues to resolve or avoid. Best practices should be established to avoid new issues and pitfalls associated with deploying PV on water. In this regard, the experience from Singapore testbed can be highly valuable in developing international standards for FPV.

Going forward, there are plenty of new opportunities. Offshore floating solar represents one such area with great potential, especially when combining the deployment with additional functionalities, such as (i) fish farming (especially as a part of urban agriculture); (ii) hydrogen generation (for conversion of electricity to fuels); or (iii) desalination (for potable water generation).

References

1. Trapani, K., & Redón Santafé, M. (2015). A review of floating photovoltaic installations: 2007–2013. *Prog. Photovoltaics: Res. Appl., 23*(4), 524–532.
2. PV-Tech. (2017). Masdar and Indonesian power giant to build world's largest floating solar plant. [Online]. Available at: https://www.pv-tech.org/news/masdar-and-indonesian-power-giant-to-build-worlds-largest-floating-solar-pl.
3. PV-Tech. (2017). Sungrow building another record 150 MW floating solar project in China. [Online]. Available at: https://www.pv-tech.org/news/sungrow-building-another-record-150mw-floating-solar-project-in-china.
4. PV Megazine. (2017). 40 MW floating PV plant in China connected with Sungrow's inverters. [Online].. Available at: https://www.pv-magazine.com/2017/05/19/floating-pv-plant-in-china-connected-with-sungrows-inverters/.
5. World Bank Group, ESMAP, and SERIS. (2018). *Where sun meets water: floating solar market report—executive summary*. Washington, DC: World Bank.
6. World Bank Group, ESMAP, and SERIS. (2019). *Where sun meets water: Floating solar handbook for practitioners*. Washington, DC: World Bank. (forthcoming).
7. Connor, P. M. (2009). Performance and prospects of a lightweight water-borne PV concentrator, including virtual storage via hydroelectric-dams. ISES Solar World Congress. (2009). *Renewable Energy Shaping Our Future*. South Africa: Johannesburg.
8. Rosa-Clot, M., & Tina, G. M. (2018). Chapter 6—concentration systems and floating plants. In: Submerged and Floating Photovoltaic Systems, (pp. 137–158). Academic Press.
9. Sahu, A., Yadav, N., & Sudhakar, K. (2016). Floating photovoltaic power plant: A review. *Renewable and Sustainable Energy Reviews, 66*(Supplement C), 815–824.
10. Santafé, M. R., et al. (2014). Implementation of a photovoltaic floating cover for irrigation reservoirs. *Journal of Cleaner Production, 66*, 568–570.
11. Choi, Y.-K. (2014). A study on power generation analysis of floating PV system considering environmental impact. *International Journal of Software Engineering and Its Applications, 8*(1), 75–84.
12. Choi, Y.-K., & Lee, N-.H. (2013). Empirical Research on the efficiency of Floating PV systems compared with Overland PV Systems. *CES-CUBE,* 284–289.
13. World Bank Group, ESMAP, and SERIS. (2019). *Where sun meets water: floating solar market report—market report*. Washington, DC: World Bank. (forthcoming).
14. Liu, H., et al. (2018). Field experience and performance analysis of floating PV technologies in the tropics. *Progress in Photovoltaics: Research and Applications, 26*(12), 957–967.
15. International Electrotechnical Commission. (1998). Photovoltaic system performance monitoring—Guidelines for measurement, data exchange and analysis.

Seascape the Landscape of Singapore, Repurposing Land in a Land Scarce Nation

Soon Heng Lim

Abstract Singapore's economy is founded on sand. From ports to airports, offices to factories, from refineries to shipyards, hotels, casinos, roads and homes, the contributors of the national GDP owe their existence to the sand on which they are erected. According to the UNEP this tiny nation state is the world's largest importer of sand. It needs more land but faces challenges: sand mining is banned. Shallow waters have already been reclaimed. The damage to the marine bio-diversity is undeniable. Sea levels are rising. This paper discusses the viability, modus operandi and merits of "floating out" land-guzzling and out-dated industries on very large floating structures (VLFS) in the sea. The plots of land they currently occupy would be better used if repurposed for 21st century cutting-edge technology and easing the housing pressures. The economic output of the land can also be increased by embracing industries that can go multi-storey. A side benefit to putting industries on VLFS is that roads are not needed for maritime transport. Land needed for road expansion is reduced. Greenhouse emission is also reduced. Rising sea levels would be an irrelevant issue. Marine biodiversity would not be decimated. This strategy is more sustainable than land reclamation. The author invites the government to take leadership in the formation of a R&D Group to flesh out this concept. The time is not far in the future when the Master Plan for this nation has to treat both land and sea as one developable continuum for work, live and play.

Keywords Singapore · Reclamation · Land scarcity · Repurpose · Rejuvenation · Restructure economy · Floating ports · Shipyards · Military bases · Refineries · Marine biodiversity · Rising sea level · Carbon capture · Algae · Hydrocarbon · Fossil fuels

Present Address
S. H. Lim (✉)
Society of Floating Solutions, Singapore, Singapore
e-mail: lsh@sixtrees.com.sg

Floating Solutions LLP, Singapore, Singapore

Keppel Offshore and Marine Group, Singapore, Singapore

C. M. Wang et al. (eds.), *WCFS2019*, Lecture Notes in Civil Engineering 41,
https://doi.org/10.1007/978-981-13-8743-2_22

1 From Third World to First World, the Role of Sand Then and Now

In a 50-year journey, after the departure of its colonial masters, Singapore propelled its way from third world to first world. Its GDP per capita has risen sharply to exceed that of any of the G7 economies.

What has sand to do with it? A lot. Much of Jurong and nearly all of Tuas was reclaimed. The entire Jurong Island, half of the Pasir Panjang coastline and all of Marina Bay as well as Marine Parade, Bedok the entire Changi Airport that was once sea and wet lands is today home to industries, parkland, ports and offices. To date, more than 150 sq. km or 25% of the original land mass is reclaimed.

This economy was literally founded on sand. Sand was cheap and environmental concern non-existent. The GDP grew exponentially.

Today, environmental concerns, deeper waters and the more than tenfold price increase of sand make land reclamation a questionable proposition.

The United Nations Environmental Program (UNEP) urge that countries ban the export of sand. In China, land reclamation is illegal. In Hong Kong, the proposal to reclaim 1700 hectares for a new metropolis and homes for 1.1 million people in Lantau Island was met with street protests.

The hardship inflicted by sand mining affect those least able to defend themselves. National Geographic March 2018 carried an article that is familiar to coastal and riverine communities. It reports in Vietnam, "sand mining poses an additional danger: It is contributing to the slow-motion disappearance of the Mekong Delta, home to 20 million people and source of half of all the country's food and much of the rice that feeds the rest of southeast Asia".

HDB (Housing Development Board) projects a land mass of 766 sq. km is necessary to house an enlarged population and economy by 2030. Singapore is short of another 46 ha as of now.

We need to develop ecologically acceptable space solutions without degrading the coastal regime. We need solutions that protect creatures that crawl and swim that form the food chain that keep us alive.

The world is incensed that the US President dump the Paris Accord on Climate Change (COP 21). We would be as guilty of the same moral turpitude if we continue to import sand knowing the damage it is doing to the world.

2 From Plastic Bottles to Icebergs and All in Between

British environmentalist and eco-pioneer Richart Sowa, created his own floating paradise with 150,000 plastic bottles in Isla Mujeres. The 8000 square-foot free-floating eco-paradise has its own garden, beach, a house where he lives, 2 ponds, and a solar-powered waterfall.

The largest man-made floating structure is the Shell Prelude, 488 m long × 74 Microsoft wide × 105 m high fabricated with 260,000 tonnes of steel. At full load, it displaces more than 600,000 tonnes. It is moored 200 km away from Western Australian shore in a harsh environment.

An iceberg, 11,000 sq. km in area has been floating for the past 18 years. A major contractor in Japan, Shimizu Corporation is selling the idea of a floating tower 1000 m high.

Everything will float, whatever its shape, size or the material's density. It just needs to displace enough water. Archimedes' buoyancy principle is Physics 101. It can be proved rigorously.

VLFS (Very Large Floating Structure) is a newly minted word. Floats, however, have served mankind for several thousand years, as ships, bridges, abodes and farms.

In some ways, Singapore is a leader. It has built and exported semi-submersibles, jack-ups, floating wharves, floating hotels that see service in extremely harsh environments of the North Seas and Persian Gulfs. The professionals are not only in the shipyards. Others include naval architects and engineers, classification society surveyors, mooring specialists, floating crane, heavy marine transport specialists and university professors engaged in teaching and research in offshore structures.

3 Clearing the Mental Block

3.1 Ports Can Float

To realise the potential of VLFS as space solutions, we need to clear a few mental blocks. In Singapore a mega port of 65-million TEU capacity is planned. In 2017, the throughput at the ports was 33.35 million up 9% over the previous year [1]. At 5% CAGR, the new port will be maxed out by 2031. By then 55 million transhipment boxes will be unloaded and reloaded on to other ships at the port. The waterways will be a maritime version of the streets of Bangkok. Ship collisions will be frequent, oil spills a common occurrence, and the city suffocated by toxic greenhouse gases and PM2.5 particles [2].

When the mega port is maxed out, do we still set aside more land to accommodate more boxes? Shouldn't land be used for housing and other activities that add more economic value than merely storing, lifting and lowering boxes?

Early ports were built on land, beside rivers and along the coast lines. That made sense since imported goods unloaded from ships go to the hinterland. This is not the case with Singapore. 85% of the goods brought ashore here do not go anywhere but await another ship to take them away. In a country screaming for land, a port onshore is a legacy of the past it cannot afford.

Ports need not be conjoined to land. Offshore floating ports make more sense. Loading and unloading large vessels out at sea also keep their pollution away from land. (Singapore, along with Shanghai and Hong Kong was listed as one of the three most polluted ports in the world in 2015.) Nearshore waters are better used for leisure, living and economic activities of higher added value. For Singapore this strategy frees up some 700 sq. km of much needed space.

Phase 1 mega port at Tuas is under construction using 222 floating concrete caissons (which are ballasted so that a part of its dead load is supported by buoyancy and the rest on a prepared seabed. It is a little too late to suggest a fully floating port for this phase. Phase 1 could remain as designed. It can serve as a backup port as well as a domestic port where boxes destined for Singapore are processed.

But plans for Phases 2 and 3 which have not started, should be reviewed. The original plan calls for a similar construction procedure: build caissons, position them and backfill the area surrounded by these caissons with sand and unavoidably denuding marine life in the seabed. The reclaimed land will need years to consolidate. Kansai Airport continues to sink 24 years after completion. "2.2 million vertical pipes, each nearly 16 inches in diameter," have not helped. It has sunk "38 feet" and will continue to sink. The Kansai problem was neither foreseen nor foreseeable. If the Japanese were caught by surprise by this phenomenon, Singapore could as well.

We will return to the mega port in a later part of this paper but for now the message is if Singapore wants to make the sea a space for higher value activities than just shipping, the port need of go offshore.

The second thing it needs to do is revamp the existing bunkering operation.

3.2 *Bunkering*

About 3500 ships called at Singapore in 2017 to upload S$18 billion worth of bunker. Yet the way bunkers are loaded has not changed for 50 years. Ships taking bunkers are moored with a single anchor and allow to swing 360° with tidal currents. This obviously take more space than necessary.

We are pleased to note that the port authority has plan to change this. The third thing that needs to move away from shore are the cruise terminals.

3.3 *Cruise Terminal*

The cruise passenger volume in Asia grew by 55% in 2016 compared to 2015 [3]. The Marina Bay Cruise Centre would be out of capacity in a few short years if the industry increases at the current rate.

The MBCC take up large tracts of prime water front land for the terminal and car parks. It need not. A floating cruise terminal moored east of St John's Island would offer the same panoramic skyline and be better place logistically.

Such a terminal could be built for a fraction of the cost. The waters there is deep enough for mega float to be floating high enough above the seabed not stress the marine life below. For mooring there are several solutions that can be borrowed from the offshore rig industry including a truss dolphin cable of taking lateral loads at that depth. (Jack up legs are designed to work in water depths of more than 100 m).

Large capacity high speed ferries can transfer or disperse passengers to several designated points on the land including Changi Airport, Marina Bay, and Harbour Front all of which are near well connected to points of interest via the mass transit system.

Several examples of floating passenger embarking and disembarking facilities for cruise passengers are on the Internet. These include the floating berths in Monaco [4] and at Juneau, Alaska [5]. Dutch Docklands [6] has futuristic design of a floating cruise terminal incorporating a very large roof surface for solar panels. The roof and side mounted panels absorb the solar heat to produce electricity which would otherwise heat the space enclosed.

3.4 The Sea Can Be Secured

The threat of pirates and terrorists out at sea are real but modern surveillance systems can make this a non-issue. Intrusions of any kind may be detected by radar, sonar or infra-red motion detection devices. AIS (Automatic Identification System) will track all vessels in the sea for their position, speed and heading. Vessels without AIS signals will be intercepted by men and weaponised drones.

Unmanned surface vessels (USV) will soon patrol the seas remotely. A 24/7 radar coverage up to 200 km radius known as Aerostat will be and added layer of security.

3.5 Mobility in a Floating City

Transport infrastructure to a city is like arteries to the body. For it to be viable, a floating city needs to have efficient ways to move people and goods and services around.

The beauty of a maritime city is that one can get from A to B without physical highways. On the sea we travel in three dimensions: on the surface, above it or beneath it. Maritime transportation leaves behind a much smaller carbon footprint land transport does.

Driverless transport systems will be safer, faster and more fuel efficient using super computers which will take data of every vessels' heading, and speed through the AIS (Automated Identification System) monitoring system to work out a safe envelope for each.

Drone taxis would be much easier to navigate over open water bodies than in a congested cityscape. Helicopters and seaplane widen up the range of possibilities for long distance commute and in cases of emergencies.

4 VLFS and Incentives to Rejuvenate Land

Singapore needs both people and space to enlarge its GDP. Shortage of people is resolved by widening the doors to foreign workers intake. Space however is more challenging. It has now for the first time ventured into polder development.

Land reclamation including polders is facing environmental and humanitarian challenges especially for small countries with limited coast lines and fairly deep waters. Singapore needs to take the next leap which is space creation with floating platforms.

Mega floating platforms can provide space to live, work and play in the sea around us. As much as a third of the 700 sq. km of our sea could be used as such.

Many prime sites in the industrial zones currently occupied by ageing, dirty and increasingly irrelevant and uncompetitive industries should be recovered and repurposed. The brown land freed up could be used for housing or the manufacturing of 21st century innovations. They use land only in the horizontal dimension but not vertically (shipyards, supply base, refineries, tank farms for instance).

Land use may be intensified by going skywards. This is possible by adopting modern hi-tech industries including the manufacturing of chips, robotics, additive manufacturing, wind turbines and electric cars or the provision of hi-tech services. Each hectare of land should yield ten times the value of goods it now produces.

This would address the concern voiced in May 2018, by the Minister of National Development in Parliament at which he said that "despite the authorities' best efforts at planning, Singapore is still severely constrained by space. If there is no more land to recycle for future public housing, it could affect future generations and their access to subsidised housing and their ability to have an affordable and quality home in Singapore".

Land is needed to build new apartments to relocate the hundreds of thousands of HDB dwellers by about 2050, ahead of the expiry of their current 99-year lease.

State planners need to ruthlessly start "land clearing." To wait for existing leases to expire may be too late. VLFS technology provides a disruptive tool for process.

In subsequent passages we shall use the term "float out" to mean relocating a manufacturing or service industry on a floating platform moored in the sea within the state boundary and therefore subject to the state's jurisdiction.

4.1 Incentives

It makes sense from an urban planning and economic perspective to incentivise industries to float out. The process does not require importing sand, nor three or four years of waiting for soil to consolidate, after pouring in hundreds of millions of dollars.

The deep water surrounding the island makes reclamation a very costly proposition. With floats deep water is not an issue.

What would prompt the stakeholder to consider a float out if his lease is has not expired? Obviously, he is going to look at the technical feasibility as well as the economics of such a move.

They need to be convinced that it is technically feasible. This is not insurmountable. The offshore oil and gas industry have made excellent inroads in the past decades of offshoring many productions systems in deep waters around the world. The stability and performance of mega platforms under wind and current loads are well understood and many mooring systems have evolved. Accreditation for the issue of seaworthiness certificates are easily available in Singapore through organisations such as Lloyds Register of Shipping, American Bureau of Surveyors, and DNV-GL.

They also need to be persuaded that economics of such a move is attractive. For this, the government needs to offer a slew of incentives. These could take the form of tax breaks, waiver of mooring charges, provision of free shuttle-ferry service (justifiable since they use less of the public roads), soft loans to the stakeholder to finance the capex for the construction of the platform.

They have to understand that a float is an asset not an expense. It has a life of more than 100 years and will have a substantial market value even if it is written down completely in the balance sheet. That value arises from the fact that unlike land-based real estates it may be repurposed and deployed. It is attractive as a mortgage. Leased land has no value; in fact, it will cost the stakeholder to reinstate the land before returning it to the land lord.

4.2 Candidates for Floating Out

Attractive candidates for floating out include transshipment ports, power stations, incinerators, reservoirs, desalination plants, shipyards, refineries, golf courses, recreation parks, airports, naval, air and army bases, data centres, hydrocarbon storage tanks, prisons, vegetable farms, and campuses for colleges and universities. These facilities occupy very large tracts of land which often are largely used in a single horizontal plane. The nature of their activities makes it difficult for them to exploit the vertical dimension.

4.3 Why Offer Incentives to Offshore?

The trade-off for offering the incentives to encourage offshoring is that the state saves billions of dollars which it would otherwise have to spend to import sand to reclaim land and to maintain it. The state also benefits from the fact that no highways are needed to service these industries when they float out.

Private sector funds may be tapped. Taxpayers money is better spent health, education and housing.

The value of the land after rejuvenation will be worth considerably more. They will generate better returns than the old industries which are under competition from countries with cheap land. Clean industries can replace pollutive ones. The pressure on our existing roads will ease as more goods and people will use marine vehicles for transport. Much of the work that goes into the construction of a floating asset need not be in Singapore. Our dependence on foreign workers to support the construction industry will reduce, easing social pressures.

What this paper proposes is not a quick fix. It is bold. It is necessary and with the wherewithal that has seen this country deliver offshore rigs to far flung corners of the world it can be done.

5 Land Guzzlers in the Little Red Dot

The Government's allocation for land use in 2030 is shown in the pie chart in Fig. 1. By then we need 6.4% more land than we have now.

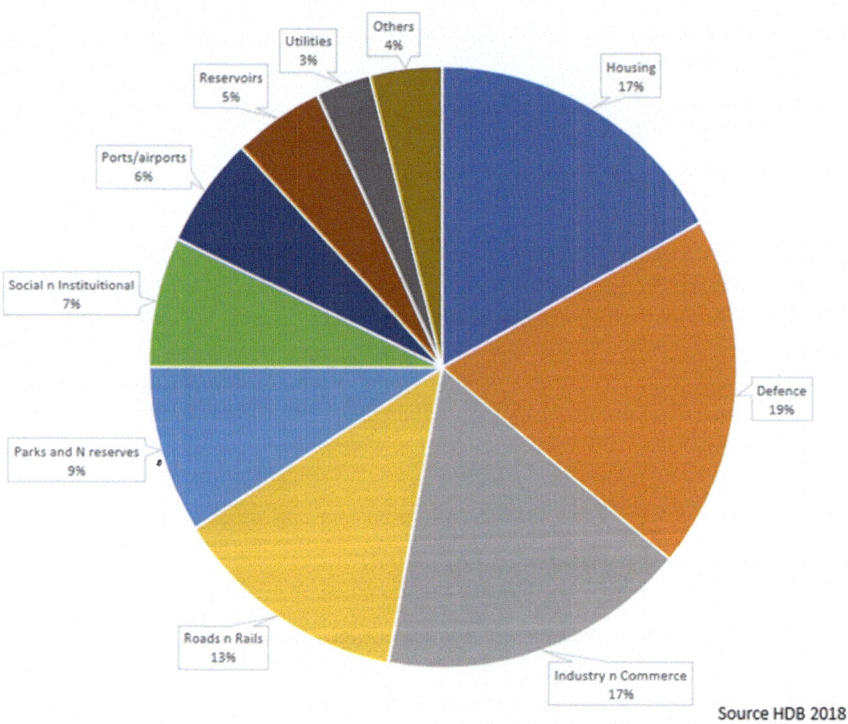

Fig. 1 Projected land use of Singapore 2030 (766 sq. km)

VLFS technology will eliminate the need to import vast quantities of reclamation sand. The attempt to reduce sand import by adopting polders has severe limitations. Polders is not feasible in the Straits of Singapore where water is deep and marine traffic is dense. Even in shallower waters north of the island, there are many natural risks and risks associated with geopolitics and shipping traffic, making polders a questionable solution. A dyke is a low-hanging fruit for terrorists.

5.1 Defense

Defense needs take up the most land. A total of 145 sq. km is provided for army, air force and navy. The army probably takes the largest share followed by the air force and navy.

5.1.1 Army Base

The army needs swathes of land for training. The land is largely rough terrain, hills, and wetlands. Armoured personnel carriers, tanks, rocket launchers, trucks and jeeps are equipment that would make daily use of these training ground. Live bullets, mortars and grenades would be used.

5.1.2 Replicating Natural Terrain

Would it be possible to replicate these terrains and features on VLFS? Yes. Quite easily. Blocks of concrete may be built in modules and may be connected to form any shape or size. The top deck may be contoured with sand, earth or granite to any shape to look like natural verdant terrain with either mock or real trees and shrubs and marsh land.

5.1.3 Bomb Proof

The main deck of the float will be heavily reinforced to take direct hits by bombs. (The author has personally been involved in a building a bomb proof pump house for a graving dock). Additionally, it will be designed with reserved buoyancy that automatically kicks in when the hull is damaged. It will have a depth that will ensure it remains substantially above water line even it loses its buoyancy and rests on the seabed. It will be capsize-proof.

5.1.4 Space Below the "Natural Terrain" Deck

The deck below could be used for storing military equipment, weapons and ammunition and meeting rooms. If desired another deck could be added below that for storage of food, fuel and water as well as space for emergency generators, batteries. Surface run offs following a rainstorm could be collected, filtered and processed for drinking. Unprocessed water could be used for watering plants.

Void spaces separated by water tight bulkheads will be fitted with water tight doors as is the practice in ships. The top deck may be reinforced to the extent required to withstand exploding grenades and mortars that may be used during military training.

A bunker could also be built such that it is submersible when desired. Being mobile, this strategic command centre may be relocated at will to frustrate the enemy's surveillance. In it will be a redundant communication back up system in case the primary system on land is disabled in conflict.

5.1.5 Protecting the Mainland from the Sea

Mass mobilisation of men and machine to and from land may be facilitated with pontoon bridges. Armoured personnel carriers, and tanks and supplies can be transported to and from mainland roll-on roll-off landing crafts. These may be supplemented by military hovercrafts. The LCAC [7] has a speed of 74 km/h and a payload of 60 tons.

Modern artillery with an effective range exceeding 70 km based on the float in the Singapore Straits, will provide cover against intrusions anywhere in the Johor Straits. Singapore will not be caught with its pants down as in World War 2, when the invading Japanese attacked from the north.

5.1.6 Singapore's First Polder

An 810-ha area at Pulau Tekong is being enclosed by a dyke to form a polder for training of army training. A polder as a space solution is vulnerable to extremists and even psychopaths with ill intent [8]. Young national service men will be put in harm's way. Remotely controlled weaponised drones can destroy the dyke easily with a few explosive devices.

Instead of the dyke a series of gravity-based structures similar to the floating berth at Monaco may be placed at the periphery (not necessarily end-to-end but with gaps in between adjacent structures to allow water to flow through). The area thus enclosed forms a lagoon which may be dredged to say 6 m below chart datum. In this lagoon the floating base may be installed. Such a solution will have a much lower life-cycle cost (because no pumping is required, and the concrete float is virtually maintenance free).

5.2 *Air Force Base*

The prime consideration for an airbase is the runway to enable fighter jets to take off and land as well as the means to keep them operationally ready. This can be accomplished on a floating air base.

The viability of floating runways has been proven many times over in aircraft carriers. For example, the USS Gerald R. Ford has a flight deck of 333 m × 76.8 m and a lower deck to service more than 75 aircrafts.

A floating runway was built and successfully tested in Tokyo. It was 1000 m in length; long enough for an F-15 fighter jet to take off with its payload.

The concept of floating air base actually goes back to WW 2. Code name the Habakkuk Project [9] it was to be a 1200 m × 180 m (21.6 ha) runway. Lord Mountbatten proposed it to war time Prime Minister Winston Churchill who was excited by the idea. The final design had a displacement of 2.2 million tons. Steam turbo generators would supply 33,000 hp (25,000 kW) for 26 electric motors.

The Habakkuk air base was to have been built on an iceberg for lack of material and time. Concrete would be a material of choice today. Concrete platforms [10] have proven their durability, endurance and robustness in offshore drilling in ice laden waters of the Arctic and Sakhalin. They are less destructible than steel and artillery damage can be repaired afloat in situ.

5.2.1 Space Below Runway

Below the runway two or three decks may be included. The first lower deck may be used to park the fighter jets that are not in immediate use away from the line of sight of the enemy as well as workshops to service and maintain the aircrafts, stores for fuel and armament. The second lower deck could be used for amenities, lockers, training rooms, auditorium and data centres.

By multi-tasking the runway on a floating base, not only is the footprint reduced, the efficiency of operations is improved as material handling is reduced and interdepartmental miscommunication less likely to occur.

5.2.2 Freedom from Airspace Constraints

Very importantly, the severe airspace constraints [11] placed on the state's fighter jets taking off from land-based runways will no longer be an issue with floating runways oriented towards the sea instead of to Malaysia (This legacy of the British rule has recently become problematic in the relation between the two neighbouring countries).

5.3 *Naval Base*

5.3.1 Faslane HMNB Clyde

The first known embrace of VLFS technology as a component of a naval base is the floating nuclear submarine jetty [12] at HMNB at Clyde in Faslane, Scotland commissioned in 2013. "The 44,000-tonne jetty has six berths and is designed to serve the Navy for the next 50 years." It is moored and restrained from lateral movements by four pipe piles. It is designed for "loading or unloading of other specialist secondary munitions such as cruise missile and nuclear torpedoes." The fact that it rises and falls in tandem with the tide was a huge operational benefit. It means mooring ropes are always at a constant tension and all shore to ship connections (gangways and services) may remain untouched round the clock.

The top deck of the jetty serves the traditional purpose of loading unloading, embarkation and disembarkation. However, below the first deck, the space is used for maintenance and stores. It is well to remember that two decks are possible only because the structure is free to float up and down. If it is attached to a pile (i.e. restrained from vertical movement), the pile would be lifted off the seabed by the force of buoyancy when the tide rises.

5.3.2 US Navy Sea Base

As far back as 1990 the US has been contemplating the use of deployable floating ocean bases [13]. In February 2018, the U.S. Navy's Military Sealift Command commissioned a second mobile afloat 'sea base' [14]. The 784-foot (228 m) Montford Point-class expeditionary sea base (ESB) commenced construction in 2015. "The ship is designed as a mobile base to carry out missions including air mine counter measures, counter-piracy operations, maritime security operations, humanitarian aid, disaster relief missions, Marine Corps crisis response, among other missions." "The vessel features an aviation hangar and flight deck that include two operating spots capable of landing MH-53E equivalent helicopters, accommodations, workspaces and ordnance storage".

A paper authored by Lim Soon Heng, C M Wang and Dirk Jan Peters titled "Deployable, reconfigurable, affordable and repurposeable naval bases of the future" was presented at the International Naval Engineering Conference 2016 at Bristol UK. The theme of the Conference was "The Triple A Navy: Active, Adaptable, Affordable." The published proceedings are available in the public domain.

5.3.3 Strategic for Rapid Mobilisation

To deter burglars, it is best to place burglar alarms outside the house rather than indoors. Likewise, the best defence envelope should be located at the maritime and

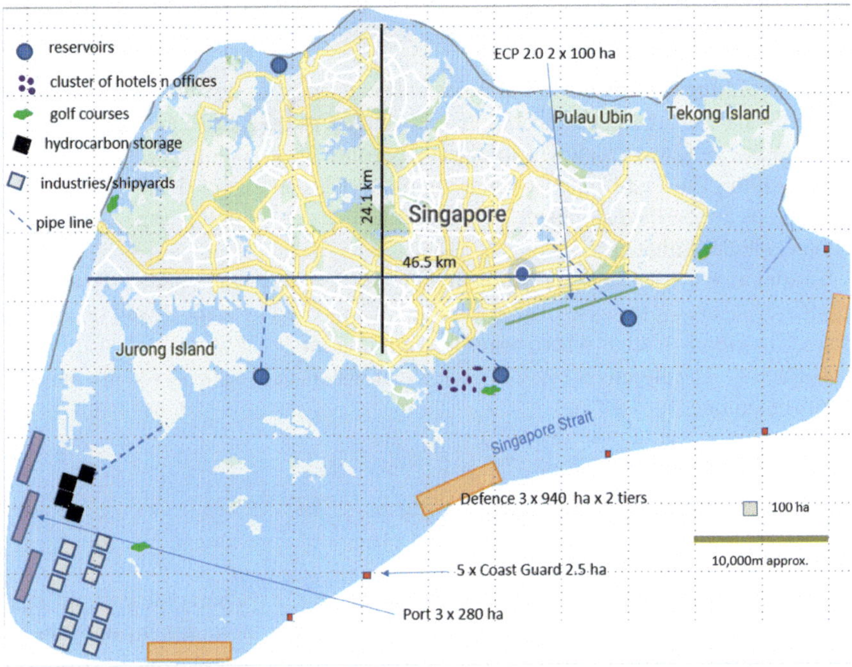

Fig. 2 Map showing location of proposed floating berths

airspace boundary of a nation. Our benign waters and very short coastline allow us to do just that without incurring exorbitant costs.

Three defence clusters in the Straits of Singapore, located at the edge of our territorial waters: one south of Jurong Island, another south of St John's Island and a third east of the existing Changi Naval Base can be more rapidly mobilised to intercept intrusions (see Fig. 2).

Unlike existing base locations on shore, the floating bases offer more flexibility to optimise the base's operational readiness. Over the years they can be readily relocated to respond to changing circumstances. When an old base need to be replaced, a new one could be built offsite while the old one is still in service. The defence capability during the swop from old to new is uninterrupted.

Swarm drone air defence systems are more easily deployed to counter air attacks from the Straits of Singapore than from the mainland. Enemy aircrafts can be brought down simply with suicide drones programmed to head for ingestion by enemy jet engines.

Each floating base is a visible marker of our national territory, just like the Great Wall of China. The recent intrusion of naval vessels from Malaysia is a wake-up call. We need to send a visible and immediate message to anyone "testing" the porosity of our maritime border that they do so at their peril. That message is best delivered by having "bastions" at our maritime boundary.

5.3.4 Unique Operational Advantage

The tides in Singapore rise and fall twice a day, at times by as much as 3 m. Floating berths have a significant operational advantage. As tides rise and fall mooring ropes slacks, gangways may be damaged or become unsafe, shore connections such as firefighting lines and potable water supplies risk being ripped off in situations where ships are berthed alongside conventional wharves. With floating berths, these problems are eliminated.

To attenuate waves, heavy concrete floating breakwaters can be added if the docking master desires it for ease of ship handling. Compared to vertical wall or rubble-mound structures in areas where water is deep, floating breakwaters are vastly more cost effective. The effectiveness of floating breakwaters may be observed in a number of videos on line.

6 Floating Out the Mega Port

It is difficult for many to wrap their head around the idea that a port of tens of millions TEU capacity can be floating. It can. Platforms of very large dimensions can be made by simply connecting smaller modules. These connections can be designed such that the assembly will behave as a rigid structure in a nearshore environment. Its hydrodynamic response to wave and wind may be determined with one of the several computational software used in the offshore industries. Wave energy may be dissipated in a number of ways if it is found necessary. In fact, the energy may also be harvest and stored in batteries kept in the lower deck of the float.

A small port in Valdez, Alaska was designed as a floating port to resolve the challenges presented by the remoteness of the location. The company that designed it US based consultant BergerBAM is now engaged in the preliminary assessment of a 906-ha floating transshipment port in the Gulf of Mexico.

Quayside container cranes, despite their imposing height and load are actually rather sturdy equipment. This is evidenced by the fact that they are commonly transported by barges across oceans, fully assembled. At their destination, these cranes are rolled off from the barge on to the quayside. Well designed, the centre of gravity is low and remain near the centre of the rail gauge even as its load moves from shore to ship.

The rolling motion of the quay is virtually imperceptible owing to the high metacentre of the float. The angle of roll can be calculated.

Fenders on the float will absorb the kinetic energy of the approaching ship. The floating quay may be designed with two "skin" walls and filled in between with polystyrene. Large void spaces are separated by water tight bulkheads.

6.1 Leverage on Local Expertise to Design but not Necessarily to Fabricate

Singapore has two sets of world class skills that may be leveraged: port management and offshore rig building. The latter includes classification societies, naval architects, shipyards, and test facilities such as the Technology Centre for Offshore and Marine, Singapore (TCOMS).

Design and Build bid documents setting out key parameters for the port platforms and equipment specifications for quayside and stacking cranes can be drawn up with the combined effort of the two sets of expertise.

To provide sufficient berth lengths it is proposed that the port be in the form of long rectangular floats. Three such floats each of 280 ha may be moored, for example, at the western end of the island. Figure 2 shows its size in relation to the surrounding islands and shorelines.

The water in this location is about 25 m, and so no dredging is necessary. The prevailing wind and waves are manageable. They would be considered in the mooring system of the VLFS. Floating structures are minimally affected by earthquakes and in fact can survive earth tremors much better than bottom founded structures. Pile failure due to liquefaction of soil does not apply.

The location need not be constrained by the suitability of the geology. It does not matter if the site has marine clay, sand or bedrock since the float is not supported by piles. The project can commence without soil investigation. There is no time lost due to soil improvement or consolidation which add to the financing cost.

Each float consists of 140 concrete modules each 200 m × 100 m which may be manufactured offshore where the price is competitive, and work is technically reliable (China, India or Vietnam are possibilities). Shipping/towing cost is marginal. GL Engineering and Construction, a Singapore contractor, built a concrete floating dock in Batam. The owner shipped it back to Hawaii [15] in a dry tow operation.

The addition of the second and third phase floaters in the years that follow would have minimal impact on the operation of the preceding port. The 200 m × 100 m modules are connected to form larger modules at a nearby staging area (to avoid congesting the active port) before final assembly at its designated position. Cranes are rolled on to the quay side or may be lifted into positioned by floating cranes (Singapore is the home port of several heavy lift ones). All bollards and fenders, pipe services and electrical cables would have been installed prior to arrival in Singapore and need only to be connected between modules.

There would not be the daily fleets of barges bringing in sand (and microscopic marine invaders) and intruding ongoing port operation as would be the case if Phases 2 and 3 of the mega port proceed as planned.

6.2 Terrorism Through Boxes

Terrorism through shipping containers [16] is little known but a major threat to many cities with international ports in close proximity. A city port that handles tens of millions of boxes from all parts of the world each year is especially exposed to illegal shipment of armaments, to the dispersion of radioactive material or biological weapons. "This is immensely disturbing, as "we are only able to scan less than one percent of the incoming containers," according to Van Hipp in *The New Terrorism and How to Defeat It*.

Separating the transshipment boxes from those inland bound proportionately reduces the risk (in Singapore case by 85%) of illegal and dangerous goods entering Singapore.

6.3 Recycling Old Ports

Singapore's first container port at Tanjong Pagar has stood empty and abandoned for more than 16 months, awaiting demolition. When the wrecker's step in, a year of dust and pollution will start to remove 800,000 sq. m of concrete deck and extract tens of thousands of piles.

With floating ports, a new one is built and when ready the old one is towed away to be repurposed for example as golf courses, condominium or be gifted to another country in need of port facilities; no dust nor noise arise in the transition from old to new.

7 Floating Out Parks and Reservoirs

More 80% of Singaporeans live in flats with 99-year lease. This is an explosive and politically sensitive issue. In about 20 years, thousands of new public housing apartment blocks have to be built to accommodate residents of old ones whose tenancy approaches expiry. With land shortage it may be necessary to also consider the float out of parks and reservoirs, the green lungs of Singapore. Floating out these is more practical than building floating houses.

A floating reservoir with a flexible containment "bag" supported by a rigid frame is a project that the author had offered as a final year engineering student's project [17]. The reservoir is modular and when several are clustered to together it can hold about one million cubic metres of water. As the density of rainwater is lower than seawater, the "bag" floats. Whether full or empty the load on the support framework is the same (virtually) and hence its draft and freeboard remain fixed.

Each cluster of floating reservoirs is connected by submarine pipes to a nearby inland reservoir to take their overflows in a heavy storm. In a drought the stored water is pumped back to the land reservoirs. With this solution about 80% of the annual precipitation would be productively used instead of the present 50%.

7.1 Floating Desalination Plant

A self-propelled floating desalination plant [18] would be a useful adjunct to the floating reservoirs. During prolonged drought it goes around the floating reservoirs to top up each of them with water that it produces with shipboard desalination plants. Additionally such a plant could be commandeered for humanitarian relief when there is a crisis such as floods or droughts.

7.2 Enhancing the Floating Reservoir Park

The floating reservoir may be complemented by multiple smaller floats (5–10 ha each) landscaped with trees, beaches and rock cliffs using disused jack ups. Bungee jumping and paragliding attractions from the jack-ups could be added attractions.

Below the water line, the sides and bottoms of the float are fitted with steel overhangs to promote the formation of habitats for corals, sea anemones to attract fishes. A world class undersea garden could be created as a major attraction.

Tenders could be invited to lease (say 50 years) mooring rights for floating golf courses, marinas, upmarket floating condominiums and holiday chalets. Investors could build these in Indonesia, Vietnam or China under supervision by the owners' representative. As in the case of ships and oil rigs, critical mechanical, and electrical items will be procured by the owner and delivered to the fabrication yard for installation.

7.3 Gardens by the Bay 2.0

The one billion dollar Gardens-by-the-Bay is an iconic landmark much loved by the locals and tourists. The decision to set aside 101 hectares of prime waterfront land for leisure activities had not been an easy one according to Prime Minister Lee when he opened it in 2012.

While we hold dear to this marvellous piece of real estate, there is a Plan B: It is a cluster four or five floating concrete islands, which could be imaginatively contoured to include beach fronts and back hills. Trees of any size can be accommodated with properly designed pits large enough for their root systems. The quintessential Supertrees, Flower Dome and Cloud Forest can be dismantled and easily re-erected; not difficult with 5000-tonnes Asian Hercules 3 (a floating crane based here. Such methods of erection are common in shipyards).

The floating version need only be a few hundred meters from the shore line and vacated plot can become a second version of the highly successful Marina Bay.

7.4 ECP Park 2.0

The other goldmine is the East Coast Park. It sits on a 185-ha prime waterfront land which is worth conservatively S$20 billion. A very attractive park with the same footprint could be built for less than S$1 billion in the vicinity as shown in Fig. 2. This would result in a net gain in asset value of at least S$19 billion. The nearby Eastern Anchorage now crowded with idle vessels waiting for bunkers or next assignment need not be there when a new bunkering system is adopted by MPA.

This idea [19] of repurposing the ECP Park was proposed by the author in an op-ed piece in 2016.

8 Floating Out Hydrocarbon Industries and Shipyards

In the following, we make the case for repurposing land for 4IR (4th Industrial Revolution) by offering sea space for the hydrocarbon industry and tapping on the capital market to create such space.

8.1 Restructuring the Economy with Floaters

Industry and Commerce take up 17%, as much as land used for housing. Finding green field sites for modern industries is increasingly a challenge in Singapore. But we must, if we are not to be overtaken by others.

A number of industries in Singapore are outdated, introduced in the 1970s when our concern was not about land shortage but a rising tide of unemployment. We need to refocus and recalibrate our priorities. It is land rather than unemployment that is the limiting Singapore's economic growth. We should be assessing the potential of an industry based on its GDP per hectare.

We need to offer an alternative to those who are heavily dependent on cheap land. That alternative would require them to vacate their land before their lease expire. It would be a win-win solution: the state gets back the land and the lessees get a cheaper alternative with little disruption in his business. The modus operandi would however require a bankable collateral. A floating estate is more bankable than a landed one because it can be deployed anywhere in the world.

A number of industries are under threat from competition from Malaysia, Thailand and Indonesia where land and labour are a fraction what they cost here.

The 80 sq. km Pengerang Integrated Complex [20] is under construction just north of the border. Within this mega complex, Petronas in consort with Aramco is investing $16 billion to develop refineries and petrochemical industries. Shipyards are popping up in Philippines, Indonesia, Vietnam and India not to mention China which in a few short years, overtook Japan and South Korea.

We need to review the policy of holding on to land gobbling industries of the past decades when land is so crucial to Singapore's future success. That review is meaningful now as the option of floating estates is at hand.

Our dependence on hydrocarbon related economic activities needs rethinking as the world gravitates towards renewable energy driven by concern for the environment, health issues, climate change and the depletion of fossil fuels. The future of hydrocarbon industries is not a positive one. It needs to be phased out in a managed way. The luxury of letting industries die naturally is not one for a land deprived country like Singapore.

8.2 Floating Refineries

It used to be that crude oil was always pumped from the ground and piped or transported by ocean tankers thousands of kilometres away for refining. That is changing. Crude oil pumped from the seabed are already processed at sea in in floating production storage and offloading (FPSO) facilities. Shell's floating refinery Shell Prelude, the biggest man-made structure will work in offshore gas fields in West Australia for the next 25 years. When the field is depleted it will relocate to another. This begs the questions: why are our refineries and petrochemical industries still sitting on thousands of hectares of land that could be repurposed for 4IR industries? Why not place them out at sea? Why not indeed when we have such sheltered waters around us?

8.3 Floating Storage Tanks

Much of the land on Jurong Island is dotted with storage tanks. Storage tanks on land cannot be erected too close to one another or too high for safety reasons and therefore land use is inefficient. It is time we consider floating out storage facilities (in addition to rock caverns). In Japan, floating oil storage facilities have been in use for several decades [21]. NUS (National University of Singapore) and JTC (a Singapore Government agency in charge of land allocation for industries) are exploring this option and we hope the relevant agencies will put this into action, after all it is a tried and tested solution.

Besides the fact that the land they occupy can be put to better use, storing oil at sea is multiple times safer. Storage tanks on land despite their separation distances and containment bunds would fail structurally when a fire occur. In 2016, a fire at

Jurong Island caused a tank to melt down like jelly. Terrorism is also a matter of serious concern as shown in the Libya incident [22].

Floating storage tanks are safe enough to be closely packed and therefore more efficient in terms capacity to space ratio. Containment booms, skimmers and dispersants are tools more effective tools to contain spillage in water than what are used for spillage on land.

8.4 Floating and Relocatable Shipyards

Singapore's marine industry owes its initial success to its location on one of world's busiest tanker trade route. Other shipyards have sprung up on the same trading route and competition is tough at the low end. In the late 1970s the industry shifted from repairs to high end ship conversion and offshore rig building, thereby staving off a downfall. However, oil prices have collapsed, and shipping traffic may one day with the polar ice cap melting away, sail the Arctic route to the Atlantic Ocean. Business may not be as usual. Shipyards rooted firmly on land pay a heavy price when business environment changes.

8.4.1 Lessons from the Demise of Two Shipyards

Harland and Wolff arguably the world's most famous shipyard, delivered the Titanic and the Olympic. At its peak, it employed 35,000 people, today it has only 500. Hundreds of millions of pounds spent by the Government to modernise its facility left no positive impact. The market shifted, leaving it stuck in the wrong place.

Back in the 1970s Japanese shipbuilder MHI (Mitsubishi Heavy Industries) constructed a massive shipyard in Tuas. In the early 1980s faced with a severe downturn in business, MHI decided to exit the business. JTC required MHI to reinstate the land back to original. There was this 400,000-dwt graving dock and a massive pump room which JTC wanted to demolish. The dock floor was 2.5 m thick. The walls were 15 m deep and the quayside about 600 m long. MHI is a classic case of a shipyard caught in economic changes and unable to extricate the hundreds of millions of dollars poured into the ground.

8.4.2 Relocating Old Shipyards in Singapore

When structural changes in an economy take place every few years assets must be nimble enough to move transnationally. Floating assets offers that advantage.

Many shipyards in Singapore are occupying land with leases that are expiring. JTC, quite rightly wants to recover the land to revitalise it for other purpose. The author has advised JTC that their strategy of multi-user yard where several shipyards share berths, docks and workshops will not work. The users are merely renters of the

Fig. 3 Floating shipyard and floating cranes

facility. They can break lease when the going gets tough. They have no long-term commitment.

The author strongly urges JTC to consider the alternative of floating shipyards where the shipyard build their own floaters and JTC offers attractive incentives as mentioned above. No new land is needed. The old land can be repurposed. Each shipyard would have full control over the use of his facility.

Workshop machinery (panel lines, blasting and painting chambers, presses, lathes and boring machines) perform just as well on floating platform as they do on land. The performance of quaysides and gantry cranes work are not affected by the fact they are on floats. Shifting large blocks may be done with SPMT [23] or with floating cranes. A floating crane can access loads all around a floating shipyard; something impossible on land (see Fig. 3).

Floating docks have been in use in many shipyards. The biggest to date is 85.6 m wide, 432 m long and 130,000 DWT in South Korea.

Workers dormitories wash rooms, toilets and amenities can be fitted in spaces beneath the main deck. Air compressors, power generators, freshwater generators and air compressors can be installed below the main deck.

A time-consuming problem with shipyards is that they need to constantly adjust any connection between shore and ship (including mooring ropes and gangways) as tide changes. This problem is eliminated.

8.5 *Floating Power Stations*

8.5.1 Power Plants Compete Unnecessarily for Land and Seafront

A thermal power plant should never be built on land if the option of building it on water exists. This is because thermal power plants need to dump heat to condense the steam into water before returning to the boiler. They also need fuel unloading

Fig. 4 Floating power plant proposed by Siemens and Sevan Marine

and storage facilities. These requirements make floating power plants more energy and land usage efficient.

A floating power plant requires no space on land for storage tanks, nor waterfront of unloading fuel and navigational access from the open sea. In Fig. 4, we see a floating power plant being refuelled offshore. This 700 MW plant is a recent proposal from Siemens and Sevan Marine to Japan.

Singapore emits 22 million tons of CO2 annually using LNG to produce some 50TWh of electricity. Currently, all 22 million tons go into the atmosphere. That mass is 3 times the amount of waste produce in the country.

Carbon capture with algae [24] is a promising green technology. The sea has an inexhaustible supply of algae which a sea based floating power plant can tap into. The cost may be offset with savings from the carbon tax which by 2030 could be as high as $15 per ton of emission.

The author has recommended to EMA (a Singapore government agency responsible among other things for planning energy strategies) to discuss the viability of using concrete floaters [25] as platforms for future power stations. A concrete platform is preferred as it is more durable (longer than 100 years) and does not require drydocking for repainting. Its heavier mass offers higher stability.

8.5.2 Senoko Power Station

Senoko Power Station deserves special mention. This station not only occupy land space, it also kills off the sea space that could otherwise be developed with floating solutions for housing or commercial activities. With the floating out of Senoko Power Station to the southern border (eastern anchorage) and the departure of Sembawang

Fig. 5 Freeing of sea space by floating out Senoko Power Station to the southern border (eastern anchorage) and Sembawang Shipyard to the west

Shipyard to the west about 80 sq. km of sea is free to be developed (see Fig. 5). The dyke and polder structure that is under construction would be unnecessary in this scheme. Connectivity to the mainland and Pulau Tekong, Ubin and Coney Island is easily achieved with floating bridges between the 8 proposed floats marked red. This space could initially be used for floating farms including fish farms and as air cargo logistic bases currently on land. If a new power plant is needed to serve the north east, it could be a floating one located south of Pulau Tekong so as not to necessitate ships entering the enclosed area for refuelling.

9 Portability and Reusability Reduce Investment Risks

Real estates by definition are fixed. Assets which are movable such as ships are known as chattels. VLFS is new class of chattels which has two unique features that enable them to push the boundaries. The two examples below will illustrate their usefulness.

9.1 The Forest City, a S$100-Billion Lesson

Malaysia was viewed by the Chinese as the safest place outside China for property development. Up until May 2018 it was. Then came a political earthquake. The incumbent Prime Minister Najib was ousted by Mahathir Mohammed, a nonagenarian and astute politician. He pulled the rug beneath the feet of the Chinese developers, notwithstanding their partnership with the Sultan of Johor [26] in whose state the project was located, as well as the backing of the Chinese Government.

The Chinese had reclaimed land close to Singapore and developed a real estate worth a hundred billion dollars targeted at the super rich from China. Mahathir realised he could make political capital by holding the project up as an example of the extravagant indulgence of his nemesis for mega projects. The project had become a pawn in the domestic politics.

The Chinese were caught; stuck with a huge project and very little leverage to negotiate a way out. In hind sight, the Chinese would surely wish they had built the entire project on floating islands instead of on reclaimed ones. Floating assets would have weakened the hands of the host country as the investors and financiers have the option to move everything across to a friendlier location.

9.2 The Golden Mile Complex

One of Singapore's few iconic buildings and the work exclusively of the first batch of locally trained architects in early Singapore, the Golden Mile Complex landmark will be rendered to dust and debris to make way for a higher density development [27]. What a shame.

If we could somehow turn back the clock and the young architects had access to experts to build a VLFS on which to build their dream, the grand old dame would perhaps be around for another fifty years, revamped redecorated it could be repurposed as a heritage hotel in Singapore or a high school in Pattaya (the complex is a hangout for many Thai workers).

VLFS's portability, reusability and recyclability enables a less destructive, more caring and sustainable use of building material. The platform on which it is built can used as a foundation on which a multi-tier vegetable farm or solar farm may be erected or simply as a playground. Such a versatility is out of question real estates in an urban environment because a building can have only a single purpose determined by the regulations applicable to the specific zone.

10 Summary

The takeaways from this paper are briefly as follows:

i. The Singapore territorial waters south of the mainland offer opportunities for space creation with floating structures to which land intensive and ageing industries may be relocated to enable the freed-up land to be used as living spaces and 21st century industries.

ii. 85% of the container boxes unloaded in Singapore are for transshipment. They should stay offshore. Singapore territorial waters may then be freed of international shipping traffic and be used for a variety of non-shipping commercial, residential and recreational activities.
iii. Singapore defence infrastructure takes up 19% of the land in Singapore. A large part of the army, navy and air force facilities can be replicated offshore with floating structures. The vacated land could be used for another city centre in the northwest to support 4IR industries.
iv. Singapore benign and deep waters provided safe haven for sailing ships in the days gone by. They are no longer needed for ships of today. Let's repurpose it.
v. Polders are easily damaged by ships psychopaths and enemy action. They are costly to construct and maintain in deep waters. They are not resilient to earth tremors and rising sea levels.
vi. The massive PIC (Pengerang Integrated Complex) and the Tanjung Langsat Port Complex emerging in southeast Johor will diminish the viability of the Jurong Island hydrocarbon complex. JTC must reinvent the island for a cutting-edge technology.
vii. Industries that should be floated out include the hydrocarbon industries, refineries, tank farms and shipyards.
viii. Hanging on to old industries, in land-scarce Singapore limits the space necessary for quality living for the future generation.
ix. The global capital market may be mobilised to create floating space in Singapore. In contrast money for land reclamation has to come out of state coffer.
x. We should endorse the UN red line on sand mining and not facilitate contractors to find ways around it. VLFS enable us to support the preservation of marine eco-systems.
xi. Floating concrete platforms has a lower life cycle cost. It is cheaper than land in many cases. It is unaffected by rising sea levels and seismic activities.
xii. Floating out assets reduces greenhouse emissions arising from road transportation.
xiii. The local expertise in the offshore rigs sector is easily adapted for this new technology. A lot of the hardware are just right for the construction of VLFS: goliath and floating cranes, dry tow vessels, pipe and steel fabrication workshops, piers and docks.
xiv. Foreign firms are readily available to support local firms when needed.

11 Recommendation

The author recommends that the Singapore Government set up a National Study Team for Floating Solutions to identify existing public and private sector real estates which could be advantageous to put offshore allowing the land they occupy to be rejuvenated and repurposed. The capital market should be mobilised to support this multi-billion-dollar initiative.

References

1. https://www.businesstimes.com.sg/transport/container-volumes-up-98-at-psa-ports-globally-in-2017.
2. https://www.sciencedirect.com/science/article/pii/S0048969717319757.
3. https://ftnnews.com/cruise-travel/33157-cruise-industry-booms-in-asia.
4. https://translate.google.com/translate?hl=en&sl=es&u=http://www.ciudadfcc.com/-/dique-flotante-de-monaco&prev=search.
5. https://ascelibrary.org/doi/pdf/10.1061/9780784479902.071.
6. http://www.dutchdocklands.com/Development/Floating-Cruise-Terminal.
7. https://en.wikipedia.org/wiki/Landing_Craft_Air_Cushion.
8. http://floatingsolutions.org/blog/2018/08/11/presidents-message-07-2018/.
9. https://en.wikipedia.org/wiki/Project_Habakkuk.
10. https://www.offshore-mag.com/1/volume-74/issue-11/arctic/concrete-drilling-platform-could-extend-exploration-season-offshore-northern-russia-full.html.
11. https://www.maphotosg.com/constraints-rsaf-air-bases/.
12. https://www.dailyrecord.co.uk/news/scottish-news/valiant-jetty-designed-berth-nuclear-2458995.
13. https://apps.dtic.mil/dtic/tr/fulltext/u2/a416198.pdf.
14. https://www.popularmechanics.com/military/navy-ships/a15392206/woody-williams-us-navy-ship/.
15. https://www.rina.org.uk/New_concrete_dry_dock_for_Marisco.html.
16. https://www.centerforsecuritypolicy.org/2018/04/12/terror-by-shipping-container/.
17. http://floatingsolutions.org/blog/2018/05/13/conceptual-design-of-modular-floating-reservoir/.
18. http://www.hyfluxmembranes.com/images/documents/brochures/fdpv-eng.pdf.
19. https://www.straitstimes.com/opinion/floating-a-novel-idea-for-the-future-east-coast-park.
20. https://www.hydrocarbons-technology.com/projects/petronas-rapid-project-malaysia/.
21. https://www.researchgate.net/figure/Kamigoto-Floating-Oil-Storage-Base-Nagasaki-Prefecture-Japan_fig2_283484673.
22. https://www.reuters.com/article/us-libya-security-oil/second-fire-shrinks-storage-capacity-amid-fighting-at-libyan-oil-ports-idUSKBN1JD0DW.
23. https://en.wikipedia.org/wiki/Self-propelled_modular_transporter.
24. https://phys.org/news/2018-11-algae-sustainably-carbon-dioxide-atmosphere.html.
25. http://floatingsolutions.org/blog/2018/11/06/presidents-message-11-2018/.
26. https://www.channelnewsasia.com/news/asia/forest-city-johor-state-government-surprised-mahathir-comments-10659392.
27. https://mothership.sg/2017/05/golden-mile-complex-called-an-architectural-cultural-marvel-in-documentary/.

Floating Solutions: The New Meaning of Mobility

Milica Simovic, Sonja Krasic and Marko Nikolic

Abstract Floating architecture has a new dimension: from completely static and permanent forms on land to becoming non-stationary. The aim of this paper is to define the new meaning of mobility when it comes to modern, floating space and to analyze opportunities it offers in the future. Through analysis of some examples of good practices, different ideas and aspects are presented. For example, floating buildings can be readily relocated when necessary, be used only periodically on specific locations or for a specific purpose, and be moved only vertically during the tide. Movable architecture is more adaptable and sustainable. The design approach changes a lot as we can now manipulate space in a more dynamic and flexible manner. Water, as an unstable environment, requires people to be more active and provokes their curiosity. Since structures are movable and can be relocated elsewhere offshore, users may need to use alternative ways of getting to them or might employ specific modes while using the space. Mobility, being the pivot of contemporary society, gives ability to movable structures to follow the pace of modern life and global tendencies, while understanding and satisfying people's needs.

Keywords Mobility · Floating structures · Water-friendly architecture · Sustainability · Architectural designing

1 Introduction

According to Macmillan's definition, mobility means the ability to move a part of your body, walk normally, travel from one place to another or tendency to move between places, jobs, social classes [1]. When we use the term "*mobility*" more specifically, in a certain context, the first associations are mostly urban or social mobility. But, when it comes to architecture, we don't use "architecture" and "mobility" in the same context, or so it has been until now, as it is not usually considered as a movable type of structure. Even though there is no such term as "architectural mobility", both urban and social

M. Simovic (✉) · S. Krasic · M. Nikolic
Faculty of Civil Engineering and Architecture, University of Niš, Niš, Serbia
e-mail: milicavsimovic@gmail.com

C. M. Wang et al. (eds.), *WCFS2019*, Lecture Notes in Civil Engineering 41,
https://doi.org/10.1007/978-981-13-8743-2_23

mobility fall under the scope of architecture. Urban mobility refers to the efficient movement of people and goods: transportation and mobility are recognized as central to sustainable development. According to definition of sustainable urban mobility [2], the term presupposes a system that incorporates economic viability, environment stability and social equity by the needs of transport and land use of both current and future generations in a sufficient manner. On the other hand, if we talk about social mobility, Aldridge [3] describes it as the movement or opportunities for movement between different social groups, and the advantages and disadvantages that go with this in terms of income, security of employment, and opportunities for advancement. Both terms describe certain aspects of mobility that directly have an impact on architecture. We need to start reconsidering architecture, and planning as well: rather than unchangeable structures with one constant function situated on a permanent location, we need to redefine the meaning of mobility within architecture. The new form of architecture, designed to float on a water body, permits movements that can completely change the way cities work. According to that, we can simply define it this way: *Floating mobility in architecture means mobility in terms of floating buildings and structures.* In order to represent it more precisely, the aim of this paper it to define a term of floating mobility in case of floating buildings (structures). This specific ability is shaped by several aspects that we want to illustrate through the analysis of examples of good practice. The focus of this research is on single objects (avoiding VLFS) in order to gather facts and generalize into relevant recommendations which can easily be applied to individual project and incorporated into existing urban tissue. This particular instance would be of great help in making the practice more common and widely applicable. Further research on the topic of floating mobility is likely to yield valuable contribution to the enumeration of possibilities in this area.

2 The Global Tendencies

The new way of thinking, when it comes to urban planning and architecture, significantly changes in accordance with global tendencies. Modern architecture's aim is to serve the contemporary society in a highly effective and sustainable way. Perhaps a more functional architecture that is movable, adaptable, transformable and capable of disengagement and reassembly could keep up with the occupants' need to indulge in multiple activities in one space, help to maintain economy and ease the overuse of energy and resources [4]. As a response to the most recent requirements, we have started developing a different direction in architecture—floating and offshore structures. The mobility of a floating structure has been exploited for industrial purposes since the latter half of the 20th century. These facilities can be constructed in a particular location, towed to site and be installed as a permanent facility, or moored and towed again to a subsequent site where the need arises [5]. This way, from a completely static form, architecture has become a dynamic form. Rising sea levels and a shortage of development sites are leading to a surge of interest in floating buildings, with proposals ranging from mass housing on London's canals to entire amphibious cities in China. The nomadic lifestyle and working patterns of our mobile society as well as other consequences of globalization require new dimensions from modern architecture.

In an up-to-date society, mobility has a huge impact on the quality of life and the prediction of economic growth, creativity, and trade, as well as it does on personal well-being. People who are able to move around easily can enjoy more opportunities. In Acharya's opinion, as corporations and communities continue to expand and contract, re-locate, emerge and vanish, their need for the capability of adaptation is increasing. There are two options usually taken into consideration when architects and urban planners discuss the social response to climate change: adaptation and migration. Adaptation is the art of survival. It is about analyzing, questioning and managing the risk of disastrous events. At the same time, it is about dealing with the increasing problems connected to extreme events and societal vulnerability. While Europe and other western countries have a financial capacity to cope with extreme events, more remote regions are much more exposed and vulnerable. Adaptation has to identify priority areas, priority needs and priority actions. The role of the development of new adaptable architecture is therefore crucial in modern urban planning. Migration, on the other hand, is very often the last chance for those threatened by the impacts of climate change or other dangerous situations. Migration also occurs when adaptation fails [4]. Now, we can have both options included in this one solution: in terms of floating mobility, buildings could simply adapt to the climate changes (follow the changing water level) or be migrated themselves (easily relocated when-and-wherever it is suitable).

3 The Concept

Nowadays, liveable cities equal mobile cities. As extensions to cities, urban planners have begun to supplement the existing urban environment by programming cultural and recreational functions into floating facilities. These enable a city to expand its ability to cater to the needs of its growing population and enhance coastal living [5]. If we consider the future of cities on a macro level, it is possible to describe them as a changeable form, composed of different segments (buildings and other kinds of structures) able to move and easily adapt to all kinds of requirements. Buildings on or in water enormously increase possibilities for relocating and manoeuvring structures that can float. Water environment also brings a completely new perspective: it changes the way we act, feel and move and it conditions specific types of social behaviour. Valk[1] claims that being literally disconnected from land has an impact on human perceptions and perspective. The moment people leave shore, they are crossing a

[1] Sikko Valk is ward-winning Dutch industrial designer, working as independent industrial designer & consultant, as well as creating products for his collections & design-label SIGU®. Designer of a floating hotel, *Good Hotel*, located in London, UK.

bridge over water, and there their break from the everyday routine begins, which is quite symbolic [6].

In the Gif im Focus,[2] Olthius[3] presented his general idea of a Blue City concept: "…Buildings will interact better with the climate of a city. It is strange that many architects still build houses that are the same for severe winter conditions and for hot summers. I think we will have seasonal houses and neighbourhoods in the future, which will change their configuration and identity along with the changing seasons. Another new concept is "meantime" cities where neighbourhoods or functions can be placed in a location. They then have to make space for new uses when their economic value no longer matches the needs of the location. This means you will be able to make space for new developments in the centre of the city without having to demolish buildings that are still functional. You just replace, re-use and re-organise to suit your needs. A common feature will be city apps—small temporary floating functions that can meet a specific need or solve a specific problem in a location: temporary parking places, floating sports facilities for a big event or temporary floating affordable housing for students. As green space is under pressure in expanding cities, we will see green spaces appear in blue cities. Floating habitats, floating forests, floating parks can all have a positive effect on the environment of a city…" [7].

4 Floating Mobility

Generally speaking, the floating building can be defined as a structure for living/working space that floats on water using a flotation system, is moored in a fixed place, *doesn't include a water craft for navigation*, and has a premises service system (electricity, water/sewage and city gas) served through the connection by permanent supply/return lines between floating and service station on close land, or has self-supporting service facilities for itself [8]. This description although excellent does not fully grasp the idea. The design considerations for a floating structure should satisfy the structural requirements that address the operating conditions, structural strength, serviceability durability and safety standards and socio-political criteria that address the aesthetics, environmental sustainability, budgetary and legal constraints [5].

By analysing the aspects of mobility which influence architectural design as well as urban planning, we may proceed to group the following properties into:

[2]Gif im Focus (Society of Property Researchers, Germany) 25 years jubilee, gif 1/2018, pdf available at (https://www.gif-ev.de/onlineshop/download/direct,396).

[3]Koen Olthuis is a Dutch architect, studied Architecture and Industrial Design at the Delft University of Technology. founder of Dutch architecture practice Waterstudio. Since 2003, he has worked on floating houses, schools, resorts, swimming pools and other projects, all of them enabled by a proprietary floating base technology. In 2007 he was chosen as no. 122 on Time Magazine's list of most influential people in the world due to the increasing worldwide interest in water developments. In addition, the French magazine, Terra Eco, choose him in 2011 as one of 100 green persons that will change the world.

(1) Relocation; (2) Permanent function-current location; (3) The transport issue (along and outside of urban settlement); (4) Flexible configurations; (5) Vertical movements; (6) Buildings that include navigation system.

4.1 Relocation

Even though mobility is not the primary feature in the story of architecture, floating buildings are possible to move during utilization, theoretically speaking. A building may be relocated in the case of contamination on site, the reconstruction of a harbor or the quay where the construction is docked, or the repurposing of an urban settlement into whose function the construction no longer agrees with, but also in the event that the building at hand must undergo a shift in function which requires a different urban area. The relocation can be planned ahead so that the given object would rest at a specific location during a particular period of time and be transported along. On the other hand, this benefit is built in the peculiar nature of floating units from the beginning; in other words, a building retains the potential for relocation, even if the construction alone is not intended to answer this purpose at the start, but remains an option in case of such needs arising at the time of utilization (as long as the location satisfies all conditions for "setting sail").

Waternet is a floating office building which belongs to the Waternet company, a Dutch public company dealing with water supply, sewerage and water management. It is located in Amsterdam (The Netherlands). The building was designed by the Attika Architekten practice, realized in 2010. This office is planned to satisfy all the requirements of the company and workers and it perfectly matches the current location and context (see Fig. 1). The building is set between the company vessels that sail out daily to clean the waters of the Amsterdam canals from garbage. It is located in the northern part of the old city harbours left by the harbour industry, although planned, but not yet transformed into a living-area. The architects envision that this area will be transformed over the course of the next five to ten years, making it an unsuitable location for rubbish-collecting boats. They planned that the building be relocated in due time [9]. Designing it as a floating structure, besides satisfying the current needs, they provided a sustainable solution for the future as well. This represents the means of easier handling of location issues.

4.2 Permanent Function-Current Location

Buildings of constant function and the possibility of relocation represent very ambitious new concept of long-term architectural and urban planning. Single or periodic-use massive objects, in terms of an independent structure or a series of interconnected elements in a colossal complex of buildings, which are used at the one and the same locality make for inordinate investments which, in turn, results in the necessity of

Fig. 1 Transporting office (left photo); Office in boatyard (middle photo); Concept sketches by Marshall Attika Architekten (right photo) (Photos: Courtesy of Martine Berendsen, Bart van Hoek and Attika Architekten) (available at: https://www.archdaily.com/585536/floating-office-for-waternet-attika-architekten/54af54aae58ece507000005e-attika-waternet_-13-jpg, accessed 8 Jan 2019)

cost rationalization. This type of concept encourages space saving as constructions of this kind take up significant amounts of land.

Following the idea of temporary utilization, G. Wolff[4] proposed the concept in her winning essay for the Inaugural Wheelwright Prize: *Floating City: The Community-Based Architecture of Parade Floats.* Her proposal is based on the study of the tradition of parade floats—an elaboration of temporary and mobile constructions that are realized annually in carnival festivals in Rio de Janeiro (Brazil), Goa (India), Nice (France), Santa Cruze de Tenerife (Spain), and Viarreggio (Italy). The float transforms the city while its enormous scale makes streets resemble rooms of a large street theatre. This research ties into contemporary interests in performance and architectural notions of mobility, temporality, spectacle, urban space, and community-based design [10]. The concept engages with modern-day architectural concerns with mobility, flexibility, modularity, art and shows the interest in community-based creative production of carnival floats. This in turn opens up a social dimension that resonates with the current preoccupation with local fabrication and maker economies.

Analogous to the abovementioned carnivals, Olthuis (see Footnote 4) considers this issue in a very similar way: he analyses the case of the Olympics in Miami. It costs a lot of time and money. Building stadiums for all these European sports is pointless because after the Olympics they are not being used anymore. So instead, it is possible to build floating stadiums, floating hotels and just lease them after the event at hand. It is better to have cultural events and museums that go from city to city [6].

The same concept, but on a single-building project, is finally applied and developed by German architects, the so-called *stadiumconcept* for the FIFA World Cup 2022. The Floating OffShore Stadium represents an innovative concept in the realm of the entertainment industry architecture—a structure that can be relocated to the seaside venues anywhere across the oceans. So far, the potential target groups for this stadium are the *UEFA European Football Championship*, the *Confederations Cup*, the *Africa Cup of Nations*, the *AFC Asian Cup*, the *North America Nations*

---

[4]Gia Wolff is an associate professor at Pratt University and is an adjunct assistant professor at Irwin S. Chanin School of Architecture at Cooper Union. In past projects, Wolff contributed to urban installations, theatre and set design productions.

Cup or the *Copa America Massive.* The stadium may be used more than once which proposes more effective ways of using the structure while it also avoids wasting resources, with such constructions standing as multi-decade investments after all. The stadium is powered by high energy-efficient diesel engines which have a double function as heating and power stations (co-generation), as well as for moving the floating stadium. Transatlantic transfer is alternatively supported by tug boats [11].

4.3 The Transportation Issue (Along and Outside of Urban Settlements)

During transportation and building manipulation, it is necessary to keep sufficient space provided for maneuvering within the urban environment. Aside from moving the construction itself, this action normally includes enough room for the front tugboats that pull the building and the back boats that serve to control the movements precisely. Tugboats navigate the object by towing and positioning it to fit the allocated place. This sort of circulation presupposes river streaming of a structure which could reconfigure, for example, to overcome stable bridge passages; a structure which could also undergo land transportation (in case the construction is built in a sheltered environment) or cruise along a wider waterfront, in which case the issue becomes less prominent (relocating across states, oceans, continents, etc.).

Good Hotel is a floating hotel located in London (UK). The building was designed by Sikko Valk and art director Remko Verhaagen, realized in 2016. The building was originally a floating jail and it was converted into a hotel and fully restored and reused on water. Having been constructed in Amsterdam, Good Hotel was hoisted onto a submersible barge and ferried across the English Channel to London. Manoeuvring the hotel, built more robustly than most land-based buildings that weighs eight million kilograms, along and outside of urban settlements needed meticulous preparation.

It was towed from the side by a leading tug ship sailing on the front and had two ships following it at the back, tied to the opposite two corners. The most challenging part in transportation was fitting 100 m × 20 m building footprint trough King George V lock, with only half a meter on either side [12] (see Fig. 2). Even though the building was massive, it was possible to materialize a long-distance transportation through the urban settlements in its entirety.

When it comes to urban-scale settlements, Ijburg is a unique pilot-project that forms a residential urban hybrid, located in Amsterdam (The Netherlands). There are two parts: the project-based part was designed by Architectenbureau Marlies Rohmer in 2011, and the other part by owners. The project-based part contains the houses developed by a private company, with a lot of attention to cohesion and architecture. In the part with private plots, the residents were granted more freedom; without the restriction of regulations regarding the architectural appearance, they were able to design and construct their floating home as desired. The restriction they had, on the other hand, was the maximum size that the building technique permits,

Fig. 2 Transporting and positioning hotel (left photo); Good Hotel now (right photo) (photos provided by Good Hotel Group) (available at: https://www.goodhotellondon.com/images/, accessed 24 Dec 2018)

as a result of which large-scale building is impossible. The main reason for this is that the concrete foundation on which the houses float may not be wider than 7 m, or else they could not pass the set of locks during transport. The houses are officially classified as immovable properties. Only a small minority of homes (mainly trailers and houseboats) are mobile. Special rules apply to them. In the case of floating houses, we are dealing with houses that can theoretically be moved, even though they are clearly not intended to do so. After the necessary preparations, and with the aid of a tugboat, a floating house can be transported to another place (it is possible even though it is not likely that people are intending to do so).

A house sails in Ijburg pulled by a tugboat and controlled by a second boat at the back. When the bridge opens and a new house sails in, it is moored at its final destination (see Fig. 3). With appropriate difficulty, this ensemble manoeuvres through a set of locks that is barely wider than the house itself and need some help from the people on the jetty, to push and pull as well. The house is then neatly docked to a jetty in the inner water, precisely at the designated location. There it is riveted onto two mooring poles. The social aspect of this experimental project turned out quite advanced. Being a part of floating community helps habitants gather and provides great connection with their local environment. It encourages more active utilization of all age categories: people swim around the houses during the summer, sail out in their boats, go ice-skating during the winter etc. [13].

4.4 Flexible Configurations

Buildings (or modules) apt to reconfiguration—folding and reshaping in accordance with the purpose at hand—whose nature grows into multifunctionality add to the constructions' flexibility, sustainability and adaptability. The multifunctional character allows for seasonal adaptability in terms of module configuration, space economiza-

Fig. 3 House sails in Ijburg (left and middle photo: Marcel van der Burg); Ijburg from the waterfront perspective (right photo: Luuk Kramer) (available at: https://www.archdaily.com/120238/floating-houses-in-ijburg-architectenbureau-marlies-rohmer, accessed 3 Jan 2019)

tion, potential needs arising suddenly, whereby cost reduction is necessary in this case as well.

The example of a massive structure that represents this concept is the world's largest floating performance stage in Marina Bay located in Singapore. The platform is designed to be a multi-purpose facility on the bay for mass spectator events, sporting activities and cultural performances (it hosts the nation's National Day Parade, the Fireworks Festival and the Water Carnival). The structure was designed by Defense Science and Technology Agency, and realized in 2007. The design of the floating platform at Marina Bay has to address the key requirements of mobility and versatility in deployment. The floating platform is designed to be modular and is constructed from the assembly of fifteen steel pontoons with a unique connecting system that allows for the dismantling and re-assembling of the floating platform. Given the modular nature of the design, the floating pontoons can be easily relocated within the bay area or re-configured into various shapes and sizes to meet different event requirements. The inherent mobility of its component parts has also been incorporated in its design function, ensuring that the structure may be re-configured for use as a mobile dock for water sports and boat shows [14].

A smaller scale project of the same concept is the Archipelago Cinema, a combination of a floating structure representing an auditorium and a floating screen, both creating a unique open-air cinema, placed in Ko Kudu Noi (Thailand). The cinema was designed by architect Ole Scheeren, and realized in 2012. Modular pieces, loosely assembled, symbolize a group of little islands that congregate to form an auditorium. A work with a strong connection to the local community, Archipelago Cinema is based on the techniques used by fishermen to construct floating lobster farms. The raft is built out of recycled materials as a series of individual modules to allow for flexibility for its future use (see Fig. 4).

The aim of this project is to consider connecting people with nature by focusing on creating more effective and unique experiences. The authors presented their general idea emotionally by emphasizing the effect nature has on the inhabitants who now interact with it through an urbanized method. The strong concept can serve as a sort of a prototype and inspiration for the future projects. Subsequent to a journey which will see the raft travel to further places as an auditorium for other film screenings on

Fig. 4 Archipelago cinema (left and middle photo: Piyatat Hemmatat); Locals constructing the structure (right photo: Courtesy of the Rocks Noi Foundation) (available at: https://www.archdaily.com/226936/archipelago-cinema-buro-ole-scheeren-film-on-the%25e2%2580%258b-rocks-yao-noi-foundation, accessed 6 Jan 2019)

water, it will eventually return to the island and be donated to its actual builders, the community of Yao Noi, as its own playground and stage in the ocean [15]. Architectural structure can travel and explore new locations, having its active life during utilization together with users and being an integral segment of the general context. At the same time, the author managed to make an economical and multifunctional solution, following and respecting the local environment. This is a level up in a manner of humanization and adaptability when it comes to architectural designing of floating structures.

When it comes to open, public space that is needed for everyday use, solutions that we need are structures whose mobility provides flexibility in making different urban configurations. The conceptual project called Parkipelagio, located in Copenhagen (Denmark), is a well illustrated idea by Blecher and Maarbjerg[5] which has a series of floating platforms with the view to add interest to the waterways. So far, the prototype is represented in the form of 20 m^2-floating platform with a single linden tree, designed for use as a public space. The author proposes this prototype island as a resting place for kayakers and swimmers, for sunbathing, fishing and for small events, and has already hosted a lecture series about the future of harbour cities [16].

The general concept comprises of nine islands, each with a different function. The idea is to combine platforms and form a network. These include a diving board, a small stage, a sail-in cafe, a mussel farm and a sauna. Islands will be designed so that they can be connected together for festivals or special events (see Fig. 5). The project is focused on the place and function of public spaces in the city—both in a local context where rapid urban development along the harbor side threatens the recreational spaces, but also in a global context where rising sea levels creates new challenges for urban environments. The islands will be dispatched on suitable locations around the inner harbor but will also find their way to more forgotten and underused corners of the harbor, catalyzing life and activity. The authors believe that the project could be adapted for use in other cities as well and help in democratizing

[5] Marshall Blecher and Magnus Maarbjerg, Australian architects working in Danish design studio Fokstrot.

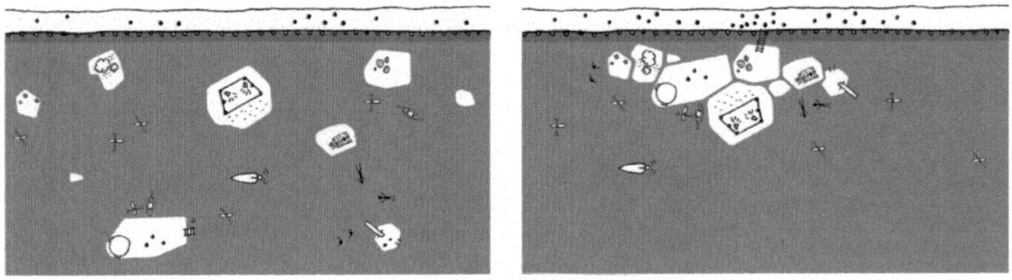

Fig. 5 Parkielagio: summer months events (left photo) and winter months (right photo) (Concept sketches drawn by Marshall Blecher and Magnus Maarbjerg (Fokstrot)) (available at: https://www.copenhagenislands.com/content/, accessed 3 Jan 2019)

harbours in relation to land residences in terms of public space while bringing some life back onto water.

4.5 Vertical Movement

The key difference between land buildings and water constructions lies in the existence of a mooring system which allows free vertical movements, thereby ensuring synchronization with the fluctuation of water masses and/or payload requirements. This type of system represents an entirely new degree of sustainability and adaptation to the conditions of natural environments. Floating buildings, therefore, gain the capacity to endure the lack of stability inherently found along the body of a river or sea due to the consequences of climate change such as rising levels, increased rainfall, flooding and daily shift of the ebb-and-tide cycles. Such arguments speak in favour of the safety that floating structures thus introduce, as the surface of the building remains constantly above water levels.

An example of vertical (floating) mobility is IBA Dock, an international floating Exhibition building (*Internationale Bausausstellung* = *IBA*) placed in Hamburg (Germany). The building is designed by Han Slawik, and realized in 2013. The superstructure is made of steel and modular in form (see Fig. 6). It resembles stacked containers on a sea freighter and also functions in the same way. This keeps the weight down and makes it possible to remove parts of the building for transport purposes so that the structure can pass under low bridges. The multifunctional building rest on a concrete pontoon, and rises and falls by around 3.5 m with the tide. The IBA organizers see the dock as a prototype for future housing construction in areas where there is danger of flooding. This experimental structure tests adaptability, especially in terms of vertical movements and aims to find the pattern for easy transport [17].

Fig. 6 Constructing modules (left photo); IBA dock (right photo) (photos: Rudiger Mosler) (available at: https://www.archdaily.com/288198/iba-dock-architech, accessed 4 Jan 2019)

4.6 Buildings that Include Navigation Systems

A heated debate is going on around the issue of categorizing those structures which share parallel functions—or more specifically: these might, on one hand, be architectural constructions which are created as unambiguous public applications, however, which also hold the feature of being motorized floating objects with transportation as their primary function, on the other hand. The problem arises when it comes to the registration of such a construction—should one label it as an architectural or as a transportation construct? Even though a floating building is characterized as a living/working space structure according to The Moon and British Columbia Standards, this does not include a watercraft with the capacity of navigation. An innovative kind of multifunctionality is rearing its head out of this particular subtype of floating buildings with navigation systems, or rather, floating constructions which had previously been rebuilt and readjusted to suit a particular function. It is understood that the given object may be utilized in the form of a public building (a restaurant, a museum, a cinema, etc*), regardless of which the users may enjoy the travel* per se *alongside the sightseeing agenda. This potential redefines the very notion of architectural mobility (as well as floating mobility, if one might even presume the term as such). The new form of dynamic certainly ensures a unique experience to its customers.*

The New Floating Cinema is a resounding on the Floating Cinema[6] pilot project, the floating structure tied to canals (rivers). It is designed by the UP Project in collaboration with Somewhere[7] in 2013. The aim is to create the structure through which artists' films, independent cinema and archive footage can be brought outside of the gallery or cinema space and presented programs of outdoor screenings, as well as on board presentations, talks, performances, workshops and events for audiences

[6]Designed by Studio Weave in 2011. An old work boat re-imagined into a cinema for intimate on-board film screenings, larger outdoor film events and other film-related activities.

[7]Nina Pope and Karen Guthrie, the film-maker artistic duo.

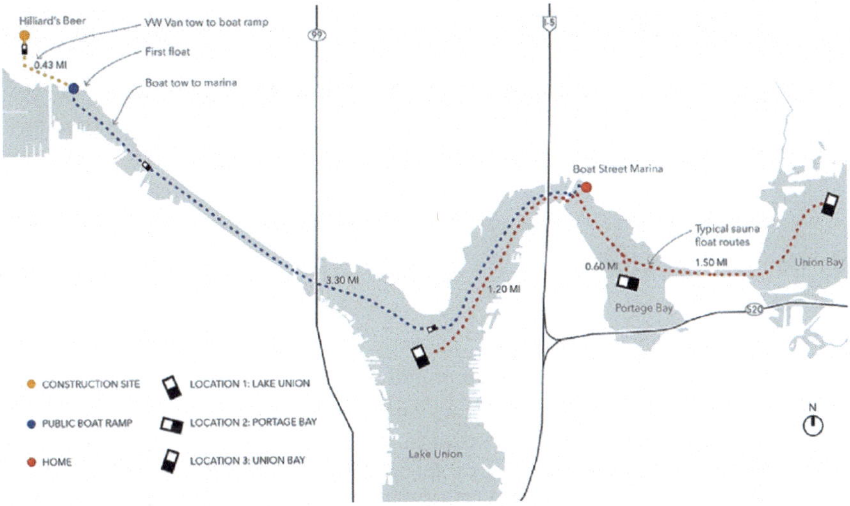

Fig. 7 Location diagram (available at: https://www.dezeen.com/2016/03/30/wa-floating-mobile-sauna-gocstudio-seattle-lakes-washington-state/, accessed 7 Jan 2019)

across the country. Powered by a hybrid engine that runs on biofuel, the new floating cinema is currently touring the London urban waterways [18].

Wa_sauna is the first floating sauna, located in Seattle (USA), designed by goC-studio and realized in 2015, situated on the lake Union. The project combines the static purpose of a sauna with the mobile platform that serves local tourism.

The 22 m^2-floating structure (up to six people) structure is powered by an electric trolling motor and visitors have complete freedom of movement while they are using it. Although the designers generally take the vessel on two- to three-hour trips, users can create their own route and stay as long as they want (see Fig. 7) [19]. The structure engages the local waterways and encourages people to use it throughout the year. It provides a place for both the community and for travelers to share a unique experience on water: visitors can enjoy the nature and take a plunge into a cold lake after warming up in the sauna.

Guggenheim Cruises is a mobile museum, designed by OfficeUS in 2014, officially located in Helsinki (Finland). Idea combines 13-hours travelling across the Baltic Sea (making a route Helsinki-St. Petersburg-Tallinn) and enjoying an experience of art. The concept aims to dismantle the necessity of a fixed-location museum. As a global freeport, the museum develops a completely new infrastructure, offering the strategic tax benefits of freeport art storage while enabling exhibitions of some of the most important pieces of modern art and design [20].

5 Discussion and Conclusion

According to previous analyses, we can classify floating mobility into two main categories: (1) buildings (structures) that need to be towed by tugboats and (2) buildings (structures) that include navigation systems (engine).

5.1 Passive Mobility

The first category may be considered "passive" mobility. Following this classification, there can be vertical or horizontal mobility, based on the orientation of movement. Vertical mobility is the ability of a floating building to follow the water level. The mooring system permits free vertical movements so the building can rise and fall together with the water (moving on the z-axis). On the other hand, horizontal mobility means ability of a floating building to be towed anywhere along the water body (moving on the x, y-axis).

In terms of *passive* horizontal mobility, we can discuss further properties: *relocation, permanent function-current location, the transportation issue* and *flexible configuration.*

Relocation—a building tends to keep the function but for some reason needs to be relocated. When the building changes its location, it is planned to be moored permanently to a new site. The advantage of floating mobility is that it serves as a precautionary measure while it is not intended for frequent movement.
Permanent function, current location—the building tends to keep the function but the absence of a firm foundation allows for frequent relocating. The building doesn't tend to be permanently moored; rather, floating mobility serves deliberately as part of the utilization strategy.
The transportation issue—considering the fact that floating buildings need some space for maneuvering together with tugboats and the preparation for sailing along a water stream. Familiarization with the urban tissue through which a structure is about to be transported is a necessary endeavor, more specifically, the assessment of potential obstacles, according to which an architectural design is modified.
Flexible configurations—the floating structure doesn't tend to keep the function (or location). It gives the freedom of maneuvering with modules to get the most out of every configuration with a high level of adaptability and flexibility.

The exceptional property of *passive* mobility is *vertical movement.*

Vertical movement—a building tends to change the cityscape frequently. From a constant point of view (a river bank or a coastline), a section of a city is changing all the time. As the building goes up and down following the water level, the view of the section alters adequately.

5.2 Active Mobility

The second category, or "active" mobility, marks self-sufficient buildings that can autonomously navigate themselves.

Buildings that include navigation systems—whether a floating platform with an engine or a vessel transformed for a new purpose, transportation itself is not the primary function. Moving while doing some other activity and enjoying the surroundings at the same time gives architecture a completely new dimension. By having this type of relatively small structures on rivers, lakes and seas, we can physically and visually activate cities a lot.

To sum up after this consideration of all the aspects of mobile architecture examples, a more wholesome definition of *floating mobility* can be given:

Floating mobility is the ability of buildings with floating foundations to move with a great degree of flexibility in terms of internal configurations, vertical movement and relocation along water bodies.

Floating mobility grants buildings the unique ability to follow the pace of modern life together with the global tendencies. It offers a huge amount of new possibilities and space for development in the area of urban planning and architecture. At the urban level, floating mobility represents the new way of thinking, much more adapted to everyday needs, incorporating a high level of social awareness. Not only can the buildings move, but the users also need to move in much more dynamic ways while using them. It provokes people's curiosity and pushes them to be more active. Floating buildings can be used for residential as well as for public purposes, they can be moved annually, once a month, on a daily-basis or only when needed. Complete freedom of construction can significantly contribute to the dynamics of the city appearance. Mobility, in general, tends to create brand-new rhythms in all fields and configurations of coastal cities are no exception. There is an overlap in the mindset growth of the current generation and the materialization of ideas the dynamic of such architecture presupposes. The process of symbiosis begins between human perspective and the way it changes along with the way the vision coming to life transfigures further thought. That means we can have dynamic cities that can understand and satisfy people's needs by changing their face day in and out.

References

1. Macmillan, E. (2007). *Macmillan english dictionary* (2nd ed.). London, England: Macmillan ELT.
2. Kayal, P., Singh, R., & Kumar, M. (2014). *Defining sustainable urban mobility, Teri-NFA* [Online]. Available at: http://www.teriin.org/projects/nfa/2008-2013/pdf/working-paper-11-Defining-Sustainable-Urban-Mobility.pdf. Accessed January 12, 2019.
3. Aldridge, S. (2001). *Social mobility-a discussion paper.* London: s.n.
4. Acharya, L. (2013). *Flexible architecture for the dynamic societies*. Toronto: Faculty of Humanities, Social Sciences and Education.

5. Wang, C. M., & Wang, B. T. (2015). Great ideas float to the top. In *Large floating structures* (pp. 1–36). Singapore: Springer.
6. DeWolf, C. (2018). Water world. *CLAD Magazine, 4*(1), 102–106.
7. Olthuis, K. (2018). *Rise of the blue city for gif im Fokus*. Interview; Interwier: Koen Toblis. January 2018.
8. Moon, C. (2015). Architectural characteristics of floating building. In *World architects congress* (pp. 1–6). UIA 2017 Seoul, Korea: SEOUL.
9. Mairs, J. (2015). *Thatched office for Amsterdam's canal cleaners is "the biggest ark in the Netherlands*" [Online]. Available at: https://www.dezeen.com/2015/01/22/floating-thatched-office-amsterdam-canal-cleaners-attika-architekten/. Accessed January 15, 2019.
10. Harvard Graduate School of Design. (2013). *Floating city: the community-based architecture of parade floats*. New York: Harvard University.
11. Sebastian, J. (2011). *Floating OffShore stadium/stadiumconcept* [Online]. Available at: https://www.archdaily.com/138162/floating-offshore-stadium-stadiumconcept. Accessed January 10, 2019.
12. Valk, S. (2016). *The good hotel* [Online]. Available at: https://sikkovalk.nl/good-hotel-amsterdam. Accessed January 18, 2019.
13. Witsen, P. P. (2012). *Floating Amsterdam: The development of Ijburg's Waterbuurt*. Amsterdam: Municipality of Amsterdam, Projectbureau IJburg, Ontwikkelingscombinatie Waterbuurt West.
14. Lim, Y. B., & Koh, H. S. (2015). Floating performance stage at the Marina bay, Singapore. In *Large floating structures* (pp. 37–60).
15. Scheeren, O. (2012). *Archipelago cinema/Buro Ole Scheeren + film on the Rocks Yao Noi Foundation* [Online]. Available at: https://www.archdaily.com/226936/archipelago-cinema-buro-ole-scheeren-film-on-the25e2%2580%258b-rocks-yao-noi-foundation. Accessed January 10, 2019.
16. Ravenscroft, T. (2018). *Artificial Island creates floating events space in Copenhagen harbour* [Online]. Available at: https://www.dezeen.com/2018/03/13/copenhagen-islands-marshall-blecher-magnus-maarbjerg-fokstrot-floating-artificial-island-harbour/. Accessed January 13, 2019.
17. Slawik, H. (2010). *IBA dock* [Online] pdf available at: http://www.slawik.net/images/bautenundprojekte/pdf/Slawik_IBA%20DOCK_projekt%20paneel_EN.pdf. Accessed January 14, 2019.
18. Borgobello, B. (2013). *Floating cinema sets in London for the summer* [Online]. Available at: https://newatlas.com/floating-cinema-london/28580/. Accessed January 10, 2019.
19. McKnight, J. (2016). *WA Sauna by goCstudio floats on Seattle's Lake Union* [Online]. Available at: https://www.dezeen.com/2016/03/30/wa-floating-mobile-sauna-gocstudio-seattle-lakes-washington-state/. Accessed January 10, 2019.
20. MacLeod, F. (2014). *OfficeUS propose floating museum for Guggenheim Helsinki competition* [Online]. Available at: https://www.archdaily.com/546245/officeus-propose-floating-museum-for-guggenheim-helsinki-competition. Accessed January 13, 2019.

The manufacturer's authorised representative in the EU is Springer Nature Customer Service Centre GmbH, Europaplatz 3, 69115 Heidelberg, Germany. If you have any concerns regarding our products, please contact ProductSafety@springernature.com

Printed and bound by CPI Group (UK) Ltd, Croydon, CR0 4YY

25/09/2026

02222847-0003